Trafficking Materials and Gendered Experimental Practices

This work is a partial representation of *Trafficking Materials and Gendered Experimental Practices* by Maria Rentetzi, a multimedia work of scholarship published in the Gutenberg-e online history series. As such, this print edition does not include the images, maps or the index contained in the online edition. The Gutenberg-e series is part of our commitment to create new kinds of scholarly and educational publications through new media technologies. Our mission is to make these works as innovative, efficient, and cost-effective as possible. We encourage reviewers and readers to also consult the complete work online at the free access site: http://www.gutenberg-e.org/ or through ACLS Humanities E-Book (HEB) at: http://www.humanitiesebook.org/series_GUTE.html

The online version of this work contains all the printed text here, in addition to digital images, artwork, audio, video, and hyperlinks that allow the reader to experience the full meaning of this scholarly work. The organizing structure, content, and design of the online work was created by the author in collaboration with a team of editors, web developers, and designers in order to give the richest meaning to the historical narrative and argument. The intellectual content of this work is designed to be read and evaluated in its electronic form. This text is not a substitute for or facsimile of the online version of this work.

Trafficking Materials and Gendered Experimental Practices

Radium Research in Early 20th Century Vienna

Maria Rentetzi

www.gutenberg-e.org

COLUMBIA UNIVERSITY PRESS

NEW YORK

Columbia University Press
Publishers Since 1893
New York Chichester, West Sussex

Library of Congress Cataloging-in-Publication Data

Rentetzi, Maria.
Trafficking materials and gendered experimental practices : radium research in early 20th century Vienna / Maria Rentetzi.
p. cm.
Includes bibliographical references.
ISBN 978-0-231-13558-0 (cloth : alk. paper) — ISBN 978-0-231-50959-6 (e-book)
1. Radioactivity—Research—Austria—Vienna—History—20th century. 2. Radium—Research—Austria—Vienna—History—20th century. 3. Women in science—Austria—Vienna—History—20th century. 4. Blau, Marietta, 1894–1970. I. Universität Wien. Institut für Radiumforschung und Kernphysik. II. Title.

QC795.34.R46 2008
539.2072'043613—dc22
2008040581

www.gutenberg-e.org

Columbia University Press books are printed on permanent and durable acid-free paper. This book is printed on paper with recycled content. Printed in the United States of America.

c 10 9 8 7 6 5 4 3 2 1

CONTENTS

ACKNOWLEDGMENTS

I began this study as a doctoral dissertation at the Science and Technology Studies program at Virginia Tech. I owe my thanks there first and foremost to my advisor Richard Burian who generously granted me his time, his ideas, and his friendship. I am also grateful to my dissertation committee members, Bernice Hausman, Gary Downey, Joseph Pitt, Peter Galison, and Aristides Baltas for their critical advice, moral support, and constant encouragement throughout my years of dissertation research and writing. To my fellow students both at Virginia Tech and at the National Technical University of Athens where I completed my first master's degree in history and philosophy of science and technology I am indebted for their friendship, long discussions, and the chance to work out ideas. I would especially like to also thank Brad Kelley, Joan Marie, Sarah Mitchell, Pei Koay, Jean Miller, Ty Brady, George Fourtounis, and Lara Scourla. Over the years I have profited immensely from discussions with, and thoughtful critique by, Bert Moyer, Peter Machamer, Alan Rocke, Tyler Smith, Gary Hardcastle, Ellen Balka, Bernhard Wieser, Doris Wallnöfer, Ulrich Dolata, Leo Slater, and Godfrey Guillamin. Thanks also go to Londa Schiebinger, Elizabeth Crawford, and Hermann Fuhrmann for their significant contributions to the revision of my dissertation. I thank Leopold Halpern, Arnold Perlmutter, Agnes Rodhe, and especially Artur Svansson for allowing me to interview them and for sharing with me their memories, family pictures, private letters, and unpublished manuscripts.

In presenting my work to several conferences and workshops, I wish to acknowledge the insightful comments and suggestions made by Annette Vogt, Ida Stamhuis, Sonia Strobanova, Annette Lykknes, Marsha Richmond, Brigitte Bischoff, Eva Kranakis, John Krige, Ann Laberge, Bill Leslie, and Margaret Rossiter.

I owe a tremendous debt to all the archivists whose help has been indispensable during my years of historical and archival research. I would like to particularly acknowledge the help of the archivist, and now dear friend, Stefan Sienell, of the main archive of the Institute for Radium Research at the Austrian Academy of Sciences. Stefan went out of his way to suggest and locate important material that I would have never been able to find otherwise. I also owe a debt of gratitude to the archivists in the following archives: the University of Miami archives; the Archives of Scientific Philosophy, Rose Rand collection of the Hillman Library at the University of Pittsburgh; the Association of American University Women Educational Foundation; the archives of the University of Albany State University of New York; the Rockefeller Archive Center; the Nobel Archives of the Royal Swedish Academy of Sciences; the Churchill College Archives in Cambridge, Lise Meitner Collection; the archives of the University of Cambridge, Ernest Rutherford collection; the Max Planck Gesellschaft archives in Berlin; the Albert Einstein archive; the Niels Bohr archive; the archives of the Escuela Superior de Ingenieria Mecanica y Electrica (ESIME); the archives of the Centro de Documentacion e Investigacion de la Comunidad Ashkenazi de Mexico; the archives of the Göteborgs Universitetsbibliotek; the Archiv der Universität Wien; the archives of the Österreichische Zentralbibliotek für Physik, and especially the archivist Brigitte Kromp who provided invaluable guidance for most of the photographic material which appears in this book. I would like to also acknowledge Ruth Freund, physicist and director of the Institut für KH Physik at the hospital in Lainz, Vienna. Although she is not an archivist, she has a remarkable understanding about the important historical material that is held at the hospital. Without her help, I would not be able to locate scattered letters, reports, and files that were not even catalogued.

A number of research grants and fellowships made my research and the completion of this book possible. I acknowledge the following institutions: Virginia Polytechnic Institute; American Institute for Physics; Konrad Lorenz Institute for Evolution and Cognition Research; and the Institute for Advanced Studies on Science, Technology, and Society in Graz. During 2003 and 2004, I was a recipient of a postdoctoral fellowship at the Max Planck Institute for the History of Science where I have returned twice as a visiting scholar for Ursula Klein's research group. My work there enabled me to turn my dissertation into this book. I am deeply indebted to Klein and Wolfgang Lefèvre for sharing their ideas with me and for providing advice, insightful criticism, and genuine friendship over the years I spent in Berlin. To Hans-Jörg Rheinberger, I owe my thanks for his warm welcome at the colloquiums of his department at the Max Planck Institute and for his interest in my work.

In 2003, the American Historical Association in conjunction with Columbia University Press awarded me the prestigious Gutenberg-e Prize, an honor that changed my professional life. The prize included the publication of a revised version of my dissertation manuscript and a generous monetary award. Most important, this made available to me an astounding network of other prize recipients, experienced editors, and the possibility to participate in informative workshops about electronic publication. Living in Greece, a country with limited resources and inadequate libraries, I have come to appreciate the importance of the AHA/CUP joint effort to make available electronic versions of new books and to publish the work of young scholars at a time when many are finding it difficult to publish first books in the humanities. I greatly enjoyed being one of the first to experience the transition from print to electronic books and I am especially grateful to Elisabeth Fairhead and Kate Wittenberg for their help in doing so. As a non-native speaker of English I also had to rely heavily on Snigdha Koirala and Jim Callan to revise my manuscript. I thank them for their patience and thoughtful suggestions.

Throughout this book, I have resorted to materials and articles that I have published elsewhere previously. I would like to thank the publishers for permitting me to use this material and wish to cite the following:

- "Women and Radioactivity Research in Vienna, 1910–1938," in *Women Scholars and Institutions*, edited by S. Strbanova, I. Stamhuis, and K. Mojsejova, 611–38.
- Prague Proceedings of the International Conference: Women Scholars and Institutions, Commission Women in Science of the IUHPS/DHS 2004, "The City as a Context of Scientific Activity: Creating the Mediziner-Viertel in Fin de Siècle Vienna," *Endeavour* 28, no. 1 (2004), 39–44.
- "Gender, Politics and Radioactivity Research in Interwar Vienna: The Case of the Institute for Radium Research," *Isis* 95 (2004), 359–93.
- "Designing (for) a New Scientific Discipline: The Location and Architecture of the Institut für Radiumforschung in Early 20th Century Vienna," *British Journal for the History of Science* 38, no. 3 (2005), 275–306.
- "From Cambridge to Vienna: The Scintillation Counter in Female Hands," *Nuncius*: *Annali di Storia della Scienza,* 2 (2004), 675–89.
- "From Experienced Experimenters to Scanning Girls: Women Designers and Users of the Scintillation Counter Before and After the Second World

War," in *Isotopes: Science, Medicine and Industry in the 20th Century*, edited by Xavier Roqué, Néstor Herran, and Jeff Hughes (forthcoming).

Finally, my greatest debt is to my parents for supporting me from the beginning of this journey. To Spiros Flevaris—the architect who taught me the difficult task of reading architectural plans, instilled in me his passion for architecture, and, luckily for me, changed my life in several respects—I owe the completion of this book.

INTRODUCTION

This is a book about the materials and experimental practices of radioactivity research. It is neither an overview of the emergence of the field in the early twentieth century nor a collection of institutional histories of different research centers. It is also not an account of women's participation in a discipline that has been portrayed as one of the research areas most highly populated by women. Instead, the book is centered on radioactive materials, instruments, and gendered skills.

In this work, I employ a gender perspective in the study of experimental practices and laboratory materials. I seek to show how experimental cultures—ensembles of scientific practices employed by gendered subjects who share a certain material and epistemic style of research—in radioactivity are constructed and reshaped by scientists of different gender. At the same time, I explore the different ways experimental practices affect men and women in laboratory sciences and I broaden the notion of material culture by focusing specifically on materials that were both commodities and objects of scientific inquiry. I attempt to understand how purified radium ended up on laboratory benches, who completed the hands-on work of its extraction and isolation from tons of residues, who designed the experiments and instruments for probing radium's properties, and who carried radium outside of the physics laboratory to the clinics and medical amphitheaters.

I am interested in the circulation of radium, the making of connections, and the pursuit of power through strategies of partnership and collaboration. The redrawing of paths of exchanges brings us to a different millennium and moves us from scientific laboratories to hospitals, from academic sites to industrial sites; it connects different disciplines such as physics to medicine and chemistry; it reflects struggles for scientific preeminence and diverse partnerships between scientists and industrialists; it even reveals efforts for professional existence.

At the end of the nineteenth century, the discoveries of x-rays and radioactivity changed the meaning of physics by reshaping its boundaries and transforming its material culture. New apparatus such as scintillation counters and photographic emulsions were developed in order to detect individual particles. Physicists were now able to replace previous macroscopic experiments with finer-grained experiments that were sensitive to such microscopic phenomena as interactions between individual particles. The shift in experimental techniques and instrumentation contributed to a new, dynamic interrelation of physics with other disciplines such as chemistry and medicine. In fact, physicists, chemists, industrialists, instrument makers, and medical technicians claimed radium for their own fields and traded, bargained, bought, and sold it.

The flow of radium as commodity—medical, chemical, or otherwise—in the early twentieth century had very diverse and important implications for both individuals and institutions. In all cases, it functioned as a device for reproducing relations and linking producers, distributors, consumers, and users. However, commodities represent very complex forms of knowledge, for example, the knowledge that goes into the production of the commodity and that which goes into appropriately using it. In the case of radium, I want to know who had the skills to prepare radium sources, brush the mixture of alcohol and zinc sulfide from the small rectangular glasses used for scintillation counting, read the photographic plates of emulsions, set up apparatus for experimental work, analyze radium samples, define radium doses for medical use, and perform experiments.

But what did it mean to do experiments? To answer this question, I explore the laborious laboratory processes and the culture that surrounded them. I study the ways instruments were designed and modified, and how materials were obtained and prepared. This book is an attempt to convey a sense of what it meant for the participants—both men and women—to be engaged in experimental practices in early radioactivity research. At the same time, keeping in mind that laboratories are gendered spaces, I show how the use of materials and instruments reveal the ways gendering of experimental skills took place.

AN EPISTEMOLOGICAL AND HISTORIOGRAPHICAL SHIFT ON MATERIALS

As Jeff Hughes has recently argued, from the 1940s to the 1980s, the historiography of radioactivity and nuclear physics had been dominated by canonical accounts that either stressed the theoretical developments and significant discoveries in the field or produced hagiographies of the important figures along with reminiscences and uncritical popular history. Even the historiography of women in radioactivity research has followed the same trend. From Marie Curie's to Lise Meitner's biographies, the main concern has been to provoke scientific and public fascination and appreciation for those women connected to radioactivity. Only during the 1990s did historians begin to rethink the history of the field in terms of the experimental practices and material culture that established and defined it as such.[1] Recent studies have been centered on instruments, scientific objects, conceptual tools, disciplinary networks, and experimental skills, following a general trend in science studies which shifted focus among other issues to scientific objects.

As early as 1989, Susan Leigh Star and James Griesemer coined the term "boundary object" in order to capture the entanglement of heterogeneity and cooperation in scientific practice. Focusing on the early history of a natural museum, Star and Griesemer analyzed the ways in which amateurs, professionals, trappers, and administrators involved in the museum made sense of their different viewpoints. Objects of their common interest, such as fossils, species, and subspecies of mammals and birds, field notes, and maps inhabit multiple worlds, simultaneously satisfying the informational requirements of each of them.[2]

In 1997, reading the history of the synthesis of proteins in the test tube from material traces, Hans-Jörg Rheinberger introduced the notion of "epistemic things"—things that embody concepts. In contrast to those historians of science who have dealt with disembodied ideas, Rheinberger emphasizes physical structures, chemical reactions, and biological functions, in general, material entities or processes that take place in the laboratory. His history is, he argued, "a biography of things, a filiation of objects, not as pictures of an exhibition, but as records of the process of their coming into existence."[3]

In 2000, Lorraine Daston compiled *Biographies of Scientific Objects*, which presented the histories of several scientific objects from ether to dreams, from mortality tables and blood pressure to cytoplasmic particles. The book is about the ways scientific objects are born, forming contingent creations out of local cultural and material resources. Four years later, "things," and this time *Things*

That Talk, gave life to another collected volume of essays, which "knit together matter and meaning."[4] The book spans a wide spectrum of "things"—from soap bubbles to paper clippings to glass flowers—and is concerned both with the ways "things" are made and with what "things" actually are.

In the last 15 years, for historians of physics and chemistry, writing the biographies of their objects of inquiry has meant writing about instruments and laboratory technologies.[5] In the life sciences, objects such as manometers, tissue slices, or even living organisms such as *Drosophila*, have been analyzed as part of the material culture of the laboratory.[6] It is only recently that materials have also attracted the attention of historians of science and acquired epistemological value. In "A Science Whose Business Is Bursting," Simon Schaffer explores the paths through which a commonplace commercial entity, such as soap bubbles, turned into a representative "thing" in nineteenth-century physics. To Schaffer, "the ephemeral bubble" becomes a vital link between the warehouse and the cabinet, giving the historian the chance to have a glimpse of both worlds.

In a different venue, Ursula Klein and Wolfgang Lefèvre have extended these lines of investigation by shifting their focus from instruments to material objects and from institutions and experimental activities to sites of intersection in the arts and crafts. They depict eighteenth-century chemistry as a science of materials and trace their shifting ontologies over the span of a century. They argue that "Eighteenth century chemists treated materials as useful commodities, perceptible objects of nature, and entities carrying imperceptible features." Such a perspective challenges traditional and powerful views of Lavoisier's role in chemical revolution.

"Seen from our new perspective," Klein and Lefèvre explain, "Lavoisier and his collaborators reaped the rewards of a century. In so doing, they introduced reforms of concepts, theories, analytical methods, classificatory structure and language. Yet, they did not initiate an ontological rupture. Chemists continued to live in largely the same world of objects before and after the Lavoisierian reforms." In the *Making of Materials*, a collected volume, Klein touches once again upon the importance of the materiality of things and the intersection of science, technology, and society. The material objects under inquiry were all materials applied in the everyday world as commodities and at the same time were objects of scientific inquiry. Thus the central view on the history of experimentation is replaced by a de-centered approach that takes into account forms of making and knowing in early modern natural history.[7]

WRITING THE BIOGRAPHIES OF MATERIALS

Anthropological and cultural studies have focused on the social life of commodities as they are the material objects that unite archaeology with cultural, social, and economic anthropology, art, and economic history.[8] Borrowing an anthropological view, science studies scholars have argued that "things" have cultural biographies since they are not only produced materially but are also culturally marked as being a certain kind of thing. "In doing the biography of a thing," as the anthropologist Igor Kopytoff puts it, "one would ask questions similar to those one asks about people: What, sociologically, are the biographical possibilities inherent in its 'status' and in the period and culture, and how are these possibilities realized? Where does the thing come from and who made it? How does the thing's use change with its age, and what happens to it when it reaches the end of its usefulness?"[9] What could it mean then to write the history of radioactivity as a biography of radium and radioactive materials, commodities which link the laboratory to the mundane world?

In ways familiar to the anthropology of material culture, what is significant about the adoption of radium as a commodity is not the fact that it was adopted but the way it was culturally redefined and put to use. Its biography reveals a wealth of cultural data depending on the path of its exchanges: the way it was acquired; the institutions that host it; how and from whom the money was assembled to pay for it; the identity of both those who produced it and those who consumed it; the ways it was manufactured, advertised, and sold; the owner's relation to those who produced and promoted it; the networks developed for its promotion; and the signs embodied in its use. Radium's biography is complicated by the many social lives it lived in an unsettled geography. From France to Austria and from Austria to the United States, from the mines to the chemical laboratory to the hospital or the corporation's sales catalogue, radium has been in constant traffic. Throughout its several lives, it has been endowed with meaning and became a sign in a system of signs. In Roland Barthes's terms, radium was transformed to a myth—maybe the most powerful one in the early twentieth century—that is, a system of communication, a mode of signification.[10]

Let us take an example. In 1921, two grams of radium went on display in Grand Central Palace in New York City. This is an immense amount considering that the production of one gram requires 500 tons of carnotite ore, 500 tons of chemicals, 10,000 tons of distilled water, and the energy of 1,000 tons of coal. The exhibit was organized by Radium Chemical Company, which first produced luminous paint for watch dials and instruments and then moved to the manufacturing of radium-containing needles and other sealed devices (largely for cancer therapy) for lease or sale to hospitals and research laboratories. Obviously, the event targeted physicians and nurses. The exhibit also included daily demonstrations of the penetration of gamma rays of radium through a three-quarter inch vanadium steel cylinder. Motion pictures and photographs in display made the audience familiar with the complex process of radium's extraction, purification, and refinery. Every night, the radium was transferred under high police protection to the house of the chemist Hamilton Foley, who was responsible for the exhibit.[11]

In this exhibit, radium clearly held powerful meanings constituted by the myths of science, medicine, industry, commerce, and state power. The public lined up to witness the accomplishments of scientific authorities and the miracle of the new element. What made it a miraculous material were its properties: its rays penetrated opaque objects, radium needles cured cancer, it was used in paint to decorate everyday life, and its ore needed tedious work to be transformed to the material in display. Faced with these several myths, the visitor needed to identify himself as a consumer, a potential buyer of the Radium Chemical Company's products.

In an attempt to unravel its different myths, radium and the experimental practices that it defined are kept at the front and center of the chapters that follow. These chapters tell the part of radium's life story that relates to the Institute for Radium Research in Vienna, and to certain gender relations in the work place. Moreover, the metropolitan setting of Vienna, a center of an empire in transition, reveals the ways in which the material was interwoven in an international scientific network and caught up in major political upheavals.

THE SCENE OF MY HISTORICAL PLAY

In 1910, a Viennese lawyer and industrialist combined resources with the Austrian Academy of Sciences and the state to establish the Institut für Radiumforschung (Institute for Radium Research) in Vienna. The institute specialized in investigating chemical and physical aspects of radioactivity and was devoted exclusively to research. Franz Exner, a prominent member of the Viennese bourgeoisie and the Austrian Academy of Sciences, became its formal director. However, his student, Stefan Meyer, a young and talented physicist, handled the supervision, planning, and administration of the institute from the beginning.

By the end of the 1910s, the physicists of the Vienna Institute were already considered leading figures in the European community of radioactivity research. Serious competitors were located in only three other research centers: Marie Curie's institute in Paris, Ernest Rutherford's laboratory in Manchester, and the Laboratorium Hahn-Meitner at the Kaiser Wilhelm Institute for Chemistry in Berlin. The Vienna Institute provides an ideal setting for studying both radium and the experimental culture of radioactivity.

On the one hand, the Viennese had access to the rich resources of uranium pitchblende of the St. Joachimsthal mines in Bohemia at their disposal and were able to acquire radium in sufficient quantities for experimental work. They even entered the small network of physicists doing research on radioactivity as the main providers of the pricey material to scientific laboratories and soon became known for their prominence in radium research.

From its inception, the institute had a very strong staff, including many women—an extremely unusual fact for the time. Leading figures such as Ludwig Boltzmann, Victor von Lang, and Exner had helped women assimilate themselves into Vienna's academic community and welcomed them into their physics laboratories. After their first admission to the University of Vienna in 1897, the number of female students increased substantially, rising from 37 in 1897–8 to 796 in 1913–14. During the same period, about 30 percent of those who were enrolled in the philosophical faculty chose the field of physics.

The Radium Institute was able to attract large numbers of those educated women and integrate them into the local and international scientific community. Especially during the 1920s, more than one-third of the institute's personnel were women. Most importantly, they were not just technicians or invisible assistants; they were independent researchers who handled radium sources for their own experiments, designed their instruments, and participated in the international scientific community.

To analyze such an exceptional institutional setting, I will focus both inside and outside the laboratory and show that these boundaries are never fixed. At the turn of the twentieth century, Vienna was an extraordinary cultural setting where the city's intelligentsia produced distinct "schools" of thought in architecture, music, painting, philosophy, and psychology. In this creative milieu, a surprisingly dense network of academic institutes and medical facilities developed on the margins of the Ringstrasse, the impressive boulevard that surrounded the old city of Vienna. Enmeshed socially and culturally in the life of the city, an area known as the Mediziner-Viertel grew out of the needs of natural scientists and physicians and was also shaped by the liberal ideology and feminists' petitions for women's acceptance into university studies, Vienna's world of intellectuals and artists, and the Kaffeehaus culture. For instance, in the area's coffee shops, female students associated with their male colleagues and exchanged ideas. Indeed, it was at this time that Viennese coffeehouses served as centers of social and epistemic interactions.

As recent studies have suggested, however, the city should be considered not merely as a location where science occurs but also as a setting that has affected the evolution of science, its methods, instrumentation, concepts, and objects of research.[12] The case of the Radium Institute reveals several aspects of such an interrelation between science and the city. The symbiosis of medical institutes with those of chemistry and physics within the Viertel provided the Viennese community of radioactivity researchers with a unique opportunity. For example, a network of scientists conducting research on the application of a variety of radioactive materials was expanding rapidly in the neighboring medical and biological institutions. On several occasions, radium was prepared at the Radium Institute and delivered to the Radium Station of the Allgemeines Krankenhaus (General Hospital) to be used in cancer therapy. The interdisciplinary nature of the field and the trade of radium offered scientists exceptional opportunities for crossing disciplinary boundaries. In Vienna, this proved to be a successful strategy for women to gain access to radioactivity research and sustain their role as active experimenters.

In addition, the fact that the Social Democratic Party gained control of the government of the city in 1919 and retained it for the following 15 years proved to be of crucial importance. A socialist program of municipal reforms was soon designed to transform the social, cultural, educational, and economic infrastructure of the capital. One of the targeted issues was gender politics, which was significantly altered by the policies and discourse of the Social Democrats. As historians of the period suggest, it is not surprising then that from 1919 to 1933, the enrollment of women at the University of Vienna increased steadily.

The unique constellation of progressive politics and supportive, politically aware personalities created the local conditions that enabled women to play a major role in the Radium Institute as well. The fact that the institute was a piece of Vienna's political history became more obvious during the 1930s. When the fascists seized power in Vienna in 1933, they deliberately put an end to most of the progressive forces of Viennese society. Especially after the *Anschluss* and the German Nazi occupation in 1938, Social Democrats, feminists, and Jews, including those of the Radium Institute, became the targets of a hate campaign and totalitarian politics.

How did women's everyday dynamic presence in laboratory life affect men's perception of their own research practices? A central argument of this book is that gender actively and constructively shaped experimental cultures in radioactivity and was not merely an external factor in laboratory life. To support this claim, the focus shifts to the study of a controversy, a favored methodological tool in the sociology of experimentation.[13]

The methodological advantage of examining controversy is twofold. First, it allows the comparison of experimental practices in diverse institutional settings, particularly when laboratories that share the same epistemic goals differ in their material cultures. By studying such differences, the historian can gain an understanding of how experimental cultures are shaped and how they affect the production of scientific knowledge. Second, instances of scientific controversies can reveal the roles that gender plays in an ongoing conflict. How, for example, are experimental skills gendered in different settings, and how are the material cultures of the laboratories transformed when more women are actively involved in experimentation? What role do materials play in the sustainability of research and research positions?

The controversy examined in this book took place during the 1920s between the Vienna Institute and Ernest Rutherford's research laboratory at Cambridge University. Although the starting point of the quarrel concerned discrepancies over the artificial disintegration of light elements, it gradually developed into a fierce battle over the reliability of instruments and techniques, Rutherford's credibility in the field, and the accuracy of various theories and experimental results.[14] Rutherford's group, all male and mainly young students of physics, used the affluent resources of Cambridge and tied their research to the reliability of the scintillation counter, an instrument deployed for counting tiny flashes of light produced on a zinc sulfide screen by the impact of charged particles. In contrast, the Vienna group, under the leadership of the Swedish physicist Hans Pettersson attracted a young generation of researchers—many women among them—who drew from the scarce sources of the institute and succeeded in gaining support of Swedish patrons and the International Education Board.

Among other things, the controversy accounted for the transformation of the material culture of the institute. Tabletop technologies such as x-ray photographic films, scintillation counters, photographic plates of emulsions, all tabletop, portable, and cheap apparatus that were easy to design and use were closely tied to the experimenters' everyday lives. Interestingly, the instruments tell us the stories of their designers and users. For instance, the transfer of the scintillation counter from the laboratory benches of Rutherford's group to the Vienna Institute is about politics of collaboration between the physicists in each group, gender assumptions concerning the instrument's use, and strategies of retaining authority in the field. In each setting, definitions of experimental tasks as skilled and unskilled were constructed based on different gender relations.

In the late 1920s, the Viennese shifted parts of the scintillation counter from one laboratory bench to another in order to save a technique that by all accounts was dying. Instead, the group at the Cavendish Laboratory was introducing the novel methods of wave mechanics to nuclear problems, altering their instrumentation and theoretical approaches. While the loyalty of the Viennese experimentalists to their material culture reveals some of their insecurity in the field, changes in Rutherford's group point to the importance of funding and resources in scientific practice. In the end, for the Viennese to lose their epistemic authority was less important than to lose funding and be "gently" excluded from the scientific community through the private resolution of the controversy.

In this book, sociological attention is centered on how the controversy was resolved and on how differently women and men responded to the rearrangements of their laboratory. Nonetheless, to tell the history of technology transfer and to focus on scintillation counters and zinc sulfide screens is also to explore the epistemology of the experiment. For instance, by sustaining the material culture of their institute, women were also maintaining the gender assumptions of their group, of who had the right to perform experiments, prepare materials, and design instruments. After the Anschluss, some of the institute's women were forced to become part of large scientific groups where access to the instruments was hierarchically prescribed. They lost control over their experiments and had to abandon their patterns of research and alter their work relations in the laboratory. How did this affect their actual research? Part of the answer lies in the shift from scintillation screens to photographic plates, photoelectric cells, and finally photomultipliers in the 1940s. It also involves the shift from radium to radioisotopes, from an expensive commodity manipulated on a laboratory bench by tabletop apparatus to a product cheaply and easily manufactured by big machines. The details of the above history will be covered throughout the following chapters.

In chapter 1, I sketch the biography of radium as a trafficking material—a material which is traded and bargained as both a commodity of the mundane world and as a scientific object—the building block of the field of radioactivity in the early twentieth century. It was the vital material around which research communities and, in the long run, a whole discipline became organized.

Chapter 2 introduces the reader to the main scene of my story—the Institute for Radium Research in Vienna—and follows the physicists' attempts to organize their research in one of the most technologically advanced institutes of their time. It is there that radium and the experimental practices that this book explores were centered. In this chapter, I pay attention to the ways the architecture of a science building shapes and is shaped by gender relations in experimentation.

Chapter 3 explores the interrelation of the science of radioactivity to the city of Vienna, emphasizing the role of the urban setting of scientific institutions to women's participation in research. In both chapters 2 and 3, radium is shown to be a trafficking material that defined individual and institutional partnerships, the urban construction and architectural design of science buildings, and the exchanges of knowledge, instruments, and expertise.

Chapter 4 moves on to the 1920s and concentrates on the role of politics to the construction of the institute's gender profile, while chapter 5 concentrates on the controversy that developed between the Cambridge and Vienna experimenters. The key player in this controversy was the scintillation counter, an instrument that was transferred from one laboratory to the other, exemplifying gender differences in experimental practices.

The topic of chapter 6 is the aftermath of the controversy, examined at two closely intertwined levels: the scientific and the political. In the backbone of the narrative is the fact that dealing with a trafficking material, the women of the institute, after losing financial support, were driven to transform the borders of their discipline. The fascist seizure of power in 1933 and the new political and social order the Nazis imposed during the late 1930s, deeply affected life at the institute. After the Anschluss, Jewish men and women were forced into exile, underlying the fact that gender and racial discrimination were entwined. With an emphasis on those women who played a crucial role at the institute, I follow their life and work trajectories into the early 1940s.

In the last chapter, I sketch the story of Marietta Blau's eventful immigration to the United States and her equally eventful consideration for a Nobel prize in 1950. Her constant struggle to survive in the new world of high energy physics provides some hints about how and why women shifted from being active experimenters to industrial designers and laboratory support staff in the 1940s and 1950s. Having lost the advantage of manipulating a trafficking material—mostly useless in the 1950s world of easily produced radioactive isotopes—and forced to function in a different experimental culture than the one in Vienna, Blau found herself working as industrial designer and far from innovative research.

A NOTE ON SOURCES

Historians are expected to rely on archives as the primary depositories of facts and decisive historical sources. As the historian of science Helge Kragh defines it, "a source is an objectively given material item from the past created by human beings."[15] Going through a list of the most important sources, Kragh includes letters, notebooks, manuscripts, scientific papers, and diaries. In the research for this book, I went through published scientific papers, letters, notebooks of finances and revenues, manuscripts, the annual bulletin of the Institute for Radium Research, as well as the almanac of the Austrian Academy of Sciences. Trips to several archives scattered throughout Europe, the United States, and Mexico gave me the opportunity to follow the trajectories of some of the protagonists of this story and reconstruct their laboratory life. Oral history and interviews filled some of the gaps about which the archival material was silent. Unless otherwise noted, translations of the above materials are my own.

Despite the fact that these common historical sources are absolutely necessary, they are not sufficient to grasp the complexity of laboratory life. I therefore additionally looked at photographs, not just as illustrations of a historical period, but also as evidence of the politics of collaboration within the institute, as proofs of material exchanges, and as indications of radium consumption and commodification. The photographs drawn from the online Oak Ridge Associated Universities' Health Physics Historical Instrumentation Museum Collection, an amazing source of radium instruments, radioactive quack cures, and medical products particularly illustrate the wide use of radium as commodity. Laboratory photographs either from the Central Library for Physics in Vienna or Agnes Rodhe's and Artur Svansson's personal archives also give a hint of how everyday life was structured at the Radium Institute.

I also used blueprints as a valuable source for historical analysis, as they express, contain, and portray gender discourse. The science studies scholar who is willing to pose questions of how gender politics and human relations define, and are defined by, spatial arrangements can find answers in the inner architecture of the laboratory and its spatial positioning in the city. Further, the map of the city provides a glimpse of existing scientific and commercial networks, of intellectual and material exchanges, and of political and academic alliances. I argue that the ways scientific buildings are arranged, interconnected, designed, and function define the ways practitioners alter, diversify, and rework the structure of their laboratory life.

NOTES

1. Hughes, "Radioactivity and Nuclear Physics" (2003).

2. Star and Griesemer, "Institutional Ecology" (1989).

3. Rheinberger, *Toward a History of Epistemic Things* (1997), 4.

4. Daston, *Biographies of Scientific Objects* (2000); Daston, *Things That Talk: Object Lessons from Art and Science* (2004), 10.

5. The literature has grown vastly. See for example Galison, *Image and Logic* (1997); Holmes and Levere, *Instrument and Experimentation in the History of Chemistry* (2000); Lenoir, "Instrument Makers and Discipline Builders" (1997).

6. See for example Clarke and Fujimura, *The Right Tools for the Job: At Work in Twentieth-Century Life Sciences* (1992); Kohler, *Lords of the Fly: Drosophila Genetics and the Experimental Life* (1994).

7. Klein and Lefevre, *Materials in Eighteenth-Century Science: A Historical Ontology* (2007); Klein, "Shifting Ontologies, Changing Classifications: Plant Materials from 1700 to 1830" (2005); Klein, *The Making of Materials* (1500–1800) (forthcoming).

8. Appadurai, "Introduction: Commodities and the Politics of Value" (1986). Anthropological studies of material culture rest on the differences between gift and exchange and concern the ways objects come to convey and condense value and cultural meanings, fashion selves, and construct identities. As Nicholas Thomas argues, "Objects are not what they are made to be but what they have become." Thomas, *Entangled Objects* (1990). See also Myers, *The Empire of Things* (1991); Thomas, *The Commodity Culture of Victorian England* (1990). Gender studies have added one more dimension to the material culture that of the ways things are gendered. See for example Martinez and Ames, The *Material Culture of Gender, the Gender of Material Culture* (1997); de Grazia and Furlough, *The Sex of Things* (1996); Kirkham, *The Gendered Object* (1996).

9. Kopytoff, "The Cultural Biography of Things: Commoditization as Process" (1986). See also Lamberg-Karlovsky, "The Biography of an Object: The Intellectual Style Vessels of the Third Millennium B.C." (1993).

10. Barthes, "Myth Today" (1973).

11. "$500,000 Radium is Exhibited Here," *New York Times*, November 14, 1921, 9:2(3).

12. Galison and Thomson (eds.), *The Architecture of Science* (1999); Dierig, Lachmund, and Mendelsohn (eds.), *Science and the City* (2003).

13. See, for example, Shapin and Schaffer, *Leviathan and the Air-Pump* (1985); Pinch and Bijker, "The Social Construction of Facts and Artifacts" (1987). Controversies have been discussed from a philosophical point of view in Machamer, Pera, and Baltas, *Scientific Controversies: Philosophical and Historical Perspectives* (2000). See also Nelkin, *Controversy: Politics of Technical Decisions* (1992).

14. Experiments on artificial disintegration involved the transmutation of one element to another by bombardment of alpha particles and the emission of long-range particles.

15. Kragh, *Introduction to Historiography of Science* (1987), 120.

1. THE BIOGRAPHY OF A TRAFFICKING MATERIAL

In 1904, an article titled "Radium and Radioactivity" appeared in *Century Magazine*, a monthly popular magazine published from 1881 to 1930. The article presents Marie Curie's personal account of the discovery of radium and radioactivity. In the article, Curie discusses in depth her arduous attempts to study the radiation of the compounds of uranium and that of known chemical elements, hoping to discover more which are endowed with atomic radioactivity. She revels that it was the chemists who supplied her with the materials she needed. "As I desired to make a very thorough investigation, I had resource to different chemists, who put at my disposal specimens—in some cases the only ones in existence—containing very rare elements."

Her next step was to examine different minerals, especially the oxide of uranium ore (pitchblende). To her great surprise, this specimen was found to be four times more active than oxide of uranium itself. The explanation was more than obvious. "The ore must contain a substance more radioactive than uranium and thorium, and this substance must necessarily be a chemical element as yet unknown."[1] Her attempts to isolate the new element would lead Marie and Pierre Curie, who joined her research shortly after, to the discovery of both polonium and radium, to the Nobel Prize in Physics in 1903, and to a second Nobel Prize in 1911, this time in chemistry.

This chapter attempts to present a cultural biography of radium. It is the biography of a material that, from the very first moment of its birth, became involved in the physical system of alpha, beta, and gamma rays and the atomic structure; in the chemical system of atomic weights, emanations, and transmutations; in the medical system of cancer treatments and radon spas; in the commercial system of luminous watches, women's cosmetics, and medical remedies; in the artistic system of luminous paintings and middle-class American culture; and in the industrial system of radium extractions, the production of luminous paint, and the beauty industry.

During the early twentieth century, radium evolved into a material of the everyday world as well as a "new research tool," in Pierre Curie's terms, of the laboratory site, bringing the above systems into close contact.[2] It traveled from laboratories to sites of medical practice and from educational amphitheaters to the physicists' and chemists' workbenches as scientific objects often do. Yet it was also an item of commodification that was transferred from mines and sites of industrial production to those of consumption.

Anthropologists have long ago argued that in order to understand the concrete, historical circulation of things one has to follow the things themselves "for their meanings are inscribed in their forms, their uses, their trajectories."[3] Besides understanding the circulation of the element that is called *radium*, I want to understand the culture that surrounded its circulation, the transformations in gender relationships that it motivated, the kind of exchanges that took place, and the forms of learned knowledge and gendered skills involved in the circulation of the material. Following around radium, one discovers what historians of science are less familiar with. In addition to being a scientific object with "wonderful and fabulous qualities,"[4] radium has been a highly valued commodity. In 1904, its price was $10–$15 per mg and just before the First World War, it escaladed to the astronomical price of $180.[5] Radium showed up as a consumer commodity in luminous watches; in women's lotions and creams; in toothpastes, cigarettes, and radium condoms; in ointments for medical use; and in food, drinks, clothing, and endless medical products. Its commodification involved complex political maneuverings, imperial commerce and colonial trade, efforts for monopolization, transformations of women's work, as well as rivalry between scientists and quack doctors over radium quantities, and the exchange of knowledge, skills, and radioactive materials among physicists, chemists, and medical practitioners.

To come to grips with radium's multiple identities as both a scientific object and a commodity and to explore the paths of its circulation, I developed the concept of a *trafficking material*. These are materials, which travel from hand to hand, from discipline to discipline, from laboratory to laboratory, or from the scientific world to the world of commodification and consumption. Their main characteristic is their ability to take on multiple identities, not because they are shared between different worlds but because they are transferred across them.[6] Motion becomes the inseparable part of their identity which has never been fixed in the first place. Physicists and chemists needed more than a decade after radium's discovery to fully identify the new element, ascribe to and understand its properties, and describe and classify them.[7]

Used in scientific laboratories, trafficking materials shift focus from instruments and experimental practices to material substances set on laboratory tables and manipulated by skilled experimenters who aim to reveal the "order of nature" in the distinct cultural setting of the laboratory.[8] How do these materials arrive in the laboratory? Trafficking materials such as radium are traded, produced, prepared, and sold, and thus provide a glimpse of how the laboratory and its experimenters are connected to the rest of the mundane world. These materials, furthermore, provide a vital link between the laboratory and the warehouse, the hospital and the academic institution, between sites of production and those of consumption. They are the objects of overlapping networks of knowledge built by different actors. Becoming part of this network it is not only a matter of gaining scientific expertise but also of using the right strategy and of possessing the power to impose oneself. For instance, trafficking materials worked as Trojan horses for sustaining women's experimental work in the field of radioactivity in the early twentieth century. Being able to prepare radium sources for medical use, counsel radiologists about radium's curative properties or perform radium measurements on ocean sediments, women physicists and chemists were able to cross the boundaries of their discipline and move to the fields of medicine or oceanography.

Trafficking materials possess not only a scientific value but also a social, economic, and sometimes a patriotic one, as we will see in the case of the Viennese physicists of the early twentieth century. They are intimately connected with the knowledge of those who sell them and of those who buy them. They take many journeys and sojourns from production to consumption and scientific exploitation, and they have been used as instruments in order to obtain something else or to restrict and control the kinds of exchanges they are involved in. Moreover, trafficking materials function as devices for reproducing certain relations, often gendered ones, between people in the laboratory, in the work place, or in the market. Thus, trafficking materials are not merely objects but are essential signs in systems of signs such as those that we have already referred to in the case of radium. Mapping the geography of radium's use, a highly complex undertaking, leads to a gender system of laboratory work and is key in understanding how the lively system of science works.

THE RADIUM OF THE PHYSICISTS

Radium was discovered in France at the end of the nineteenth century, but it was strongly linked to Röentgen's discovery of x-rays in his laboratory in Würzburg, Germany. In late 1895, Röntgen announced that the passage of an electric discharge from an induction coil through a partially evacuated glass tube produces what he called x-rays. The emitted radiation was able to penetrate not only the black paper Röntgen used in his original experiment but thick-material objects as well. In January 1896, Henri Poincaré presented Röntgen's discovery to the French Academy of Sciences, posing the question whether any naturally fluorescent or phosphorescent substance could emit penetrating radiation similar to that of x-rays. To Antoine-Henri Becquerel's satisfaction, that proved to be possible in the case of uranium.[9]

The same year, Becquerel observed the darkening of a photographic plate in contact with uranium crystals and described the invisible radiation emitted by uranium. In March 1896, during a meeting of the French Academy of Sciences, Becquerel presented his findings, which were then published within ten days. Shortly afterwards, based on intensive experimentation, he concluded that the emissions were not related to visible fluorescence but to a specific property of uranium.[10]

At the time, Marie Curie was working on her doctoral thesis under Becquerel's supervision. She conducted her research on the rue Lhomond at the Ecole municipale de physique et de chimie industrielles (EPCI), in a damp storehouse turned into a physics laboratory. *Le hanger* (the shed), as the laboratory was known, was directed by her husband, Pierre Curie. Although modest and short of apparatus, the lab provided Marie with enough space to perform her experiments. After the discovery of the new radiation, Marie Curie decided to work on Becquerel rays or uranium rays as they were called at the time. Her aim was to investigate the properties of uranium radiation and a necessary first step was to develop an accurate and reliable method of measuring radiation. Pierre and his brother Paul-Jacques Curie, both prominent physicists working on piezoelectricity and physics of crystals, provided the necessary apparatus.[11]

A method based on photographic plates was unassailable for demonstrating the existence of the new radiation but insufficient for measuring its intensity. The density of the exposure on the film, however, could be used as a crude measure of the intensity of the radiation, but it was not accurate enough. Curie aimed for something better. "Instead of making these bodies [the compounds of uranium] to act upon photographic plates, I preferred to determine the intensity of their radiation by measuring the conductivity of the air exposed to the action of rays."[12] The ionizing property of the radiation had already been used for the case of x-rays to measure their intensity. Despite their wide usage, instruments such as the gold leaf electroscope and the spinthariscope designed by William Crookes were not accurate and precise enough to be employed for the measurement of uranium rays. The need for new instrumentation became more pressing with the discovery of the new radiation. Pierre and Paul-Jacques Curie had already devised an electrometer in the early 1880s based on the piezoelectric effect of quartz crystals. The apparatus measured small quantities of electricity in absolute terms.

In the light of the new and slowly emerging field of radiation physics, Marie employed the Curie electrometer in her research, equipped it with an ionization chamber, and transformed it into a reliable tool for ionization measurements. Carrying an electrometer from crystal to radiation physics, Marie demonstrated that the intensity of the radiation was proportional to the amount of uranium. The new method proved to be superior to the photographic plate method. As Marie conducted research on a number of other substances, she soon discovered that only thorium possessed properties similar to those of uranium. Her hypothesis was that radiation was an atomic property unrelated to its chemical structure. Additionally, pitchblende, raw ore containing uranium, was more radioactive than the amounts of included uranium could explain. Obviously, she concluded, there had to be a new radioactive element in the pitchblende. Her experiments with synthetic chalcolite supported her hypothesis and prompted Pierre Curie to involve himself fully in the study of radioactive substances.

Both Marie and Pierre had approached the problem through physics and had been trained and integrated in that culture. Pierre worked on crystal physics and Marie had been working on the magnetic properties of various minerals under the supervision of the physicist Gabriel Lippmann.[13] Deeply devoted to research, Pierre concentrated on the study of the physical properties of the radiation while Marie performed the radiochemical analyses. Between the two, Marie crossed the boundary of her discipline by using methods from chemistry to analyze pitchblende samples while Pierre kept his identity as a physicist rigid. As Davis argues, "If he [Pierre] tended to concentrate on the physics aspect of the work of radioactivity and she on chemistry, this would seem to have been a matter of personal preference."[14]

I argue that it was more than a simple preference. As Helena Pycior documents, Marie started to work on radioactivity in December 1897 and it was not before late March 1898 that Pierre joined her.[15] Forced by the subject of her research, by the time Pierre got involved, Marie had already integrated chemical methods in the study of the new substances. Her experiments with synthetic chalcolite, the study of all chemical compounds of uranium and thorium, and the testing of all known chemical elements and certain minerals all occurred in this early period. Thus, the core of her own research program required her to continue working on the isolation of new elements.

In July 1898, after using a combination of electrometric methods and chemical analyses, the Curies discovered polonium. Since they were not members of the French Academy of Sciences, they were not allowed to present their results in the weekly meetings of the academicians. It was Becquerel who presented the discovery on their behalf on July 18.[16]

The emergence of the Curies' new research challenged the unity of traditional chemistry. French chemists and academicians soon became uncomfortable, feeling that they might lose control over their discipline. To chemists, outsiders like the Curies seemed to be encroaching on their resources and disrupting disciplinary boundaries. None of the first researchers was a chemist. Marie had primarily studied physics, receiving the Diplôme de licence ès sciences physiques (1893), and a year later, a *licence* in mathematics as well.[17] Pierre had received his *licence* in physics from the Sorbonne in 1877 and in the early 1880s, collaborating with his brother, discovered the phenomenon of piezoelectricity. At the time of their early research on radioactivity both were working in a physics laboratory, training engineers, and "recruiting students from the *écoles primaires supérieures*."[18] Becquerel, who came from a family with a strong tradition in physics, attended the Ecole Polytechnique in 1873. Two years later, he was appointed as a demonstrator at the Polytechnique and then professor of physics in 1895. That year was significant for him because he was also appointed as a professor of physics at the Museum of Natural History, a position already held by two previous generations of Becquerels. In 1889, he was elected to the Academy of Science in recognition of his work in physics.[19]

THE CHEMISTRY OF THE IMPONDERABLE

It was probably because of Becquerel's membership in the Academy of Sciences that French chemists did not overreact. They simply insisted that before the new element could be given any official status, it had to be successfully isolated, its atomic weight had to be measured, and its spectroscopic characteristics analyzed.[20] The chemist Gustave Bémont, Pierre's close collaborator and director of the chemistry laboratory next door in EPCI, joined the team and the spectroscopist Eugéne Demarçay was also enlisted as a collaborator. Bémont offered his expertise on the tedious chemical analysis. In December 1898, painstaking measurements and studies of the properties of radiation led the Curies to the discovery of a second element, the one they named *radium*. Becquerel was once again their representative in the academy. Shortly after, a joint publication by the Curies and Bémont appeared in *Comptes rendus*, announcing the discovery.[21]

It is no coincidence that the publication immediately following that one was Demarçay's "Sur le spectre d'une substance radioactive" (On the spectrum of a radioactive substance"), where he analyzed the spectrum of radium.[22] Marie coined the term *radioactivity* to name the research in radiation physics, rejecting the term *hyper-phosphorescence* used by J. J. Thomson in England as misleading for the nature of the new radiation.[23]

A Nobel Prize in Physics, awarded jointly to the Curies and Becquerel in 1903, affirmed the importance of the new emerging field. In his presentation speech, H. Törneblach, the president of the Royal Swedish Academy of Sciences, referring to the discovery of radium and the findings of Ernest Rutherford and William Ramsay on the release of helium by radium, admitted that they were "discoveries that are bound to be of great importance for the physicist and for the chemist alike."[24]

In the years that followed radium's discovery, there was a constant interchange of ideas, practices, techniques, and new knowledge between physicists and chemists. The contribution of the chemists proved to be essential in deciphering the mysteries of the new element. The French chemist André Debierne, the *chef de travaux* at Marie's laboratory after Pierre's death in 1906, suggested a number of chemical techniques and facilitated the work of physicists employed in the lab. Many young researchers flooded Curie's laboratory, working on the chemistry of the new science, while she was busy perfecting her methods of radium extraction and detecting new radioactive elements. From 1907 to 1914, 58 people worked at Curie's institute, most of them foreigners who came to learn at the source.[25]

In the Anglo-American scene, the situation was similar to the one in France. As Lawrence Badash describes, because some of the experiments required chemical separations of radioelements, the physicist Ernest Rutherford "secured the services of a young demonstrator in the chemistry department, named Frederick Soddy."[26] At the time, Rutherford was still at McGill University in Montreal, Canada, working on the nature of radium emanation. Harriet Brooks, Rutherford's first graduate student and one of the first women in the field, joined their research and in 1900, she successfully identified radium emanation as a radioactive gas with lower atomic weight than radium. The following year, she moved to the Cavendish Laboratory in England to work with J. J. Thomson and later to Curie's institute where she conducted research with André Debierne as a *travailleur libre*.[27]

The traffic at Rutherford's laboratory, however, "was uncertain" as John Heilborn points out and "brought only a half of dozen people into the laboratory during Rutherford's tenure." Among them was a young German chemist, Otto Hahn, who arrived at McGill in 1905 after spending two years working with Sir William Ramsay in England and with an important discovery in his curriculum vitae. Hahn had identified a new radioelement that emitted thorium emanation and which he labeled *radiothorium*.[28] Meanwhile, Soddy had moved to the Chemical Institute of University College in London to join Ramsay's research. Only when Hahn returned to Berlin and was joined by the Austrian physicists Lise Meitner in 1907 did Germany start to be considered one of the most important radioactivity centers in the world.

The transfer of knowledge and its carriers—the physicists and chemists who conducted pioneering radium research—was taking place along with the transfer of the materials themselves. Before Rutherford moved to McGill, he was a fellow in the Cavendish Laboratory. A day or two before he left England in 1898, he ordered uranium and radium salts to be sent to his new address and he was also among the few lucky scientists to receive radium substances from Marie Curie as a gift.[29] Hahn also did not arrive in Montreal empty handed. With Ramsay's consent, Hahn had taken with him the radiothorium he separated from samples of barium chloride and the actinium that he and a collaborator had earlier proved to be identical to the new element that the German chemist Friedrich Giesel had named *emanium*. Hahn's radiothorium and actinium salts also followed him to Emil Fischer's laboratory in Berlin.[30]

Chemical work required large samples of scarce radium and thus materials mattered even more to the chemists than to the physicists. At the time, one of the most important suppliers for small quantities of radium was Friedrich Giesel, manager of the quinine production at the laboratories of the Braunschweig quinine factory Buchler & Co. Giesel used his chemical expertise for the extraction of radium from uranium ore and, starting in 1901, his factory became the first radium supplier. The Austrian Stefan Meyer, the Curies, and even Dmitrii Mendeleev in Russia received their first samples from Giesel.[31] Rutherford was also able to order pure radium bromide from Giesel, taking advantage of a 300-dollar endowment offered by Sir Williams MacDonald, a significant patron of McGill University.[32]

Badash alone refers to an amazingly rich traffic of materials in 1913 when the concept of isotopy was at stake. The German chemist Kasimir Fajans sent his student Max Lembert from Karlsruhe to Harvard to work at the laboratory of the atomic weight expert Theodore Richards, an authority in the field. Soddy, who was also involved in research on isotopes, used Ramsay as the mediator to approach Richards as well. Lembert and Richards conducted control experiments on the atomic weight of ordinary lead to be compared with radioactive samples. The trafficking of radioactive samples was impressive. Fajans provided lead previously obtained from Giesel in Braunschweig; Richards got samples from Ellen Gleditsch, the Norwegian chemist who was then working with Boltwood; Ramsay sought supplies from the Cornwall pitchblende and the British Radium Corporation, a new company that established a factory for producing radium bromide under Ramsay's own directorship.[33]

Despite the fact that some chemists were eager to probe the mysteries of the new radioelements, the more traditional ones saw radioactivity not only as a physicists' intrusion into their field, but more importantly as an attack on the world view and the basic doctrines of their discipline. When in 1896, Mendeleev proposed the periodic table as a pedagogical way to systematize the existing chemical elements, his primary ordering was by atomic weight or, in his terms, the "elemental weights." Each element had a stable, fixed place in the system depending on its mass. Additionally, it was immutable without any possibility of becoming another element and had no substructure. Up to the end of the nineteenth century, Mendeleev's definition of an element prevailed in the chemical community. In his persuasive account of Mendeleev's work, Michael Gording argues that among the three phenomena that brought chemistry under attack, destabilized the entire discipline, and forced Mendeleev to defend his system was radioactivity.[34]

A key step that changed the notion of a chemical element and reshaped chemistry was taken in 1902. Studying the nature of radioactive emanations from thorium, Rutherford and Soddy came up with the theory of atomic disintegration. They concluded that a primary radioactive substance undergoes a series of atomic transmutations; that is: radioactive atoms give birth to a series of atoms of smaller and smaller weights, emitting alpha rays. Therefore, by losing the weight of an alpha particle, an atom would necessarily change species; in other words, one element would become another. This meant that Mendeleev's basic notion of the chemical element was under direct attack. In contrast to the traditional view of a fixed element, a stable state in radioelements could be attained only when the final product was not radioactive. As Rutherford and Soddy argued, "radioactivity is at once an atomic phenomenon and accompanied by chemical changes in which new types of matter are produced. These changes must be occurring within the atom, and the radioactive elements must be undergoing spontaneous transformations . . . the changes in question are different in character from any that have been before dealt with in chemistry."[35]

The Rutherford-Soddy theory prompted a couple of remarkable reactions. "What right had physicists to tell chemists their atoms could disintegrate!" exclaimed the American chemist Herbert Newby McCoy at the news of the disintegration theory.[36] But while McCoy decided to restore the pride of the chemists by showing that there was indeed a definite connection between uranium and radium, with the second being the daughter product of the first in a decay series, Mendeleev sought to dismiss the transmutation theory as similar to the claims alchemists had put forward long ago.

The fact that radioactivity was the result of a spontaneous transformation of elements was indeed seen to be a new chemical "alchemy." In 1905, Mendeleev was invited to attend the World's Fair in St. Louis, but a cataract surgery prevented his visit. "Precisely there," he explained to Clemens Winkler, "I intended to put forth my opinion about the semi-spiritualist state into which they [radioactivity researchers] are now trying to enmesh our science. It behooves us to stop it while we can still act."[37]

On June 6, 1905, when Pierre Curie visited Sweden to deliver the late Nobel award lecture he was unable to present in 1903, he made clear that the "consequences for chemistry of the properties of the radioactive substances are perhaps even more important" than they are for physics. He was not wrong. To Pierre, the disintegration theory was "more far-reaching than the ordinary chemical transformations: that the existence of the atom is even at stake, and that one is in the presence of a transformation of the elements." Marie Curie definitely affirmed radium's impact on chemistry in her own Nobel lecture in 1911. Radium's discovery marked "incontestably" as Curie claimed, a new epoch for chemists and a kind of chemistry that we might well call "the chemistry of the imponderable."[38]

Indeed, the establishment of radium's chemical individuality forced the revision of the notion of a chemical element, allowing the possibility for transmutation, for the existence of atomic substructure and for elements' movement in the periodic table since some of them did not possess a fixed and stable position. Besides revising the chemical theories, new experimental practices and techniques were introduced. Since the presence of radium could often not be detected by the balance or by the spectroscope, novel electrometric methods had to be developed.

The concept of isotopy and the displacement laws put forward by Soddy and Fajans in 1913 completed the disintegration of the traditional notion of an element and of chemistry as well. According to the rule of an alpha emission, a loss of an alpha particle moved a radioactive element two places to the left in the periodic table. The corresponding rule for a beta emission meant a shift to the right in the periodic table; with the loss of a beta particle, an element moves one place to the right. As a result, a box in the periodic table could hold more than one element. This resulted in the fact that the radioelements produced from the emissions had the same chemical properties as their parents, although they possessed different atomic weights. Thus, the most fundamental effect of radioactivity on chemistry was the redrawing of the periodic table in such a way that could fit the radioelements' changing the primary ordering that was now not the atomic weight anymore but the atomic number of the elements.

THE RADIUM INDUSTRY IN FRANCE AND THE AUSTRIAN INVOLVEMENT

From the discovery of polonium in July 1898 to that of radium in December of the same year, it took the Curies a little more than five months to overcome the academicians' concerns about the nature of their discovery. While the French Academy insisted on the isolation and spectroscopic corroboration of polonium, the Curies needed to process enormous amounts of uranium pitchblende in order to extract a few milligrams of polonium. The only available uranium mine was the one in St. Joachimsthal, Bohemia, then part of the Austro-Hungarian Empire. The mines had a monopoly on the uranium used in industry to color the famous Bohemian glass and porcelain.[39]

Since the Curies conducted research in their old, miserable laboratory with insufficient funds and apparatus, they had no chance of approaching the mines directly to ask for a donation of pitchblende. To enlist the support of uranium producers, they had to choose an indirect way and approach those who spoke their parlance. The place to turn to was more or less evident: Pierre and Marie wrote directly to the Kaiserliche Akademie der Wissenschaften (today, the Austrian Academy of Sciences), asking the Austrian academicians to help them obtain 30–100 kilos of residue from the Bohemian mines. "The purpose of this research is exclusively scientific," they emphasized, and thus the academy "would arrange that the administration of the Joachimsthal facilitate our research."[40]

The Austrians had already acquired a reputation as scrupulous and progressive front-line experimenters since the discovery of x-rays. Immediately after Röntgen observed the effect of the new rays and presented his findings to the Würzburg Physical-Medical Society, he sent copies of the report, together with x-ray photographs, to several scientists in Europe. Among them was Franz Serafin Exner, a leading Viennese physicist who received the material on January 4, 1896, including nine x-ray pictures. Amazed by the new phenomenon and persuaded by the impact of observing unexpected images revealing inner parts of the human body, Exner immediately reported the discovery to his colleagues and prompted his advisee Eduard Haschek to work on this topic. Three days later, he also reported the event to the Chemical-Physical Society and his brother, the physiologist Sigmund Exner, broke the news to the Society of Physicians.[41] Ernst Lecher, a young assistant professor from Prague, attended Exner's presentation. Immediately afterwards, Lecher published a report in the *Wiener Presse*, edited by his father.[42] *Die Presse* and the rest of the Viennese press devoted long articles to the discovery, including x-ray photographs.

Around the end of 1896, Sigmund Exner, deeply impressed by the use of x-rays in medicine and physiology, presented an apparatus for the localization of infected areas to the Viennese Society of Physicians. Within the next months, the Viennese clinics anticipated the use of x-rays for medical purposes and various physicians began to use them as a diagnostic tool and to treat skin diseases, tumors, and leukemia.[43]

The Curies thus believed they had good reason to seek the assistance of their Austrian colleagues in acquiring the pitchblende for their research. They were not wrong. Eduard Suess, the director of the academy, sent Pierre a telegram assuring him that the mines would provide uranium-ore residues for free. In his reply on November 19, 1898, Pierre praised "the liberal spirit of Suess's government." A month later, the Curies discovered radium, and in appreciation of Austrian's help, they sent an enriched radium sample, a valuable gift, to the Vienna Academy.[44]

As Marie Curie later admitted "to try to isolate the supposed new element was a great temptation." In their attempt to separate radium from bismuth, both contained in pitchblende, Marie and Pierre found that by crystallizing the chloride of radioactive barium from a solution, they obtained crystals that were more radioactive and richer in radium than the chloride which remained resolved. Repeating the crystallizations several thousands times they were able to obtain pure chloride of radium.

The Curie method, the so-called fractional crystallization, was very time consuming, and demanded laborious operations and gigantic amounts of pitchblende to finally yield a few decigrams of radium chloride per ton of ore. Such a procedure clearly demanded the facilities of a factory with enough space to deposit the ore, treat it, and purify it. "Fortunately for us," Curie claimed, "the curious properties of these radium-bearing compounds had already attracted general attention and we were assisted in our research." Among those assisting were the Société Central de Produits Chimiques, which undertook on the Curie's behalf the treatment of the St. Joachimsthal pitchblende in exchange of part of the produced radium.[45]

Curie's statement supports Xavier Roqué's argument that her involvement with the radium industry was as essential to her career as her scientific pursuits. In 1904, she and Pierre saw the prospect for the financial support of their institute in Emile Armet de Lisle's attempts to market radium salts. The son of a chemical industrialist, Armet de Lisle extended the family's quinine business located in Nogent-sur-Marne outside Paris, by including the manufacture of radium salts. With the Curies' scientific success, he anticipated the opportunity to create a profitable business. He foresaw the possibility of supplying radium compounds to Curie's laboratory as well as the medical potentials of radium.

Armet's factory, Sels de radium (Radium salts), soon succeeded in establishing a steady production of radium salts which were then tested and certified by Curie. The factory also functioned as a resource for some of those who wished to visit Curie's laboratory but lucked financial support.[46] A few years later, in collaboration with the Bank of Radium of Paris, Curie sent Henry Change, an expert in radioactive minerals, to Colorado to purchase the largest carnotite deposit in the United States. After the final negotiations in 1912, Chagneux forwarded to Paris the papers transferring the property to Curie.[47] The fact that historians and biographers have failed to take notice of such connections and Curie's attribution of immense value to the radium industry indicates the lack of concern with the material culture of science.[48]

TRAFFICKING MATERIALS AMONG AUSTRIAN INDUSTRIALISTS, GOVERNMENT ADMINISTRATIONS AND SCIENCE PRACTITIONERS

While French physicists and chemists focused intensely on radium extraction and purification and explored the industrial dimensions of radioactivity research, Austrian scientists did not want to play the role of mere providers of radium for their French colleagues. Their concerns were scientific as well. The physicist Stefan Meyer, a student of Franz Exner, had already expressed a strong interest in probing radium. In the year of its discovery, Meyer acquired a small sample of radium from Giesel in Braunschweig, to attempt to measure the magnetic permeability of radium.

According to Meyer's assumption, magnetic permeability depended upon the atomic weight of the elements. Combining the skills of a physicist and a chemist, Meyer extended his research to some 200 inorganic compounds of all the known elements, including radium and polonium. He soon teamed up with the physicist Egon von Schweidler to conduct experiments on the magnetic properties of the radiation of radium and polonium. Those experiments led them to the discovery of the magnetic deflection of the "Becquerel rays" which become known as beta rays, emitted by radium. They were also able to distinguish them from those emitted by polonium (alpha rays).[49]

In 1901, given the interest of the Viennese physicists to radium research, Suess and other members of the mathematical and natural sciences division (*mathematisch-naturwissenschaften Klasse*) of the academy were persuaded by Exner to set up a commission for "the intensive study of the radioactive substances."[50] The academy agreed with the Ministry of Agriculture to purchase the necessary raw material from the mines in St. Joachimsthal.

At the time, physicists were housed in a makeshift institute at Türkenstrasse, a street closely located to the University of Vienna. However, the building proved to be inadequate for conducting radium research. The limited equipment was not the only obstacle. Theoretically, out of 23,000 kg residue, only 12 grams of radium could be extracted.[51] Such gigantic amounts of pitchblende required a huge amount of space—as the French had already noticed—experienced chemists, and the facilities of an industrial-like laboratory which the small, improvised Physics Institute was not designed for. [52]

It was Karl Auer von Welsbach who eventually offered to the commission the appropriate space in his factory, located in Atzgersdorf near Vienna. Welsbach had studied chemistry and physics at the Technical University of Vienna, worked at the University of Heidelberg as a lecturer on inorganic chemistry, and later served as *Assistent* at the second Chemistry Institute in Vienna.[53] His patent for a new method to produce incandescent mantle out of 99 percent thorium oxide resulted to the commercial production of mantles in 1886. Welsbach directed the production of the impregnated fluid at Vienna's Chemical Institute while the Pintsch Company in Berlin manufactured the burners. In 1887, he acquired the factory Würth & Co. for chemical-pharmaceutical products in Atzgersdorf and moved the entire chain of production there, creating an affluent gas-lamp industry by improving his patent in the early 1890s.

At that time, Ludwig Haitinger was the gas-lamp industry's manager.[54] Haitinger had studied chemistry at the University of Vienna and worked as an organic chemist before he accepted a position at Welsbach's industry in 1887. He had already conducted research on uranium, rare-earth elements, and the technical improvement of the gas lamp at the Chemistry Institute. Although trained as a scientist, Haitinger's role soon shifted to that of the industry's manager. One of his new tasks was to arrange the exportation of Welsbach's gas lamp to the United States. Three years later, he quit his position and returned as an adjunct to the Chemistry Institute. The technician and chemist won over the industrial manager but only until 1892, when Haitinger returned to the factory as director.

Combining the scientific and industrial cultures with expertise in chemistry, both Welsbach and Haitinger appealed to Exner and the Austrian Academy as the right persons in the perfect position. Welsbach owned a suitable, spacious, industry-like laboratory for the extraction of radium and was an experienced, innovative chemist already familiar with radioactive elements, having worked with thorium. Haitinger carried over his skills as an experimenter from the Chemistry Institute to Welsbach's industry along with a dexterity with financial issues back to the domain of science.

In 1901, Suess and Lang drafted a letter to Welsbach on behalf of the Austrian Academy. They proposed to involve him in radium research by using his laboratory either under his or Haitinger's directorship. "Such a research," as Suess and Lang explicitly stated, "cannot be carried out by ordinary aid."[55] Obviously, the work on radium introduced a unique and novel cooperative tone into experimentation as chemists and physicists were forced to collaborate with industrialists. Shortly after Welsbach's agreement, the academy's Commission for the Investigation of Radioactive Substances was finally formed, chaired by Exner, and Suess, with Boltzmann, Lang, and Welsbach as members.[56] The Austrians, like the French, had finally succeeded in assuring radium's traffic between their own science and industry. In 1903, the Austrian Government placed an embargo on the export of uranium and its residues, marking their power in regulating radium supplies and industry worldwide.[57]

After ensuring adequate space for the treatment of pitchblende, the next step for the Austrian Academy was to ensure ample amounts of radium for research. On January 15, 1904, Suess and Lang addressed the Ministry of Agriculture, asking for 20,000 kg residue. As they argued, "For many years, no phenomenon has affected the scientific world in such an extraordinary way as the observation of the strange appearances of radium, a substance that is according to phenomena, an inexhaustible source of light and heat, and thus seems to contradict fundamental assumptions of today's physics." The same day, Suess and Lang emphasized the importance of radium research to the Ministry of Culture and Education. To strengthen their argument, they focused on the Nobel Prize awarded to the Curies for their discovery of radium and the crucial role that Austrians played by providing the radioactive material. They did not fail to mention that the French Academy had economically supported the Curies with 155,000 francs.[58] Between the lines, Suess and Lang implied their own demand for financial support.

While they emphasized the need for a "parallel, up-to-date line" of research in Vienna, comparable to that in Paris, Suess and Lang also negotiated low prices for the pitchblende with the Ministry of Agriculture. Both research groups, in Paris and Vienna, needed 4 grams of radium, which meant at least 10,000 kg of residue apiece. Since the annual production of the mines was only 6,000–7,000 kg of residue, it might have been more than two years before such quantities were available. The negotiation was not an easy one. To persuade the government, the Austrian Academy affirmed that the radium would be used for strictly scientific reasons and that the research was innovative and absolutely important.[59]

The deal was eventually closed in March 1904. The mines provided the academy with 10,000 kg residue in two parts and received 8,040 kronen in payment. Welsbach's industry was well compensated for the space provided and for the chemical elaboration of the residues. In 1904, Welsbach received 3,064 kronen and after the extraction of the radium, he received an additional 6,121 kronen. Part of the payment came from a donation that Haitinger made to the academy in the form of a prize for scientific research in his father's memory.[60]

Working painstakingly for two years, Karl Ulrich and Haitinger extracted 4 grams of radium bromide out of 30 tons of pitchblende by transferring laboratory methods to the factory and extending their practices to industry. At the same time, they shifted the culture of science from an enterprise focused on teaching and modest experimentation to a modern, relatively expensive research practice. Eventually, the extracted radium constituted the main radioactive material used at the Institute for Radium Research. "With it," Meyer later recalled, "had been laid down the basis for systematic research in this field in Austria," and, as Exner acknowledged, it elevated the institute to one of the richest centers in radioactive materials.[61]

The Curies, of course, were regular customers for pitchblende. In addition, the Royal Society of London placed a similar request on May 1, 1904. The English were ready to buy the portions of residues that were not set apart for the Curies.[62] Three years later, while both William Ramsay in London and Rutherford in Manchester were involved in radioactivity research, the Austrian Academy of Sciences loaned 350 mg of radium bromide to the English scientists after their persistent requests. Highly respected as he was, Meyer played the role of the mediator. The collaboration of Ramsay and Rutherford soon proved to be unfortunate since the first insisted on keeping the radium for one and a half years, making only radon available to the latter. A second loan in 1908 of 170 mg of radium as 300 mg of radium bromide was sent exclusively to Rutherford for his own experiments.[63]

While English and French scientists had already embarked fully on radioactivity research, Emil Warburg, president of the Physikalisch-Technische Reichsanstalt, a physics institute in Berlin, was trying to transform his laboratory into a leading scientific center. As David Cahan notes, by the start of Warburg's presidency in 1905 the institute was not able to meet the demands of contemporary science. Radioactivity was a promising field and Warburg knew the place to start. "I now intend to carry out several works on radioactivity," he wrote to Exner in 1909, "and allow me the following question, whether it is possible to let us have a loan of a small quantity of radium bromide for this purpose." As it is documented in one of the academy's reports, the Commission for the Investigation of Radioactive Substances "had already provided radium to a large number of institutes in France and Germany" first for free and then for moderate prices.[64]

By the end of the first decade of the twentieth century, radium had become a trafficking material used by both chemists and physicists, threatening the disciplinary unity of both and blurring their boundaries. For each discipline, radium carried a different identity as it did for the various institutes involved in radioactivity research. The discovery of radium provided the Curies with the prestige they both were lacking when they worked at the EPCI. They also acquired new space and apparatus to continue their experiments. Warburg used radium to transform his institute in Germany into a leading scientific center. For the English also, research on radium was a way to express their aim to maintain their leadership in physics. For the Austrians, however, radium meant much more.

After the painstaking work of separating radium from pitchblende, Ulrich gained the directorship of the St. Joachimsthal mines, acting as the key link between the Bohemian industry and his colleagues in Vienna.[65] Auer von Welsbach, with a serious commitment to both science and industry, was challenged by the opportunity to both contribute to fundamental research and seek financial benefits from new industrial achievements. By using radium as a bargaining tool, the ministries of agriculture and education of the dual Austro-Hungarian monarchy located in Vienna exercised their political strength over the other half of the empire and, after 1903, Austrian academicians such as Suess and Exner proved their administrative and scientific talents by serving as mediators among the Austrian ministries in charge, their international colleagues, and the directors of the Bohemian mines.[66]

Exner and the Austrian physicists foresaw in radium research the emergence of an exciting scientific field. Thus, to simply administer the radium sources in St. Joachimsthal and play the role of radium merchants was not part of their goal. As Suess and Lang argued, they had "a kind of moral obligation to the whole scientific world" to pursue research on radium.[67] Although not explicit, radium actually meant much more to them. As we will see in the next chapter, radium eventually became the vehicle for obtaining a new physics institute appropriate to the prestige of the Viennese physicists.

RADIUM IN THE CLINIC

The discovery of radium brought Austrian physicists into a kind of disciplinary interchange that they had not anticipated. Up to the end of the nineteenth century, although physicists and chemists had been offering their knowledge to medical and pharmaceutical students in the big academic lecture halls, their cooperation ended at the doors of the infirmaries and surgical rooms. Physics and chemistry laboratories were widely open to physicians for educational reasons, but the medical clinics and operating rooms were mainly closed to physicists. However, when it came to using radium for therapeutic purposes in important medical centers in Vienna's fin de siècle, the only way to bring it into the medical facilities required opening the doors to physicists and chemists.

The introduction of radium into the operational rooms and cancer centers, involved the intrusion of physicists into the realm of medicine and the constant trafficking of radioactive materials between physics laboratories and sites of medical practice.

Friedrich Giesel, the German industrial chemist, was the first to observe that radium had biological effects as well. By deliberately applying radium barium bromide to his arm, Bequerel confirmed Otto Walkhoff's observation that radium causes skin wounds. Bequerel and Pierre Curie repeated the experiment and immediately published their results. A picture of Curie's injured forearm after his self-exposure to radium appeared in a French newspaper, bringing the news to the general public.[68] The above findings intrigued the medical practitioners. Given the effects of radium on human tissues, many physicians became excited about the possibility of using radium in the treatment of cancer. Pierre was the first to substantiate the interrelation of physics to medicine through radium. He offered a tiny amount of the valuable element to Henri Danlos, a dermatologist at the Saint-Louis hospital in Paris, who put it to good use by treating some cases of lupus.[69]

The development of radium therapy in France was astonishingly rapid. Without delay, the radium industries financed the creation and operation of three institutes specializing in this activity. In 1906, Armet de Lisle financed the establishment of the Laboratoire biologique du radium in Paris and three of the most important French radiotherapists, Louis Wickham, Jacques Danne, and Henri Dominici, became associated with the lab. The Laboratoire included research labs for pathology, physiology, physics, and a combined clinical therapy and medical research lab. De Lisle supplied radium from his own industrial refinery in Paris and received a portion of the fee charged to each patient.[70]

In 1912, the physician and industrialist Henri de Rothschild founded a radium therapy department in his hospital, also in Paris. The same year Jacques Danne initiated the development of the Laboratoire d'essais des substances radioactives in Gif, outside Paris. The apogee of the institutionalization of radium therapy was reached in 1914 with the establishment of Curie's Institut du radium which included two laboratories. One laboratory was dedicated to physics research and the other to biological and medical research. The combination of physics and medicine both housed in the Institut du radium, facilitated perfectly the transfer of materials, instruments, experts, and knowledge. At the same time, industry was closely bound to physics, chemistry, and medicine. [71]

France had been the pioneer in radium therapy, but it was not the only one. In the following years, a considerable number of radium therapy centers were founded throughout Europe. In 1908, the first center for cancer was established in Stockholm at the university clinic of the Karoline Institute, where the gynecologist James Heyman developed his method of radium treatment of cancer of the uterus. On August 1, 1910, the Radiumhemmet (the Radium Home), a makeshift clinic started by two Swedish doctors, was also founded in Stockholm based on a small donation of 40,000 kronor and 120 mg of radium supply. By 1917, the rapid success of the clinic forced the City Council to provide generous funds and a new well-resourced building.[72] In 1910, a similar center was established in Heidelberg by the physician Vinzenz Cerny and in New York at the Memorial Hospital.

In 1911, King Edward VII of England, following the previous examples, sponsored the Radium Institute in London after his own successful treatment for a lesion on his noise. In Manchester, Rutherford's own city of residence, William Milligan an otolaryngologist, established a radium laboratory at the Manchester Royal Infirmary in 1914. Rutherford was asked to offer his advice on radium sources while his laboratory monitored the radium supplies and their quality for the newly established medical facility.[73]

At the General Hospital in Vienna, a radium department was set up with half a gram of radium available in 1912. A year later, a radiological institute was founded at the Vienna Lainz hospital where Heyman's methods were soon used for treatments of cancer. In Denmark, the first radium medical center was established in 1913, with King Christian X as its patron. Interestingly, he bought the valuable element using the money dedicated to the purchase of silver wreaths for King's Frederic VIII's recent death.[74]

During the 1910s, biologists, physiologists, and clinical practitioners around the world explored the effects of radium on a number of different human tissues, the nervous system, the eye, as well as on plants and animals. Working on radium, Viennese physicians tied their culture as medical practitioners to that of physicists. They were forced to learn the physics of the new element, to study the way radiation is absorbed by the human body and its dosimetry, and to find the appropriate ways to measure it. Sigmund Exner brought radium research to physiology by studying its effects on animal tissues. Alfred Exner and Guido Holzknecht were the first Austrian physicians to probe the pathology of radium dermatitis. Holzknecht was already recognized as a pioneer in diagnostic radiology and, in 1902 he developed the first instrument to measure x-rays. His device, the chromoradiometer, attracted the interest of the international community of physicians and was tried as a dosimetry apparatus for radium as well. In June 1903, Holzknecht joined Gottwald Schwarz, later director of the Röntgen station at Elisabeth-Spital, in a study of the possible atrophy of the optical nerves due to radium. Soon after, their article appeared in the *Mitteilungen* of the Society of Physicians.[75]

To physicians working on radium therapy, the most astonishing feature of their experimental treatments was their success rate. This success meant that physicists started to dominate the health field. The early methods of radium therapy required the close cooperation of physicists and physicians. The devices were crude and consisted of two main kinds: a) the flat applicators (radium plaques) which were flexible or rigid and designed for external use and b) the spherical ones used in endocavitary applications. They contained a quantity of radium proportionate to the area for treatment. Later on, the enormous cost and lack of radium supplies forced doctors to consider the use of radium emanation instead. Physicists, responding to physicians' needs, designed more sophisticated devices, such as steel needles containing capillary glass tubes filled with radon and treatment was based on needle puncture.[76]

Given the lack of a suitable method for standardizing radium preparations, the expanded use of radium in cancer therapy necessitated closer relationships between physicists and physicians. Physicians were in desperate need of accurate methods for radiation measurement. Their major concern was the precise description that could ultimately lead to successful duplication of their work. It was the physicists who had the expertise and the instrumentation to do so.

In 1910, during its meeting in Brussels, the Congress of Radiology and Electricity decided to appoint a committee for the preparation of an International Radium Standard. Mme. Curie was asked to prepare the original standard and the Austrians agreed to produce a similar one to be compared for accuracy reasons. Having the monopoly of the Bohemian uranium mines, the Austrian physicists were in charge of radium supplies in Europe and regulated the radium prices. For example, by 1908, the Austrian Academy of Science had already provided radium to a number of hospitals such as the general hospital and university clinics in Vienna and the university clinic in Krakau.[77]

However, the traffic of the material was both ways. On the one hand, hospitals' demands for radium increased rapidly with the establishment of radium laboratories within their facilities and thus also increased the dependency of medical practitioners on the physicists. On the other hand, given radium's astronomical price, the physicists were glad to reuse the spent radon needles received from hospitals for their own research. It was becoming apparent that physicists and medical practitioners were more and more in a mutually dependent relationship. It is indicative that in the process of organizing the second Congress on Radiology and Electricity, Rutherford admitted to Meyer that "we have very few medical representatives in the list, and this is invidious." However, on June 29, 1914, while Rutherford was arranging the last details for the meeting, he did not anticipate the outbreak of World War I in August 1914.[78]

RADIUM AS A MEDICAL COMMODITY

In early-twentieth-century American society, radium was deployed in an environment of strong racial discrimination. As early as 1904, a chemist at the University of California designed a series of experiments using radium to "bleach the negro" and turn his skin white. Radium combined with x-rays was tested for its effects on the coloring cells of the body. The case made headlines: "burning out birthmarks, blemishes of the skin, and even turning a negro white with the magic rays of radium, the new mystery of science."[79] The radium craze combined with the eugenics movement of the 1920s in the U.S. made possible articles dealing with issues such as "sterilization in the interest of race betterment" to sully radiology journals and present radium as a commodity ready to serve the medical world.[80]

The link between radium's "magic" power and its commodification as a medical item was reinforced by doctor's public claims of its therapeutic effectiveness. As C. Chase reported, over 80 articles on radium therapy had appeared in the North American literature by 1906. Medical doctors created lengthy lists of conditions that apparently benefited from radium treatment and famous scientists such as Frederick Soddy suggested radium as a great therapeutic agent.[81] Worried by physicians' concerns about radium, Rutherford wrote to Stefan Meyer: "I hope you will be able to reserve a large quantity [of radium] for experimental purposes, otherwise I am afraid it will all go into the doctors hands." [82]

The potential of radium in medicine and as a cure for cancer drove doctors to investigate the new element. Exemplary was the suggestion of Sir Alfred Pearce Gould, the famous cancer specialist at Middlesex Hospital in London, to purchase radium as a gilt-edged investment for hospital funds. At a hospital board meeting in November 1913, Gould urged his colleagues to buy radium as there was no more permanent way of investing money for the future. He added that "as to the use of radium I must speak guardedly. It is of the greatest importance in cancer case."[83]

Industrialists seemed to have had a similar view about radium's potentiality as a valuable commodity. The Standard Chemical Company, one of the largest U.S. commercial producers of radium, was established in 1911 by Joseph Flannery, a successful Pittsburgh businessman. Although he had no scientific or technical knowledge, Flannery set up a lucrative radium extraction plant in nearby Canonsburg, Pennsylvania. He sought to refine radium from ore mined, sorted, sacked in, and transported from Colorado mines. A barium-radium chloride solution was then transferred to a refinery laboratory in the Oakland section of Pittsburgh. The refinery employed 15 technicians in addition to 150 workers in Canonsburg who performed the first radium process. Bankers' investments, chemical technology, and Flannery's business genius were all decisive. Production relied on around

15 chemists whose laborious work and precision laboratory analysis yielded just one gram of radium every five weeks. Its price, however, was 120,000 dollars, a value at the time greater than the weight-based cost of the Hope diamond.

As a final commodity radium lived several lives designed by employees who were engaged in sales, clerical, and management activities at the company's offices in Pittsburgh.[84] Radium showed up in quite a few medical products designed and promoted by the Standard Chemical Company. Containers for small radium needles made with a mix of nickel, copper, and zinc (but not iron or lead which could have been the safest for transporting radium) were placed in a small velvet-lined leatherette case, carrying the company's stamp. Later, steel cases replaced the older ones and were designed to store radium plaques. Lead-lined boxes containing radium salts, radium needles, radium solutions for drinking and intravenous use, and radium compresses for pain relief proved the "unquestionable value of radium." Standard radium emanators for treating two people and even a portable office-style emanator for treating one person were designed, produced, and sold by the company. The emanator was an unusual device intended to be placed overnight in water so that a ready supply of radioactive water would be available the next day. Its functioning principle was based on the fact that radium emanation or radon when inhaled had therapeutic properties. During the 1920s, several other companies such as the Radium Life, Inc. of Los Angeles, the Radium Company of Colorado, the Radium Chemical Company of New York City, and the Canadian Radium and Uranium Corporation controlled the radium market and supplied hospitals with all kinds of radium applicators, needles, and radon seed implanters and capsules, all medical technologies used in cancer therapy.

The constant trafficking of the valuable commodity among industry, science, and technology was best elaborated in France. Radium was what Marie Curie and the industrial chemist Armet de Lisle certainly saw as the most important new commodity. While Curie used industry to serve her physical and chemical investigations of the new element, de Lisle used science and technology to extend the radium market to medical applications. "Indeed he became one of the most important protagonists of radium therapy in France" writes Soraya Boudia.[85] de Lisle's radium factory became the major supplier of medical technologies in France during the 1910s. Several apparatus were advertised through the journal *Le Radium*, mainly devices for the internal treatment of cancer.

In Austria, medical consumption of radium was intimately connected to curative baths. In June 1904, Heinrich Mache, Exner's student and *Assistent*, conducted the first study of the water at the famous health spas in Gastein. Bubbling air through bottles completely filled with the Gastein water, Mache obtained emanation that he soon concluded was due to radium minerals in the earth's crust. Within the next three months, he completed his experiments, giving a much fuller account of the amount of radioactive emanation in the water. Soon thereafter, Meyer joined Mache and they expanded their investigations to other springs such as Carlsbad, Marienbad, Teplitz-Schönau-Dux, Franzensbad, and St. Joachimsthal. Their results varied, but the St. Joachimsthal water proved to contain the largest amount of emanation found in any spring water. Given that the radium content in the ground was high, Mache's earlier conclusion proved to be right.[86] The presence of radium emanation, later called *radon gas*, in the spring water explained its therapeutic properties. In 1904, Mache proposed a measurement unit for the concentration of radon in water that took his name.[87] Both the scientific world and the popular culture embraced radioactive emanations as miraculous remedies, gifts to humanity from the foremost geniuses of an inventive age.

As the *New York Times* reported in 1911, public opinion forced the government to assume control of the exploitation of these natural sources and establish a curanstalt, a therapeutic institution accessible to all classes and open year-round. The Austro-Hungarian monarchy subsequently asked for the supervision of Mache and Meyer. The curanstalt was operated by pipes drawing the water directly from the mines. The water was filtered without losing anything of its radioactivity and directed to the baths. A luxurious bath was then built and inhalation apparatus were designed. Visitors came from near and far to soak in the supposedly therapeutic water and inhale the air. Patient prescriptions were given in Mache units. As Meyer claimed later, concerning the medical uses of radium, "Austria was in this respect the best country by means of research." [88]

The interest in radium for medical use was so great that radium preparations and apparatus to produce radioactive water quickly became commercially available in Vienna. For example, the Radiumwerk Neulengbach, a commercial supplier of radium preparations, opened a store near the Physics Institute. They advertised their products as being as radioactive as the natural springs and effective against a long list of medical problems such as neuralgia, neurosis of the digestive organs, and malignant neoplasm.[89] Radium had become firmly rooted in everyday human life as anyone could go to the local store to buy a radium emanator, a radium ointment, or a radium compress.

RADIUM AS A COMMODITY

"When we [Mache and Meyer] visited it [the city of St. Joachimsthal] in 1904, the city was up to the spirit of the time. There was a variety of 'roasted meet with radium,' (*Radiumbraten*) 'radium soap,' 'radium cigars,' 'radium cigar holders,' 'radium pastries,' etc." So wrote Stefan Meyer, describing his visit to the city of the uranium mines in Bohemia. "One owner of a guesthouse called it *Radium-Gaststätte*. This irritated the competitors. Hence another one called his own *Zur Emanation* and he ordered the printer to imprint this name on the labels of cigar holders. He wanted to distribute them as a way of advertising his guesthouse. But the word was strange to the printer and thus he printed the most familiar to him, *Zur Imitation*."[90] Maybe the Bohemian printer was unfamiliar with the word "emanation," but by the mid-1910s, hardly a person either in Europe or in the United Sates was unfamiliar with the word "radium."

A very rapid expansion in manufacturing and consumer products that contained radium was one of the most dramatic phenomena of the early twentieth century. Radium *cleanser,* for example, was a sprayer that supposedly killed flies, mosquitoes, and germs as well as cleaned furniture, paintwork, and porcelain. Radium Enamel, manufactured by J. L. Prescott Co. of New York, was a radium product for shining metal surfaces, and Radium was a boot polish made in England by Radium Polishes Ltd. In Germany, chocolate bars containing radium were sold as a "rejuvenator." Radium-brand creamery butter, keyholes and chains, and toys and military equipment coated with radium flooded the market. A trade radium preparation called *Nirama* was used as a fertilizer to increase the speed of growth in potatoes, flowers, and other plants. The miracle of science was being revealed in mundane things. At the 1904 banquet of the New York Technology Club, well-respected New Yorkers toasted with liquid glowing radium cocktails. Radium roulette, a New York rage, appeared the same year. Participants gambled and played with wheels, balls, and chips all layered with luminescent radium paint.[91]

The radium products that Meyer saw in St. Joachimsthal were another indication of the public's radium obsession that was decisively shaping both the European and American cultures during the 1910s and 1920s. Even the radium pastries in St. Joachimsthal bakeries contained minute quantities of radium. Saved envelopes describing the production of bread with radium water by Hippmann-Blach, a known bakery in the city, attest to the habits of the day.

Radium-brand cigars were produced by the A. Batschari tobacco company in Baden, Germany, and later also by the Polus tobacco company in Balerna, Switzerland. Radium was added to toothpastes such as Doramad, a German brand produced by Auergesellschaft in Berlin. Advertising and product labeling persuaded believing consumers. The toothpaste label indicates that "its radioactive rays" strengthen teeth and gums, providing "a new, pleasant, mild, and refreshing taste." In the brochure for the promotion of Doramad, under the title, "What is biologically important?", a joyful woman with a shiny smile stares at her reflection in the mirror.

Both women and men were the targets of radium advertising but for entirely different products. Men of that era were introduced to radium condoms and were promised a better sexual life. "The best radium finest seamless male pouches," moreover, "guaranteed for one year" appeared in the U.S. around the 1920s, and Radium Nutex, another condom brand came out sometime later. The most impressive product attracting "weak discouraged men" that could now "bubble over with joyous vitality" appeared around the end of the 1920s and was manufactured by the Home Products Company of Denver in Colorado. Vita Radium Suppositories promised to restore men's sex power because, as the advertisement argued, "a man must be in a bad way indeed to sit back and be satisfied without the pleasures that are his birthright." Soluble radium was added to a cocoa butter base in the form of a suppository and was introduced into the rectum in order to stimulate "the weakened organs that needed its [radium's] vitalizing aid.[92]

In the 1920s, men were also advised to use the Radiendocrinator, which was "the last word in scientific manufacture," as the American Endocrine Laboratories assured the customers of this aphrodisiac product. The 150-dollar gold-plated Radiendocrinator consisted of seven radium-soaked pieces of paper about the size and shape of a credit card, covered with a thin piece of clear plastic and two gold-wire screens placed in a beautiful, dark, embossed leatherette case. It was fashioned by William Bailey, a Boston swindler who promoted his product as appropriate for "sexual rejuvenation." This scam item was supposed to restore and enhance the male sex drive when it was worn as an athletic strap under the scrotum during the night.

However, Bailey's most amazing product of all was Radithor, a 1920s radium tonic which contained more than 2μCi (74kBq) of radium per bottle. Produced by concentrated radium and mesothorium fluid diluted with distilled water, Radithor was packaged in half-ounce bottles and sold by the case of thirty. Between 1925 and 1930 more than 400,000 bottles were consumed by men who wished to restore their virility. One of the first victims of Radithor was Eben MacBurney Byers, a wealthy steel manufacturer from Pittsburgh and well-known sportsman and playboy who died of radium poisoning in 1932. At the recommendation of his doctor he averaged three bottles of Radithor a day for two years. [93] His case was decisive in the passage of the Food, Drug and Cosmetic Act in the U.S. Congress in 1938 which mandated premarket approval of all new drugs and cosmetics.

As women's sexuality was not yet at the top of any industry's interests, early advertisements that targeted women promoted mainly radium cosmetics. The information that caught the attention of many women as early as 1904 was the claim that "radium and x-rays [can be] used to beautify." Radium was viewed as a godsend panacea that was able to safely remove even birthmarks.[94] During the 1920s, beauty was engineered in the intersection of fashion, science, and business. Social trends and "scientific" innovations fueled by appropriate forms of mass advertisement resulted in the production of a series of contraptions and quark medicines that women willingly used to captivate the male eye. Extravagant claims for the aesthetic benefits of radium products linked women's beauty to the uses of radium lipsticks, face creams, and powders. Beauty products such as Radior soaps, face creams, and lotions were guaranteed to contain "a definite amount of Actual Radium—nature's greatest aid to Beauty." Already in 1918, several Radior toilet requisites were produced by Radior Co. Ltd. of London and sold in Fifth Avenue shops in New York. Since there were no Food and Drug Administration (FDA) regulations for premarket approval of new cosmetics at that time, Radior products jumped right into the sales market. Full page advertisements in major New York newspapers argued that, "It is easy to prove the superiority of 'Radior' preparations. *Try them*. See how smoothly the vanishing cream rubs in." The 5,000-dollar guarantee the company promised to the unsatisfied customer was not enough to relieve women's radiation burns, scarring, and skin cancer that soon appeared.[95]

In the early 1930s, this time in France, a Parisian doctor, Alfred Curie, and a pharmacist, Alexis Moussali, epitomized the beauty trend in the industry and launched a cosmetic brand Tho-Radia "for pharmaceutical products, beauty products and perfumes." A publication in *La Revue des specialties* that appeared immediately after its launch, argued that the cream, with a base of thorium and radium, had great success in Paris with promising curative and beautifying properties. The company's advertisements spoke of a "scientific cream" which was promoted and sold to the entire country thanks to the public enthusiasm. Besides the cream, Parisian women had a variety of choices among powders, lipsticks, and toothpastes, all Tho-Radia products. [96]

While mass advertisement shaped women's consumerism and suggested beauty innovations, new technology created the means. In the late nineteenth century, more than 1,000 patents were used for new corset designs, a technological enhancement that supported women's bust from below. In the early 1930s, the radioactive corset was added to the rubber, electric, and magnetic corsets. It was called the *Shilouette Radiante* and was manufactured by Shilouette, a German company with subsidiary companies in London and Paris. In 1937, the partner from the Paris office obtained a patent for the new radioactive product and Shilouette launched in London as well. Its immediate success was due to the fact that it had a "stimulating, even rejuvenating influence on the cells of the human body, aids fatigue, warms the body, and helps rheumatic pain." [97]

The combination of scientific authority, mass advertising, and commercialization of women's beauty products enforced the use of radium as a highly valued commodity. Radium dial painters of the 1920s, the women who applied radium paint on watch dials in several factories in the United States, had direct access to the material and grabbed the chance to use the sensational product at no cost. While artificial phosphorescence was a novelty at the time, women applied radium directly to their buttons, fingernails, faces, and eyebrows during their daily shift. "And I can remember, too," revealed one of the dial painters to a reporter of the *New Yorker* "a lively Italian girl at the plant who painted her teeth with the stuff [radium paint] whenever she had a date, so that they'd shine in the dark."[98] The luminous paint, however, sentenced the painters to a slow and painful death as all of them pointed their brushes between their lips to give the brush an exceedingly fine point and speed up their work. Radium dial timepieces, clocks, and especially watches were popular with both the civilian and the military populations, so the women's work was important to both the U.S. army and navy and those corporations involved in the production. While the radium industry was boosted, women's daily radium ingestion resulted in painful deaths and the most infamous case of radium poisoning in the history of industrial occupation accidents. Commerce, war, and the strong belief in science and progress helped turn radium into a valuable thing. Moreover, commodity culture selectively effaced cases of radium beauty horrors and radium poisoning from the mass memory.[99]

TRAFFICKING MATERIALS, CONFLICTED IMAGES

Trafficking materials have the precarious status of having an identity in constant transition. Throughout the early twentieth century, radium embodied several, sometimes conflicted, images to those who produced it, traded it, studied it, and consumed it. In the eyes of the scientists, radium represented a strategy to organize their work place, a means to make a name in the competitive world of science, or a medium with which to reveal a piece of the natural world. To industrialists, radium coincided with the stability and prosperity of their business or was connected to their ambitions of being part of the modern world filled with radium refreshments, aphrodisiac radium products, toiletries, and cigars. To the physicians, radium became inseparable from their identity as progressive practitioners; it forced them to cooperate with physicists and technicians and to trade with industrialists and companies producing medical radium artifacts. To the public, radium reflected a myth, the powerful, often gendered, myth of the interrelation of science, technology, and medicine. Mass advertisement and popular images of radium enforced that myth. Take for example the following story.

In Vienna on May 30, 1935, a female journalist from *Neues Wiener Journal* paid a visit to the Radium Institute to see "seven young Viennese women disintegrate the atom." Although she understands nothing of radium research the journalist has an interest or a "preference" for it, because, as she puts it, radioactivity is a science where every month, there is a new discovery one reads about in the newspaper. Indeed a few months later the *Neues Wiener Journal* published an article on American efforts to light the streets of the cities and highways using radium.[100] The anonymous journalist further argues that radioactivity is "a science that was established by a woman and thus it is romantic. . . . romantic for the amateur."

In this clear association of romanticism with the presence of women in the field, one can track the public image of radium: a material associated with "the personality of its discoverer, a courageous French lady who was more than a match for a generation of male physicists." Curie's myth coupled with the 1920s myth of radium as a powerful commodity in the service of women, resulted in a positive popular image for the whole discipline. As a romantic endeavor, radioactivity suited women who "remained women despite their work," and they never lost their femininity, as the journalist discovers. Wondering around the institute's laboratories, she comes to a room with the name of a Frau Doctor on its door. The place looks like a witch's kitchen on the stage of *Faust*. Vessels, glass tubes, and apparatus that are out of her reach were laid around. "On a hook on the wall is hanging an elegant Trotteur-hat and an English, very stylish lady's coat. From this, I see that the Frau Doctor, who disintegrates the atoms and searches for radium in this room, must be a very nice woman; despite the glass tubes and vials, she follows the fashion" the journalist affirms with surprise.[101] Similar to the case of Joliot-Curie who had visited the Vienna Institute with her husband a couple of months previously, those women seem to fit a similar model; they are equally women, mothers, partners, and scientists.

However, as we will witness in the chapters that follow, radium research was far from a romantic endeavor and far different from its public image for most of the women who worked at the Radium Institute, especially toward the end of the 1930s. The point is that radium was never one object, shared among different intersecting worlds. It was always already a different material embedded with different meanings, depending on the hands that held and manipulated it.

NOTES

1. Curie, M. "Radium and Radioactivity" (1904), 461.

2. Curie, Pierre. 1905. "Radioactive Substances, Especially Radium" Nobel Lecture, June 6, 1905, http://nobelprize.org/nobel_prizes/physics/laureates/1903/pierre-curie-lecture.html.

3. Appadurai, "Introduction: Commodities and the Politics of Value" (1986), 5.

4. Himstedt, "Radioactivity" (1906).

5. Hahne, Frank. 1989. "Early Uranium Mining in the United States" in World Nuclear Association: Energy for Sustainable Development, http://www.world-nuclear.org/reference/usumin.html.

6. I am directly referring here to Leigh Star's and James Griesemer's notion of boundary objects, objects that are shared among several intersecting worlds and have different meanings in each of them, sustaining though a structure common enough to make them recognizable means of translation. (Leigh and Griesemer, "Institutional Ecology, 'Translations,' and Boundary Objects" (1989)). The notion of boundary object, although proved to be extremely useful in several STS studies, does not capture radium's shifting identities between different worlds and the exchange of knowledge and expertise that goes hand in hand with the trade of the material. The focus in the case of the boundary object is on the boundaries that are traversed. The focus here is on the actual trafficking of materials and the culture that supports the traffic.

7. In her 1911 Nobel lecture, Curie claimed that "Although radium has so far only been obtained in very small amounts, it is nevertheless true to say, in conclusion, that it is a perfectly defined and already well-studied element." The opposite holds for polonium because at the time, scientists were unable to separate it given its low proportion in minerals and its short half life. (Curie, Marie. "Radium and the New concepts in Chemistry" Nobel lecture, December 11, 1911 http://nobelprize.org/nobel_prizes/chemistry/laureates/1911/marie-curie-lecture.html).

8. I owe my focus on materials to Ursula Klein and her interesting work on "The Making of Materials," a work in progress which involves a workshop co-organized with Emma Spary at the Max Planck Institute for History of Science in Berlin in 2004 and a forthcoming edited volume, which includes papers by Pamela Smith, Seymor Mauskopf, Augusti Nieto-Galan, Barbara Orland, Ernst Homburg, Christoph Bartels, and Marcus Popplow. My sincere thanks to Klein's generosity for sharing with me the papers of the workshop.

9. Kohl, "Von den 'Becquerel-Strahlen'" (1997), 489. Dutreix and Dutreix, "Henri Becquerel" (1995) 1870.

10. Becquerel "Sur les Radiations Invisible" (1896), 501–03. For more on Becquerel's experiments, see: Kohl, "Von den 'Becquerel-Strahlen'" (1997) 488–491; Peh, "The Discovery of Radioactivity" (1996), 627–630. Blaufox, "Becquerel" (1996), 145–154; Dutreix and Dutreix, "Henri Becquerel" (1995), 1869–1875.

11. Mould, "The Discovery of Radium" (1998); Peh, "The Discovery of Radioactivity" (1996).

12. Curie, M. "Radium and Radioactivity" (1904), 462.

13. Mould, "The Discovery of Radium" (1998), 1233; Pycior, "Marie Curie" (1997), 31–50.

14. Davis, "The Research School of Marie Curie" (1995), 323.

15. Pycior, "Reaping the Benefits" (1993), 304.

16. Mazeron and Gerbaulet, "The Centenary of Discovery of Radium" (1998), 207. The original publication of the Curies' discovery appeared in *Comptes rendus*, "Sur une substance nouvelle radioactive" (1898), 175–78.

17. Mould, "The Discovery of Radium" (1998), 1233.

18. Davis, "The Research School of Marie Curie" (1995), 324.

19. Blaufox, "Becquerel" (1996), 145.

20. For the reaction of the French chemists, see Peh, "The Discovery of Radioactivity and Radium" (1996), 628.

21. Curie, M. and P. and Bémont, "Sur une nouvelle substance," (1898). See also Peh, "The Discovery of Radioactivity," (1996).

22. Mould, "The Discovery of Radium," (1998), 1238.

23. Pycior, "Reaping the Benefits" (1993), 305; Davis, "The Research School of Marie Curie" (1995), 328.

24. Törneblach, H. R. Presentation Speech, The Nobel Prize in Physics, December 10, 1903, http://nobelprize.org/nobel_prizes/physics/laureates/1903/press.html.

25. Davis, "The Research School of Marie Curie," (1995), 333.

26. Another chemist and Rutherford's close friend, Bertram Boltwood, centered his work on the chemistry of radioactive elements, assisted later by the Norwegian chemist Ellen Gleditsch. Boltwood was actually the first to obtain a university position in radiochemistry in 1910. Badash, *Radioactivity in America*, (1979), 15, 64.

27. The German physicist Friedrich Ernst Dorn first noticed in 1900 that radium was giving off a gas which he named niton, the radium emanation later known as radon. On Brooks, see Rayner-Canham, M. and G. *Hariet Brooks: Pioneer Nuclear Scientist* (1992).

28. Shea, *Otto Hahn and the Rise of Nuclear Physics* (1983).

29. Feather, *Lord Rutherford* (1940), 63. Heilborn, *Ernest Rutherford and the Explosion of Atom* (2003).

30. Shea, *Otto Hahn and the Rise of Nuclear Physic* (1983), 5,9. In France, Debierne had discovered actinium and Giesel claimed to have found a different element, emanium. The question was whether the two substances were identical or not. Hahn and Otto Sackur were asked by Ramsay to resolve the controversy. They proved that the substances were actually identical.

31. On Giesel, see Hahn, "Friedrich Giesel" (1928). Especially on Mendeleev's work on radioactivity, see Gordin, *A Well-Ordered Thing* (2004).

32. Feather, *Lord Rutherford* (1940), 105.

33. Badash, *Radioactivity in America* (1979), 205. On Ramsay and the radium industry in England see also Harvie, "The Radium Century" (1999).

34. Gordin, *A Well-Ordered Thing* (2004).

35. Nye, *The Question of the Atom* (1986), 456.

36. Badash, *Radioactivity in America* (1979), 69.

37. Nye, *The Question of the Atom* (1986), 430; Gordin, *A Well-Ordered Thing* (2004), 226.

38. Curie, Pierre. "Radioactive Substances, Especially Radium" Nobel lecture, June 6, 1905, http://nobelprize.org/nobel_prizes/physics/laureates/1903/pierre-curie-lecture.html; Curie, Marie. "Radium and the New Concepts in Chemistry," Nobel lecture, December 11, 1911, http://nobelprize.org/nobel_prizes/chemistry/laureates/1911/marie-curie-lecture.html.

39. Kaiserliche Academie der Wissenschaften in Wien, Konzepte, no. 20 015, AÖAW.

40. The Curies to the Austrian Academy, 1898, no. 25 081, AÖAW.

41. Karlik and Schmid, *Franz Serafin Exner,* (1982), 85; Fischer, *Geschichte der Gesellschaft der Ärzte* (1938), 155.

42. Julian, "The Early Days of the X-ray Revolution" (1996), 39. See also Lesky, *The Vienna Medical School of the 19th Century* (1976), 304.

43. Fischer, *Geschichte der Gesellschaft der Ärzte,* (1938), 155. Karlik and Schmid, *Franz Serafin Exner* (1982), 86; Alth, *50 Jahre Strahlentherapie Lainz* (1981), 4; Kogelnik, "The History and Evolution of Radiotherapy" (1996), 221.

44. Pierre Curie received Suess's telegram via Michel Levy. Between 1898 and 1899, the Curies received 1.1 tons of pitchblende from the Austrians. By 1906, the amount went up to 23.6 tons. The first amount was supplied at no cost and the ones followed were sold for especially low prices. See Kaiserliche Academie der Wissenschaften in Wien, Konzepte, no. 20 015, AÖAW; Pierre Curie to Suess, November 17, 1898, AÖAW; Meyer, "Die Vorgeschichte" (1950), 1; Reiter, "Stefan Meyer" (2001), 110.

45. Curie, M. "Radium and Radioactivity" (1904), 462; Curie, Marie "Radium and the New Concepts in Chemistry" Nobel lecture, December 11, 1911, http://nobelprize.org/nobel_prizes/chemistry/laureates/1911/marie-curie-lecture.html; Roquè, "Marie Curie and the Radium Industry" (1997), 270.

46. See for example the case of Ellen Gleditsch. In 1916, Curie prompted her to come to France and work at the factory, hoping that "He will offer you good conditions." Gleditsch followed Curie's advice and took Debierne's position at the factory while he was serving at the front. Sharing her time between the factory and Curie's laboratory in several other occasions, Gleditsch was able to pursue her research. (Lykknes, *Ellen Gleditsch: Professor, Radiochemist, and Mentor* (2005), 134).

47. "Seeks Radium in Colorado" *The New York Times*, s12, 1:6, September 12, 1912.

48. Roquè, "Marie Curie and the Radium Industry" (1997), 268.

49. Reiter, "Stefan Meyer" (2001), 109–10; Meyer, "Die Vorgeschichte" (1950), 1; Meyer, "Magnetic Properties of the Elements," (1899); Meyer, "Magnetic Properties of Inorganic Compounds" (1899). Meyer had studied physics and chemistry at the University of Vienna and was able to combine both in his early research on radioactivity.

50. Suess and Lang to Auer von Welsbach, July 10, 1901, AÖAW.

51. Meyer, "Die Vorgeschichte" (1950), 10.

52. The French, for example, carried out the purification of the raw material in the Central Company for Chemistry Production (*Société Central de Produits Chimiques*) a spacious factory where the chemist André Debierne could handle the enormous amounts of pitchblende. Only after the first extraction was further work completed at Curie's laboratory. As Soraya Boudia points out, Debierne "had to adapt laboratory techniques and perfect industrial treatment methods." The company provided chemical products and the staff salaries. In return, they obtained a share of the extracted radium salts for marketing. The forced collaboration of chemists and physicists with industrialists, as Boudia emphasizes, played an important role to the construction of

the radioelement industry in France. (Boudia, "The Curie Laboratory" (1997), 250–51; Mazeron and Gerbaulet, "The Centenary of Discovery" (1998), 207).

53. For Auer von Welsbach scientific discoveries, see Auer von Welsbach Museum, http://www.althofen.at/AvW-Museum/Englisch/biographie_e.htm. Also "Carl von Welsbach, the Inventor of the Incandescent Light Mantle," (1912).

54. "Haitinger-Feier," *Chemiker-Zeitung*, (1930), 182.

55. Suess to Auer von Welsbach, July 10, 1901, AÖAW.

56. Reiter, "Stefan Meyer" (2001), 111.

57. Up to that time, the mines in St. Joachimsthal were the main and only considerable source of pitchblende in the world. Echoing Boltwood, Lawrence Badash argues that, the embargo reflects Austrians' displeasure with the commercial instead of the mere scientific use of radium. Badash, *Radiaoctivity in America*, (1979), 140.

58. Suess and Lang to the Ministry of Agriculture, January 15, 1904, AÖAW; Suess and Lang to the Ministry of Culture and Education, January 15, 1904, AÖAW.

59. Suess and Lang to the Ministry of Agriculture, January 15, 1904, AÖAW.

60. On March 7, 1904 Suess and Lang informed Pierre Curie of their agreement with the Ministry of Agriculture. Given the shortage of the residue, they agreed that the two research groups, the Austrians and French, would alternate receiving one ton each at a time On March 22, Lang and Suess informed the administrators of the St. Joachimsthal mines of the above procedure. The mine's administrators would send alternatively one ton of pitchblende to Atzgersdorf, Vienna and one to Paris. Further financial issues were clarified. Between 1904 and 1905, the Curies paid 9,540 kronen to the mines for the residues. See Suess and Lang to Pierre Curie, March 7, 1904, AÖAW; Suess and Lang to k.k. Berg- und Hüttenverwaltung, St. Joachimsthal, March 22, 1904, AÖAW; Meyer, "Die Vorgeschichte" (1950), 9.

61. Haitinger and Ulrich, "Berich über die Verarbeitung," (1908); "Haitinger-Feier" *Chemiker-Zeitung*, (1930), 182; Meyer, "Das erste Jahrzehnt," (1920), 1; Exner, "Le nouvel institute" (1910), 244.

62. The Royal Society to the Austrian Academy, May 1, 1904, AÖAW.

63. Rona, *How it Came About* (1978), 21.

64. Cahan, *An Institute for an Empire* (1989), 8; Warburg to Exner, February 12, 1909, AÖAW; Kaiserliche Academie der Wissenschaften in Wien, Konzepte, 25 015, AÖAW. For example, a list of radium recipients included the Chemistry Institute in Krakau and Eduard Riecke's institute at the University of Göttingen (1908 report of the Ministry of Public Affairs, n. 904, AÖAW).

65. Meyer, "Die Vorgeschichte" (1950), 14.

66. For characterizations of Austria as the center of the monarchy's power, see Okey, *The Habsburg Monarchy* (2001), 196. When the Austro-Hungarian Empire collapsed, Ulrich lost his position as director at the St. Joachimsthal mines and moved to Vienna as a guest at the Institute for Radium Research (Meyer, "Das erste Jahrzehnt" (1920), 11).

67. Suess and Lang to the Ministry of Culture and Education, March 14, 1908, AÖAW.

68. Walkhoff published a brief description of a skin reaction to radium in October 1900. Interestingly enough, he published his report in a journal of amateur photography. This fact calls attention to the interrelation of physics, medicine, and photography. (Walkhoff, "Unsichtbare, photographische wirksame," (1900). See also Dutreix, Tubiana and Pierquin, "The Hazy Dawn of Brachytherapy," (1998), 223.); Bequerel and Curie, "Action physiologiques," (1901). As Mould correctly emphasizes, Pierre Curie and Bequerel planned their experiments and they were not accidental (Mould, "The Discovery of Radium" (1998), 1241). Radium had already started to occupy the interest not only of physicists and physicians but also of the general public. For example, in 1903, the American inventor Thomas Edison warned the public of radium's possible damaging biological effects in an interview in *Penny Press* (see Mullner, *Deadly Glow* (1999), 16).

69. Tubiana, Dutreix, and Pierquin, "One Century of Radiotherapy" (1996), 231; See also Viol, "History and Development of Radium Therapy" (1921), 21.

70. Hayter, *An Element of Hope* (2005), 13.

71. On the radium therapy in France see Vincente, "Genesis of the Pavillion Pasteur of the Institut du radium in Paris"(1997).

72. Berven, et al., *The First Fifty Years* (1965).

73. Fox, "The History of Radium" (1998), 115–124.

74. Hayter, *An Element of Hope* (2005), 13; Sell, "The Development of Radiotherapy in Denmark During 100 Years" (1995); Alth, *50 Jahre Strahlentherapie Lainz* (1981), 4–5; Luger, *70 Jahre Krankenhaus* (1977), 3; Mould, *A Century of x-rays* (1993), 136.

75. Exner, S. "Einige Beobachtungen" (1903); Exner, A. and Holzknecht, " Die Pathologie" (1903). Alfred Exner was the first to treat a tumor of the esophagus by forcing a catheter that held the radioactive source through the esophagus so as to achieve improvement in swallowing (Exner, A. "Über die Behandlung," (1904), 4–96). See also Fischer, *Geschichte der Gesellschaft der Ärzte,* (1938), 156. In 1902, Antoine Beclère, an outstanding pioneer of radiology in France, visited Holzknecht in Vienna to discuss dosimetry issues. In 1905, Holzknecht took over the leadership of the Röntgen Laboratory of the second Medical University Clinic of Vienna, later known as the Holzknecht Institute, elevating radiology to an exact science and Vienna to the center of radiological research. Five years later, Holzknecht joined Freud's psychoanalytical society after experiencing psychological anxieties. More than two decades later, in 1931, Holzknecht treated Freud's tumor in his institute. On Holzknecht see Angetter, *Guido Holzknech,* (1998), 16–17; Tubiana, Dutreix, and Pierquin, "One Century of Radiotherapy" (1996), 228.

76. Tubiana, Dutreix, and Pierquin, "One Century of Radiotherapy" (1960), 231. Also, for illustrations of early radium therapy equipment and methods, see Mould, *A Century of x-rays,* (1993), 27, 131. See also http://www.orau.org/ptp/collection/brachytherapy/brachytherapy.htm.

77. Report of the Ministry of Public Affairs drafted in 1908 n. 904, AÖAW. In 1904, Suess was interviewed by Louis Leasley from London. On his return to London, Leasley asked Suess for a radium loan in order to pursue "hospital work" (Leasley to Suess, February 18, 1904, no. 25 083, AÖAW).

78. Rutherford to Meyer, June 1, 1914, AÖAW; Rutherford to Meyer, June 29, 1914, AÖAW.

79. DiSantis, "Radiologic History Exhibit" (1991). For more on advertisements concerning racial issues, see Massing, "From Greek Proverb to Soap Advert" (1995).

80. Wiesloch, 1916. "Sterilization in the Interest of Race Betterment" (1916).

81. Chase, "American Literature on Radium and Radium Therapy prior to 1906" (1921); Viol, "History and Development of Radium Therapy" (1921); Soddy, "Radium as a Therapeutic Agent" Medical Record" (1903).

82. Ernest Rutherford to Stefan Meyer, October 22, 1910, Rutherford's Papers, Manuscript Collections, CUL.

83. "Urges that Radium Be Nationalized" *New York Times*, November 30, 1913, III, 4:1.

84. Mullner, *Deadly Glow* (1999), 21–29; The Historical Instrumentation Collection, Oak Ridge Associated Universities, http://www.orau.org/ptp/collection/Miscellaneous/photoalbum/photoalbum.htm

85. Boudia, "Radioactivity and Metrology" (1997), 251.

86. Mache, "Über die im Gasteiner Wasser" (1904a); Mache, "Über die im Gasteiner Wasser," (1904b); Mache and Meyer, "Über die Radioaktivität der Quellen" (1905).

87. Mould, *A Century of x-rays* (1993), 170.

88. "Controls its Own Radium" *The New York Times*, November 30, 1911, N22, 22:2; Meyer, "Die Vorgeschichte" (1950), 3.

89. "Re Präparate," newspaper clip, undated, AÖAW.

90. Meyer, "Die Vorgeschichte" (1950), 11.

91. For radioactive products, see The Historical Instrumentation Collection, Oak Ridge Associated Universities, http://www.orau.org/ptp/collection/quackcures/quackcures.htm; *Radiation and the Popular Imagination*, Radiology Centennial Inc., formerly available at http://www.xray.hmc.psu.edu/rci/contents_8.html and archived at http://web.archive.org/web/20060504024805/http://www.xray.hmc.psu.edu/rci/contents_8.html (accessed 7 November 2007). See also Baker, "Industrial Uses of Radium: Possibilities in Agriculture, Medicine, and Other Directions" (1915).

92. For most of the advertisements and products presented here, see http://www.orau.org/ptp/museumdirectory.htm.

93. "The Radiendocrinator" http://www.orau.org/ptp/collection/quackcures/radend.htm; Mullner, *Deadly Glow* (1999), 109–118.

94. Photograph VIII-09, "Beautifying x-rays", *Radiation and the Popular Imagination*, Radiology Centennial Inc. Formerly available at http://www.xray.hmc.psu.edu/rci/ss8/ss8_1.html; archived at http://web.archive.org/web/20051125035151/http://www.xray.hmc.psu.edu/rci/ss8/ss8_1.html (accessed 7 November 2007).

95. "Radium and Beauty," *New York Tribune*, November 10, 1918, Magazine Section, p. 12; Photograph VIII-10, "Beauty Shop Horrors" *Radiation and the Popular Imagination*, Radiology Centennial Inc., formerly available at http://www.xray.hmc.psu.edu/rci/ss8/ss8_1.html and archived at http://web.archive.org/web/20051125035151/http://www.xray.hmc.psu.edu/rci/ss8/ss8_1.html (accessed 7 November 2007).

96. As Soraya Boudia argues, it is not entirely certain that Alfred Curie actually existed. The authors of the article draw attention to the fact that Alfred's last name signified scientific authority and thus those involved in the commercialization of Tho-Radia probably aimed to secure commercial success for the new product. See Lefebvre, and Raynal, "De l' Institut Pasteur à Radio Luxemburg: L' histoire étonnante du Tho-Radia" (2002); "la curieuse histoire du Tho-Radia" http://monsite.wanadoo.fr/tho-radia/index.jhtml. For a history of women's beauty trend, see Riordan, *Inventing Beauty* (2004).

97. Riordan, *Inventing Beauty* (2004). See also http://www.double-d.com/coffeebreak/companies/silhouette.html and http://homepage.ntlworld.com/davesplace/strange.htm.

98. Lang, "A Most Valuable Accident" (1959). See also Clark, *Radium Girls* (1997).

99. Rentetzi, "Women Radium Dial Painters as Experimental Subjects or What Counts as Human Experimentation" (2004).

100. "Strassenbeleuchtung durch Radium" *Neues Wiener Journal*, November 10, 1935.

101. "Sieben junge Winnerinnen zertrümmern Atome . . ." *Neues Wiener Journal*, May 30, 1935. I owe my thanks to Brigitte Bischof who brought this article to may attention. Bischof, "The 'Marie Curie' Syndrome" (2003).

2. DESIGNING (FOR) A NEW SCIENTIFIC DISCIPLINE

In the mid-1910s, the physicist Victor von Lang, director of the I. Physikalisches Institut in Vienna, and Eduard Suess, director of the Austrian Academy of Sciences, intensified their attempts to establish a new institute appropriate for the study of radioactivity.[1] Up to that point, radioactivity research had been hosted in Vienna's Physics Institute, inside a shabby building with temporary laboratories on Türkenstrasse, a side street close to the university. A "very primitive, converted apartment house" and "a miserable space," as Lise Meitner described it, the institute was interspersed among the neighborhood's residences and shops.[2] Although it included lecture halls and laboratory space, it was insufficient for both teaching and research. In the big lecture hall, there were no desks. The students had to write on their knees. The floor was so rickety that it quivered whenever someone crossed the room, affecting apparatus sensitive to motion.[3] For those who envisioned Austria as "the centre of research on radioactivity," such a makeshift institute was inappropriate for these new "exact studies."[4]

Besides being exact, this new science was profoundly interdisciplinary. As Lawrence Badash points out, "radioactivity was something of a hybrid between physics and chemistry." Its data came from an impressively diverse number of scientific disciplines.[5] Tracing the changing cultures of theory in nuclear science from 1920 to 1930, Jeffrey Hughes similarly argues that "in a field [radioactivity] which drew its practitioners from such a wide range of backgrounds, including chemistry, physics, geology, and medicine, the variety of interpretative practices was unusually large."[6]

Arne Hessenbruch has recently traced the connections of radioactivity to technology by focusing on Ernest Rutherford's 1901 experiment on radiation energy. As Hessenbruch surmises, "Who precisely were the many different constituencies and individuals involved in the early history of radioactivity? There were physicists, chemists, and medics, each having a different perspective."[7] The cultures engaged in radioactivity research, however, were not only scientific. Soraya Boudia and Xavier Roque have convincingly argued that Marie Curie and her colleagues deliberately organized the radium industry in France and defined radioactivity as a field of enquiry that impinged both on science and on industry.[8]

At the same time, those involved in radium research came to consider themselves as a distinct disciplinary community. As Rutherford reported to Bertram Boltwood about the International Congress of Radiology and Electricity held in Brussels in 1910, "we had a rather good section and practically all the radioactive people were present." Self-references to the "radioactive people" or the "radioactivists" certainly signaled newly acquired identities, acknowledged even by K. B. Hasselberg, the president of the Swedish Royal Academy of Sciences.[9] Yet, scientists at the crossroads of well-established disciplines such as physics, chemistry, and an emerging field such as radioactivity experienced the encounter as a transition. As Rutherford admitted, "Really, I was startled at my transformation at first but afterwards saw that it was quite in accord with the disintegration theory."[10] How else might a self-defined physicist describe his feelings at the award ceremony for the 1908 Nobel Prize in Chemistry? Linked to the conventional map of physics and chemistry, radioactivists such as Rutherford were at the same time persistently remaking the map of science, fitting radioactivity within its borders.

Recent histories of radioactivity, departing from a linear narrative focused exclusively on theories, have explored the material culture of the discipline and its social practices at the intersection of academia, industry, and the modern state.[11] Attention has been paid to instruments, to experimental cultures, to exchanges of materials and techniques among disciplines, and, recently, to the gender politics that characterized different institutional settings.[12] Here, I introduce urban history and the architecture of science buildings as further important contexts for the history of radioactivity. I investigate what it meant to the Viennese physicists to acquire a new institute for radioactivity research and design a building, the first of its kind, for such a multifaceted field of inquiry, taking into account not only the malleable present but also the future of their discipline. How did architecture and the urban positioning of the institute affect the identities of its scientists? Examination of these questions sheds light on the mostly unexplored history of radioactivity in Vienna.

THE BIOGRAPHY OF AN INSTITUTE

During the first decade of the twentieth century, the institutionalization of radioactivity was essential to those who wished to legitimize their epistemic claims about new radioactive elements and their cultural position on the map of science. International rewards certainly contributed to this choice of direction.[13] Three Nobel Prizes had already been awarded to researchers in the new field. In 1903, Henri Bequerel shared the Nobel Prize in Physics with Marie and Pierre Curie. In 1908, Ernest Rutherford was awarded the Nobel Prize in Chemistry, and in 1911, Marie received her second Nobel Prize, this time in chemistry. The publication in January 1904 of *Le Radium*, a special French journal devoted to radioactivity research, was yet another attempt to legitimize research on the newly discovered radioactive elements.

The first editors, Jacques Danne and Henri Farjas, aimed to popularize radioactivity. Six months later, in July 1904, the journal shifted toward a more scientific audience, altering its content. The new volumes included scientific sections on radiophysics, radiochemistry, medicine, radiotechniques, and even photography in relation to photochemistry. Danne remained the secretary of the journal while a new committee oversaw the publication. The names of those involved—Becquerel, Antoine Béclère, René Blondlot, André Debierne, the Curies, and Rutherford—reveal efforts to bestow prestige on the field.

However important they are for assuring credibility and enlarging epistemic authority, journals and prizes cannot be compared to the advantages that might be provided to researchers by a specialized research center. Indeed, buildings of science play the double role of configuring the identities of those involved in scientific production and symbolically displaying the identity of the scientific processes that take place inside. It is indicative that during the first decade of the twentieth century, radioactivity researchers were distributed across laboratories in Europe and North America but had no single specialized institution devoted exclusively to the new field. Systematic attempts by scientists to obtain a designated space for their practices were related to their coordinated efforts to make disciplinary space within the sciences for their new enterprise.

Rutherford, the foremost authority on radioactivity in the Anglo-Saxon world, based his early work at McGill University in Montreal. The director of the physics laboratory there, John Cox, provided Rutherford with a lavish laboratory which, nonetheless, specialized not in radioactivity research but in training students in medicine, engineering, and industrial chemistry.[14] In Paris meanwhile, Marie and Pierre Curie were working at rue Lhomond in the Ecole municipale de physique et de chimie industrielles inside a damp storehouse turned into a physics laboratory.[15] After the announcement of the Nobel Prize in 1903, Pierre hoped that "all this noise will not perhaps have been useless, if it gets me a chair and a laboratory."[16]

A modest laboratory was soon set up, but only at the science faculty in rue Cuvier, where Pierre taught a physics course to those preparing for the *certificat d' études*. Eventually, in 1909, a decision was made to found the Institut du radium, a specialized institute for radium research in Paris. However, it was not until 1914 that the new Curie Institute, including two laboratories, one for physics research and one for biological and medical research, were ready for work.

In 1907 meanwhile, Rutherford had already moved to Manchester, where he once again located his work at the university physics laboratory without gaining an autonomous research centre for his experiments. As he admitted to Bertram Boltwood, "The laboratory is good, but there was not much in the radioactive line." He added that the main advantage to being back in Europe, however, was the "hope to be able to raise a good deal [of radium]."[17]

The precursors of the Curies in institutionalizing radioactivity research were the Viennese physicists. Since the end of the century, radioactivity had attracted the attention of Vienna's physicists through both its promise of scientific development and because, having access to the Bohemian radium and uranium mines in St. Joachimsthal, they were often asked to play the role of radium providers to the wider scientific community. Despite these advantages, bureaucracy and lack of money delayed the foundation of a specialized radium institute for years and embarrassed the Viennese scientists during international meetings.[18] The solution came from Karl Kupelwieser, who offered a munificent sum for the establishment of the Radium Institute in Vienna.

Although one of the most interesting and influential personalities of the late Habsburg era, Kupelwieser has not received much historical attention. He was born in 1841 into a bourgeois Viennese family. His father, Leopold Kupelwieser, was a famous painter and professor at the Academy of Fine Arts in Vienna. Karl's mother, Johanna Lutz, was related to the president of the senate, Theobald Rizy. Both parents were members of an artistic circle including the painter Moritz von Schwind, the musician Franz Lachner, and the composer Franz Schubert.[19] Karl studied law at the University of Vienna, learned to play the piano, and showed literary talent. At the age of 28, he married Bertha Wittgenstein, the sister of his best friend Paul Wittgenstein. This marriage marked the beginning of a strong relationship between the two families, as well as the development of Karl into an astonishingly powerful and affluent industrialist. His brother Paul offered the directorship of a rolling mill in Bohemia to Karl Wittgenstein, later father of the famous philosopher Ludwig Wittgenstein. Within the next five years, the wealth of the company increased enormously thanks to Wittgenstein's organizational skills. As an astute lawyer, Karl Kupelwieser bought most of the company's stocks and two additional smaller enterprises, and served as the legal adviser of the consortium.

By the end of the nineteenth century, Kupelwieser had amassed a huge personal fortune with which he supported several social and scientific projects. His charity work included financing a residence for blind, deaf, and mute people in Vienna and a tuberculosis hospital for children and adults. In 1907, he also patronized the Biological Station in Lunz.[20] His son, the biologist Hans Kupelwieser, became its first director.

In 1908, Kupelwieser's role as a patron of science was further boosted by a generous donation to the Academy of Sciences for the foundation of the Institut für Radiumforschung (Institute for Radium Research). In a letter to the academy, Kupelweiser clearly indicated that he wanted to "prevent the shame" of his country, caused by letting others "snatch away" the privilege of conducting radium research.[21] He notified them that he was willing to place a contribution of 500,000 crowns at the disposal of the academy for the construction and maintenance of an institute to be dedicated to research of the physical and chemical properties of radium. It was no surprise that his offer was considered a patriotic act, which would foster the country's progress and increase the empire's prestige.[22]

There was indeed a sense of "national pride" in establishing the institute. The institutionalization of the field gave Austrians the chance to put their own radium supplies to good use. This, for example, threatened Bertram Boltwood, as became clear in a letter to Rutherford: "I see that someone has given a lot of money for a Radioactive Institute at Vienna and I am afraid that the wholesome business will drive the small dealer like me to the wall."[23]

During the next two years, with his private initiative, Kupelweiser achieved what Exner and his colleagues had not managed to do despite their persistence for over a decade. The negotiations with the sloppy bureaucratic system of the Austro-Hungarian monarchy and its ministries were Kupelweiser's expertise. Being a lawyer, he knew the tricks. His donation set the officialdom into frenzied action. Since 1894, the administrations of the Ministries of Education and Finance had been in constant negotiations about the site where the physics institute should be established. Already on March 14, 1908, Suess and Lang had written the Ministry of Culture and Education, supporting their case. "The new building for the Physics Institute is an essential condition for the possibility of radium research, since the current one is absolutely useless for such exact studies." [24]

When Kupelweiser offered his donation five months later, both physicists and the state unhesitatingly seized the opportunity immediately. On June 5, 1909, a protocol for the erection of the institute was granted by the state and signed by those involved in the construction. The list of signatories reveals the overlapping communities and interests involved: Eduard Suess signed as the president of the Austrian Academy; Franz Exner represented the physicists; Zdenko Skraup, director of the II Chemisches Laboratorium, signified the presence of the Viennese chemists in the project; a number of engineers and city planners stood for the state; and Eduard Frauenfeld was the architect.[25]

Following Kupelwieser's specifications, although the institute was established in a separate building, it was located between a physics institute and a chemistry institute. The proposed site was a spacious state property at the intersection of Boltzmanngasse, Währingerstrasse, and Strudlhofgasse in the ninth district of Vienna. The physicists chose the quietest corner of Boltzmangasse and Strudlhofgasse to host their institute. On the corner of Währingerstrasse and Boltzmanngasse, the chemists designed their own facilities. The Radium Institute was located between the two buildings, facing Boltzmangasse.

In contrast to the Radium Institute—a semi-autonomous research centre of the academy—the physics and the chemical institutes were subject to the University of Vienna and were to be used for both educational and research purposes. By the time the Radium Institute was completed and ceremonially opened on October 28, 1910, the building of the Physics Institute had already begun and it was finished two years later. The advent of the First World War disrupted the ongoing construction of the Chemical Institute, which was partly completed in 1920 but didn't fully open its doors until 1924.[26]

The physicist with direct interest in radioactivity research was Franz Serafin Exner, member of the Austrian Academy of Sciences since 1885 and *Professor Ordinarius* at the University of Vienna since 1891. Soon after the discovery of radium, Exner played a major role in persuading the academy to establish the Commission for the Investigation of Radioactive Substances. He was deliberately trying both to ensure adequate amounts of radium for research and to enlist supporters in establishing radioactivity as a strong field in Vienna.[27] His influence in the Viennese scientific community was distinctive. Following the reorganization of university physics in Vienna and the creation of three separate institutes in 1902, Exner held the directorship of the II. Physikalisches Institut and proved to be the most important experimental physicist of his generation. In 1907, he reached the highest point of his academic career by becoming rector of the University of Vienna and participated in a number of scientific commissions, decisively shaping science in Vienna. During the first decade of the twentieth century, he was the main force in negotiations between the Austrian Academy and the Curies to provide the latter with considerable amounts of pitchblende for their scientific investigations. He was also Rutherford's closest contact in the Austrian Academy, guaranteeing his radium supplies. Scientifically, Exner was acknowledged not only for his original work on atmospheric electricity, color theory, spectral analysis, and radioactivity, but also for the wide circle of students he mentored.[28]

Thus, in 1910, there was no doubt that the establishment of the institute, made possible by Kupelwieser's donation, was also the result of Exner's attempt to legitimize the field of radioactivity in Austria. Having enormous prestige in the Viennese and international scientific communities, Exner was able to make important decisions on the research agenda, personnel, and research students' work, as well as the entire organization of the institute. He celebrated the building of such a specialized research centre as a symbol of scientific authority. He saw it as having the potential to bring Vienna to the forefront of radioactivity research and probably as a personal victory after twenty years of frustrations and difficulties in the institute at Türkenstrasse. It is significant that his only publication concerning the establishment of the institute appeared in French, in the journal *Le Radium*, the main forum of the international community of the "radioactivists." He proudly described the institute as "unique of its kind," with "the most modern facilities in the world." With the generosity of a confidently satisfied director, he assured the community that "the Academy of Sciences will be glad to see the largest possible number of scientists working there."[29] Exner was, however, already 64 years old and far from being in the most creative period of his scientific career. Under these circumstances, Stefan Meyer seemed best to fit the role of the active director (*Leiter*) of the institute, while Exner retained the position of its official director (*Vorstand*).

A key figure in the institute's history, Meyer belonged to a family of Jewish intellectuals. He studied physics at the University of Vienna under Exner and Ludwig Boltzmann. His ascent in the university hierarchy was remarkable. In 1900, he became *Privatdozent* at the University of Vienna and in 1906, he succeeded Boltzmann as head of Vienna's Institut für Theoretische Physik, a position he held for a year. In 1908, he became *Ausserordentlicher Professor* and Exner's assistant. He expressed an early and strong interest in radium research and when the International Radium Standards Committee was founded in 1910, Meyer was appointed its secretary, an indication of his scientific success. It was this dynamic involvement in the field which prompted Exner to entrust Meyer with the scientific organization of the institute, including its design and the purchase of instruments and furniture.[30]

CONCERNS ABOUT NOISE AND VIBRATIONS

As Christoph Hoffmann has argued for the German situation, "An ordinary physics institute at the end of the nineteenth century embodies in bricks and mortar, stone and lime precisely this command: disturbances have to be avoided."[31] No case better illustrates physicists' concerns about disturbances than the Physikalisch-Technische Reichsanstalt, Berlin's Imperial Institute of Physics and Technology. At the end of the 1890s, the city's street company decided to lay electric streetcar tracks in front of the Reichsanstalt, which might interfere with the institute's electromagnetic research.[32] The decision generated a serious controversy between the physicists and the streetcar company, which lasted for six years. It ended only when the company agreed to take measures to minimize disturbances for the institute.

Concern about disturbance was not restricted to the Germans, but also troubled Austrian physicists. They were just as uneasy with the heavy traffic of electric streetcars on Währingerstrasse and just as worried about external factors which would interfere with their attempts to measure radium preparations precisely or to perform spectroscopic analysis. Their installations' stability had always been important. When the physics institutes were still at Türkenstrasse, the physicists' main complaint was the shaking of the floor and the disturbances caused by vibrations. "Whenever someone was ironing in the neighboring house," Benndorf recalled, "the needle of the magnet was going here and there and we very often experienced this nasty situation."[33] There was no solid spot where an instrument could be set up nor where magnetic measurements could be carried out. In 1896, the directors of the three physics institutes, Exner, Ludwig Boltzmann, and Viktor von Lang, sent a letter to the deacon of the philosophy faculty, complaining about the intolerable situation in their laboratories:

> Any precision measurements whatsoever are impossible from the start due to the constant shaking [of the building], not only during the day but even at night, finer measurements cannot be carried out, in part due to the traffic on Währingerstrasse and due to the wind, in part due to the constant currents in the adjacent houses. Moreover, the spaces are extremely impractically arranged and far too small for scientific purposes, such that often due to the presence of the observer in the room, the temperature there rises in a troublesome way. Even the local middle schools have far more suitably arranged laboratories than the university.[34]

For the physicists, it was out of the question to put up with similar circumstances in their new location. This was the main reason they refused to accept either of the other two sites originally under consideration for the natural-science district.

The first proposed site was on Währingerstrasse, where the Volksoper was finally built. The second was the old Gewehrfabrik, the gun factory at the corner of Währingerstrasse and Schwarzspanierstrasse, where, in 1904, the Institute for Physiology was established.[35] Both sites were very close to the city and within the university quarter. According to the physicists, however, they seemed insufficient for the natural-science district. The first was too close to the street, while the second was next to the Anatomical Institute, which was equipped with high-voltage current. The induction current it generated affected the functioning of the electromagnetic apparatus, which made the physicists' research impossible.[36]

In order to control vibrations, noise, and electromagnetic disturbances, Viennese physicists chose the location across the Josephinum as one which might produce a less disruptive environment for research. The proposed location was a spacious state property of 12,065 square meters across from the Josephinum known as Bäckenhäusel.[37] Historically, the building belonged to a baker in the fifteenth century, and from 1648 to 1784 it served as a small hospital. Later on, it was converted into a residential building. Just before Vienna's physicists decided to build their institute at this spot, the building hosted a tobacco administration and storeroom.[38] After tearing down the old Bäckenhäusel, a complex of three buildings went up in its place. Traffic was redirected further away from the new building by adding an extra lane on both sides of the street. Instead of passing close to the institute's building, the streetcars traveled along the middle lane of Währingerstrasse. In this way, mechanical vibrations through the ground were eliminated. In addition, trees on the pavement in front of the buildings and inner courtyards protected the small natural-science quarter from street disturbances. By locating the Radium Institute between the Chemistry Institute and Physics Institute, Exner and Meyer also aimed to eliminate the interference of external factors with research. In collaboration with the architects, they chose to build a relatively small edifice with three stories and about 20 workrooms facing tranquil Boltzmanngasse, thereby avoiding the busy traffic of Währingerstrasse. Concerned with disturbances, they moved the façade two meters further from the street to avoid the noise of the already infrequent traffic on Boltzmanngasse.[39]

RADIUM: AN UNPREDICTED NEW INHABITANT

While they were able to take care of external factors, such as noise and mechanical vibrations, the physicists were not yet fully aware of radioactive contamination as a new kind of disturbance. As Meyer admitted, "It was a new concern; we were setting up something untested; experiences of radioactive 'contamination,' mutual interference, and so on were for the most part absent."[40] Unfortunately, they located two storerooms for radioactive materials along with the accumulator (a 469-volt battery) and a transformer in the cellar. Workrooms were grouped on the other side of the cellar, separating the area of research from that of the apparatus and the radioactive sources. Double walls on the external sides reduced the amount of radiation exposure.

According to Meyer, it became apparent that such an arrangement was a mistake since the radioactivity was spread throughout the building and was inhaled daily by the personnel. It was not until the mid-1920s that they moved 1 gram of radium to a room on the top floor of the building and surrounded the solution with lead bricks. A better resolution was chosen for the stronger radioactive materials. They built a special room in the cellar with thick concrete walls and a heavy steel door to seal the storage of strong preparations and eliminate contamination.

It was probably because of radiation precautions that the institute's design broke with the German architectural stereotype of hosting the director under the roof of the research centre. In Vienna's Radium Institute, there were no residential apartments either for Exner or Meyer. The presence of radium, a new and unpredictable inhabitant, forced changes in the common architectural canon.

The institute's first *Ordentliche Assistent*, Viktor Hess, later admitted that a second architectural mistake was to locate the radiochemical laboratory of the institute on its ground floor. Inside the front entrance, one would encounter on the left two rooms for chemical work, a darkroom, and a space for precise measurements of radioactive materials. In the hallway, showcases housed glasses for chemical analysis and other devices. Performing all the radiochemical tasks in the ground floor "was unfortunately a wrong choice," Hess recalled. He added, "The upper floor or a special building for radiochemistry would have been a better solution, because one could not avoid working with large amounts of preparations there."[41]

While mechanical vibrations and noise had always been familiar disturbances that could interfere with experiments, radiation hazards were a viciously malevolent disturbance that scientists were neither ready to face nor willing to admit. Elisabeth Rona, one of the female experimenters who worked at the institute and who specialized in polonium preparations, seemed to be one of the few who had a clear sense of how hazardous radiation could be. In the autumn of 1926, she spent a few months in the Curies' laboratory in Paris to learn from Irène Curie how to prepare polonium sources.[42] It was there that Rona first experienced the danger of radioactivity. Marie Curie and she were attempting to open a flask containing a solution of a strong radium salt when a violent explosion scattered glass all over the laboratory. It was only by mere luck that they were not injured or highly contaminated.[43]

When Rona returned to the Vienna Institute at the start of 1927, her experience with radium hazards led her to insist on taking precautions. The entire building was by now of course highly contaminated. Although hoods were used and the rooms were fairly well ventilated, the walls of the chemistry laboratory on the ground floor were especially contaminated. Rona recalled that "separated by a narrow corridor was the instrument room containing Geiger counters for beta counting and parallel plate condensers for alpha counting."[44] That spatial arrangement resulted in the contamination of instruments that finally had to be carried to the neighboring Physics Institute. Rona's own room was also one of the most contaminated. As she discovered, it was there during the 1910s that Otto Hönigschmid carried out his atomic weight experiments. The unprotected Hönigschmid often shook the solution by hand in order to homogenize the radium solution.[45]

After her accident in Paris, Rona was more cautious. In her work with colleague Gustav Ortner to open sealed tubes with radium salts, Rona insisted on using gas masks. But even in the late 1920s, Meyer was unwilling to take her request seriously:

> He laughed and tried to assure me that no danger was involved. However, I was not convinced and bought two gas masks with my own money. When we tried to open the first tube, it exploded, and the same thing happened with the second, scattering radioactive material all around. The gas mask saved us from severe damage. The basement room was closed permanently because it was impossible to get rid of the contamination.[46]

Women's individual bodies occurred as sites of experience, sites where radium and its penetrating radiation literally left their tracks in women's cells. Radiation affected the health and the everyday practices of all workers exposed, but women had a particular concern over the effect of the radiation in their reproductive systems. Unfortunately, there are no statistical data for that time concerning the effects of radiation on women's reproductive system or on the health of the institute's personnel as a whole.

Only slowly did scientists become aware of the hazards of radium. When Meyer published his book on radioactivity in 1927, he described different types of hazards, including his own accident. Emptying and transferring radioactive liquid from one bottle to another, his fingers became contaminated and, after the first symptoms of radiation burns appeared, the muscles atrophied. He hoped that "mentioning this will caution those who work with radioactive matter." Yet it is interesting that he did not enforce safety rules in his own institute. Rona admits that he "was probably honest in believing that no danger (at least immediate danger) existed because changes in the blood system, a sensitive organ for radiation, are slow." At the same time, he characterized as "hysterical fear" Kara-Michailova's claims that whenever she entered a room at the institute where 1 gram of radium was kept, she had trouble swallowing and had swelling of her tongue.[47] Ironically, in 1968, Kara-Michailova died of cancer induced by her long exposure to radioactivity.

Although Meyer and Exner tried to plan for their future research in radioactivity, the radium hazards were unpredictable at the time. As it proved, so were disciplinary changes. In the 1930s for example, the shift from radioactivity to nuclear physics required changes in instrumentation and the material culture of the institute. Nonetheless, it was not until 1966 that Berta Karlik, director of the institute after the Second World War, finally managed to restructure the building "with the help of a young, very competent architect."[48] By pulling down a few walls, Karlik arranged to fit in a Cockroft-Walton accelerator and add an additional floor. The shift from the electromagnets and the accumulators of the 1910s to the big accelerators of the 1950s had indeed been inconceivable.

THE URBAN SETTING OF THE INSTITUTE

According to contemporary tradition, the director of a prospective institute provided plans and oversaw its design and construction. At the start of 1909, Meyer and Exner thus worked on the *Raumprogram*, literarily the "room program," of the building.[49] They soon gave Frauenfeld and Berghof, a Viennese architectural firm, a sketch of the complex of the Physics Institute and Chemistry Institute as well as the Institute for Radium Research. The extended area around the new natural-science district was known as the Mediziner-Viertel, the physician's quarter.[50]

Devoted exclusively to research, the Institute for Radium Research offered possibilities for *Praktikum*, laboratory research positions for doctoral candidates to complete their thesis requirements. In the years that followed the institute's inception, a number of research students together with the permanent personnel probed the chemical and physical properties of radioactive elements. The research carried out during the first decade of the institute's operation eventually led to at least two Nobel Prizes and involved a significant number of young creative researchers with backgrounds in fields such as biology, physiology, chemistry, geology, and physics.

A surprising condition imposed in 1910 by Kupelwieser restrained the institute from radium research on medicine and from medical experiments on living organisms. For such work, the researchers depended on the Radiumstation of Vienna's Allgemeines Krankenhaus, near the institute.[51] Inderdisciplinary exchanges of this kind were facilitated by the fact that the institute was Austria's official radium standard-holder, acting as information center for scientific questions related to radioactivity. It also supplied the hospital's Radiumstation with radium for medical use.

Archival references and publications illustrate how these interdisciplinary activities were tied to the buildings, and the growing network of radioactivity researchers was drawn from physics and medicine. First, these activities included research students commuting between the different institutes. The space of the Josephinum, which was located at Währingerstrasse 25, housed the surgical and medical faculty. It was used for a variety of courses in experimental physics for medical students and in experimental chemistry for pharmaceutical students taught by physicists and chemists. Also, courses on microscope theory, inorganic chemistry and theoretical and physical chemistry at the Physics Institute were much attended by medical students.[52] Occasionally, the institute's research students pursued medical studies in the neighboring institutes. As Marietta Blau described her work at the end of the 1910s, "I conducted theoretical studies and at the same time was a research assistant at the Laboratory for Medical Radiology at the Holzknecht Clinic, where I studied medical physics."[53] The clinic was offered at the Allgemeines Krankenhaus.

Second, trade of radioactive materials as well as transfer of apparatus and expertise extended the list of activities in the Viertel. In 1911, a protocol for trade in radium preparations and a provisional plan for the function of the Radiumstation at the Allgemeines Krankenhaus were signed at the Radium Institute.[54] Shortly afterwards and on several occasions, radium was prepared at the Radium Institute and delivered to the hospital's station, which was equipped with the latest apparatus for cancer therapy.[55]

Besides those working at the Radium Institute, a growing network of scientists researching the application of a variety of radioactive materials was expanding in the neighboring medical and biological institutions. For example, Goldschmidt, from the Physiological Institute, collaborated with Rona at the Radium Institute on the study of polonium in the cure of leukemia.[56] E. Knaff-Lenz, a Vienna University pharmacologist, studied the influence of radioactivity on humans. The map of the Mediziner-Viertel soon reflected the same modes of scientific collaboration and radium trading as did the list of committee members of the Radiology and Electricity Congress, planned by Meyer and Rutherford in 1913. Unsurprisingly, besides the physicists of the Radium Institute, the Vienna delegation included the radiologist Guido Holzknecht, the roentgenologist Robert Kienböck, the physician Hans Horst Meyer, and the director of the Radiumstation at the Allgemeines Krankenhaus Gustav Riehl. Knaff-Lenz was suggested as secretary of the congress's medical committee.[57]

The point is that the proximity of science institutes located in the Mediziner-Viertel mattered seriously to the science produced and the scientists involved in its production. Within the boundaries of the Viertel, personal communication took place that facilitated interdisciplinary exchanges. Along with the flow of ideas, there was a reciprocal flow of people across Währingerstrasse and within the institutes. The few women in science became visible. As we will see, this helped them to become surprisingly well integrated within the Viennese scientific community.

Most of all, the architectural interconnectedness of science buildings provided a sense of stability to the scientists of the Viertel for their collaborations. Hosting a trafficking material, the Institute for Radium Research and its scientists depended on interdisciplinary collaborations and, thus, were in need of stable routes of scientific exchanges.

THE ARCHITECTURE OF RADIOACTIVITY

Locating radioactivity in the city and addressing safety issues in the institute's design might principally have concerned Meyer and Exner, but their patron was preoccupied with the aesthetics of the new discipline. In his donation, Kupelwieser imposed the interesting condition that the institute ought to be built in a "pleasant architectural form," without defining what that should be.[58] Vienna's architecture was then in transition; it was a period of "creative unrest."[59] From the construction of the Ringstrasse—the boulevard with its monumental Renaissance and Baroque-style buildings surrounding the old Vienna—architecture had moved through a period of radical change to the Moderne movement.[60]

The most representative architect of the period, Otto Wagner, paid great attention to the traditional architecture of the city and adopted the Viennese style of weight and monumentality. Yet, he also certainly broke with this past, establishing his Nutzstil, the functional style, arguing for artistic simplicity.[61] His motto—necessity is art's only mistress—was reflected in his buildings and exemplified Vienna's architecture during the first decade of the twentieth century.[62]

"The form of the Radium Institute appeared to meet the artistic demands of the time," as Rainer Pawkowicz notes.[63] Deeply influenced by Wagner's simplicity, the architecture of the institute, though it demonstrated nothing of

Wagner's authentic architectural skill, adopted the principle that "nothing unpractical can be pretty."[64] The form of the building smoothly followed its function. In the original plans, the façade of the institute combined elements of architectural historicism with an austere structure. The ornamentation was discreet with simple motifs above the windows, a narrow balcony at the second floor, and external ornamented pilasters. When the institute was actually built, the façade was simpler with less ornamentation and pure lines, just a simple arch above the entrance and plain motifs above the windows. A few architectural elements such as the balcony and the akrotiria on the four upper corners of the roof signified the transition from Historismus to Moderne.

On the one hand, the exterior of the building made a clear reference to the modernity of its occupants, appropriate to the new discipline they were establishing. The architectural style of the Radium Institute was especially revealing in comparison to the Josephinum, the last Baroque building constructed in Vienna between 1783 and 1785, which stood opposite the natural-science quarter. Besides its heavy ornamentation, the courtyard of the Josephinum, which was also a science building, created a strong separation between the public and scientists. In contrast, the façade of the Radium Institute did not incorporate a visibly intermediate space, but extended into the street, reorganizing boundaries between those who had the right to cross the building's threshold and those who were excluded. Social and political changes in fin de siècle Vienna had for example drawn more women to the sciences, including radioactivity. The designers of the institute can be seen as responding to changes in access to science which were already occurring in Viennese society and the city's university.

On the other hand, the interior of the institute could be understood as a combination of a modernizing and a comforting functionalism. All spaces were equipped with modern facilities that the Türkenstrasse physicists would probably have found hard to imagine. Windows provided natural light to laboratories and office spaces from all sides while thick roller blinds protected experiments and experimenters from the unpleasantly hot summer sun. During cold Viennese winters, gas central heating created a pleasant working atmosphere, relegating the chilly winters at Türkenstrasse to the status of a nightmare. Each floor was equipped with gas and with two chemical stoves, a new technological development. The electrical company Stögermayr designed the electrical system, providing each floor with direct, alternating and accumulator current, and with currents of different strength and voltage for different types of research. Delicate, simply ornamented lamps were installed in the institute, illustrating the value of functionalism combined with artistic pleasure. Meyer's office was expensively furnished by the well-known American Trade Company.[65] There were telephones in each working space, making the institute one of the most luxurious, technologically advanced, and prestigious science buildings of its time. In this respect, Vienna's Radium Institute resembled most leading science institutes built in Europe around the same time.

The different message the institute conveyed, mainly through its exterior design, was that a new science demanded new architectural forms and aesthetics. Heinrich von Ferstel's heavy ornamentation and Renaissance designs of the natural-science institutes in fin de siècle Vienna did not fit a modern world. Rather, it was efficiency, economy, and the facilitation of science that Kupelwieser, Exner, and Meyer had in mind when planning the institute. Compared, for example, to the Kaiser-Wilhelm-Institute for Chemistry that opened in 1912 in Berlin and included a department for radioactivity research under the directorship of Otto Hahn and Lise Meitner, the architecture of Vienna's institute followed a much simpler form. Although the original conception of the exterior of the Kaiser-Wilhelm-Institute was planned with a conical roof on the turret, when it was finally built, there was a cupola instead, a decorative touch that seriously concerned the scientists.

Despite the suggestions of Emil Fischer, Germany's leading chemist, that "simplicity is characteristic of all experimental science . . . therefore the workshops of experimental research should also be free from any splendor but equipped with all the means of advanced technology," the architecture of the Kaiser-Wilhelm-Institute demonstrated the difficulty of well-established disciplines such as chemistry to break with traditional architectural forms of imperial Germany.[66]

On the contrary, in Vienna, the architecture of the newly created natural-science quarter reconfigured the relationship between the traditional disciplines of physics and chemistry and the newly emergent field of radioactivity, allowing them to coexist while drawing boundaries and protecting distinctions. As an indication of the common interests that the communities of physicists and radioactivists wanted to pursue, the Radium Institute's blueprints show a bridge at the third-floor level connecting it to the Physics Institute. "The bridge to the Physics Institute was not only a material connection," as Meyer pointed out, "but also an ideal one; friendly relationships with the neighboring institute were necessities of life."[67]

During the 1920s, the contacts proved to be exceptionally important for both sides. In 1922, Swedish physicist Hans Pettersson arrived in Vienna with his wife, the chemist Dagmar Pettersson, both intending to work on the artificial disintegration of light elements. He and his Radium Institute collaborators soon seriously challenged assumptions about the nature of the atomic nucleus put forward by Ernest Rutherford and his colleagues at the Cavendish Laboratory in England. The need for specialized personnel led workers on other projects to shift to Pettersson's group, attracted young researchers to the institute, and drew others from the Physics Institute next door.

Among those who joined the group were female physicists such as Marietta Blau, Elisabeth Rona, Hertha Wambacher, Stephanie Zila, and Elvira Steppan. In response to the continuing controversy with the Cambridge group, the Viennese mobilized the facilities offered by the better-equipped neighboring institute, which housed more reliable measuring instruments. Indeed, the physicists next door had an impressive library, experimental apparatus and, most important, uncontaminated rooms, all lacking at the Radium Institute.

In his formal report to the International Educational Board in 1928, Pettersson repeatedly emphasized that:

> As these investigations [on artificial disintegration] must necessarily be divided into one "contaminated" and one "uncontaminated" part, the first for preparing and measuring the intense radioactive sources, the second for detecting their extremely minute effects on the shape of the rare atomic fragments, the work would not have been possible had not the Radium Institute happened to be in such close and excellent cooperation with the adjacent I and II Physics Institutes, the directors of which have also put a considerable part of their apartments and equipment at our disposal.[68]

The same advantages were recognized by Rona who recalled that Geiger counters and parallel plate condensers had to be moved to the Physics Institute because their own institute was highly contaminated. Above and beyond material benefits, the bridge offered the science practitioners in both institutes the opportunity to exchange ideas and to learn about their colleagues' latest research. At a symbolic level, the bridge supported feelings of membership in the same physics community, while it gave stability to particular collaborations.

Although he was director of the II. Physikalisches Institut, Exner remained the official director of the Radium Institute until his retirement in 1920, and retained his offices in both buildings. The bridge made material his and his colleagues' dual affiliation. On the other side of the building, the proximity of the Chemical Institute became essential after the end of the First World War for it facilitated the already existing collaboration between physicists and chemists yet maintained the rigidity of their disciplines' boundaries.

Disciplinary boundaries were certainly projected onto the institute's architecture, but to understand how the boundaries between the traditional disciplines and the emergent field of radioactivity were established and enforced, we need to follow the practices of the institute's experimenters. For example, although Meyer and his colleagues were required by the Austrian Academy of Sciences to publish their work in its *Sitzungsberichte*, they also published in the institute's *Mitteilungen*, a separate annual edition which collected all the papers of the institute's researchers. Partly because of its practical orientation toward research purposes, this distinct publication on radioactivity legitimized the new field and shaped its boundaries. Most scientists nevertheless continued to publish extensively in prestigious physics journals, such as the *Zeitschrift für Physik* or the *Physikalische Zeitschrift*, but they also contributed to the *Zeitschrift für Physikalische Chemie*, to the *Zeitschrift für Technische Physik*, and to *Strahlentherapie*, a journal of medical physics.

DECIPHERING THE INTERRELATION OF GENDER, RADIOACTIVITY, AND ARCHITECTURE

Marelene and Geoffrey Rayner-Canham have recently brought our attention to the fact that women were drawn to radioactivity research in the early twentieth century. Identifying three different European research schools on radioactivity—the French, the English, and the Austro-German—the Rayner-Canhams argue that women "seemed to play a disproportionately large share in the research work in radioactivity compared to many other fields of physical science."[69] In the case of Vienna, Peter Galison has fairly characterized the institute as "a Mecca" for women in radiochemistry and radiophysics.[70] Most of these women were fluent in major European languages and moved easily from Vienna for short periods to the laboratories of the Curies, Lise Meitner, Otto Hahn, or Rutherford. They participated as full colleagues in the research of the Radium Institute and played an important role in the broader scientific community.

Attention to internal arrangements and spatial layout proves to be an important tool in identifying gender relations in the laboratory workplace and in deciphering exceptional gender assumptions about accessibility in Vienna radioactivity research. The Radium Institute's architecture gave physical meaning to conditions not just on who might enter the building but also on who might gain access to the new discipline. To explore the space where women and men were involved in radioactivity research is therefore to look for exclusions and inclusions based on gender, and for ownership of space and for gender assumptions in architectural design.

Before architects began to draw detailed plans of the institute, Meyer and Exner had already defined for them the exact space requirements for laboratories, offices, and support rooms. Sharing a collegial ethos in physics work, they placed a high value on a welcoming environment for their research personnel, combining spaces for socializing and spacious research laboratories. Their design referred not only to the peculiar interdisciplinary nature of the new field by combining chemical and physics laboratories but also to the fact that both men and women were involved in radioactivity research.

The building was characterized by its central symmetry around the axis of the staircase, which in its slim form suggested its function as a means for vertical traffic. In each intermediate floor, the staircase had large open landings that gave enough room to accommodate sanitary installations, including one bathroom between the ground floor and the mezzanine. The bathroom was designed for the needs of those who worked with radium preparations or polonium sources and were often contaminated with radioactive powder. It was not at all unusual for the researchers to take a shower or wash dirty laboratory uniforms. Besides those in the cellar and the top floor, the remaining landings provided access to two small toilets, one on the left and one on the right, most probably designated for men and women. The fact that not every floor had two separate sanitary installations may reflect the smaller number of women in the institute.

Was the existence of toilets for women a common architectural feature of science buildings at the time? Sophie Forgan mentions that even today in physics and engineering departments, one has to go a long way to find women's toilets.[71] In this respect, Vienna's Radium Institute was indeed an exception and differed, for example, from another radioactivity research centre, the Physics Institute in Heidelberg built around the same time.

In 1908, the physicist Philip Lenard, director of the new institute in Heidelberg, submitted to the state the building's *Raumprogram*. The architect Friedrich Ostendorf provided the detailed designs. In addition to the Physics Institute, the plans included a radiological institute with three departments: radiophysics, technical radiology, and medical radiology. Financial disagreements between the state and the institute's designers delayed the process until July 1909. Yet the final blueprints had to be modified again in order to meet the approval of the state solely because they did not include separate toilets for women.[72]

As the early twentieth century saw increased female participation in science, architects, city planners, and the state apparently had to accommodate this newly developing body politics. The existence of separate sanitary installations suggests that women were a common feature of the Radium Institute's scholarly life. Not only were both Meyer and Exner familiar with the fact that women participated in physics as students and researchers, they also encouraged such participation. In 1893, when women were not yet admitted to the University of Vienna, Exner was among those who signed a petition supporting women's right to higher education.[73] The accounts of many of the women who later worked at the institute further suggest that Meyer was a central, encouraging figure in their laboratory work. Thus, it is not at all surprising that Meyer and Exner acknowledged women's bodily needs in the institute's everyday life, anticipating the state's inspectors.

A second architectural element of the institute's building reflected its remarkable arrangement for keeping the institute and its laboratories in order by accommodating special housekeeping personnel. Inside the front entrance, one would first reach the mechanic's apartment on the right. The apartment was directly linked to a room where a housemaid cleaned the devices used in chemical procedures.[74] Clearly, the institute's designers were trying to create comfortable conditions for the research staff, who were now concerned only with their own scientific projects. A similar setting in the Kaiser-Wilhelm Institute for Brain Research in Berlin around the same period had a notable result. As Annette Vogt has documented, the personnel responsible for the housekeeping of the institute were also responsible for cleaning the apartments of all the scientists and technical assistants who resided near the institute. This arrangement, Vogt argues, created unique conditions for the women who worked at the institute by freeing their time from everyday home tasks and laboratory cleaning, thus enabling them to play an exceptional role in scientific research.[75] So given the fact that they did not carry the burden of laboratory cleaning, a stereotypical role for women in science, women researchers in Vienna's Radium Institute enjoyed in this respect similar work conditions to that of their male colleagues.

Focusing on the architecture of laboratory and office spaces in the institute reveals that there were several other respects in which men and women made use of the space in a more or less equal way. Among the peculiarities of the institute's design was that the workrooms on each floor were all small but interconnected, providing wider access to other practitioners while maintaining the feeling of working in a semi-secluded, personal space. One was able to enter each room from at least two different doors, and interconnectedness was the inevitable result. Doors served to articulate the flow of scientists and probably that of small portable apparatus in a variety of workspaces with different functions. Men and women had access to all of these rooms regardless of their gender. As the pictures of Rona and Georg Stetter illustrate, what mattered was their scientific expertise, their knowledge of operating a specific apparatus, and their need to perform an experiment.

New scientific demands shaped and transformed spatial arrangements, especially after the Second World War. Pettersson's experiments on the artificial disintegration of light elements signaled a shift from radioactivity to nuclear physics in the institute's research agenda. Introducing scintillation counters, photographic emulsions, and Shimizu–Wilson ray track apparatus, he altered not only the materials used at the institute but also the space and culture that surrounded them. The new experiments required larger collaborative groups. Stools and chairs replaced the long benches. Plane tables were arranged around the wall with one of their short sides attached to it. Although they served as individual desks, these tables had enough empty space around them to accommodate more coworkers when needed. In addition, lack of space in the institute forced specific transformations in laboratory design, which continued to allow workbenches to be allotted to each researcher by assigning more than one scientist to a given room. Photographic material suggests that there was an interesting symmetry in the arrangements of space between men and women at the institute.

Textual evidence for women's ownership of laboratory space complements the suggestive photographic material. According to Elisabeth Rona's autobiography, when she entered the institute in the early 1920s, she was assigned Exner's former office due to a shortage of office space.[76] From the Pettersson–Karlik correspondence we also learn that Karlik "had locked the room so that only a 'Funktionär' could have entered,"[77] and later that Przibram "could hear the pump running in *my room* all the time."[78] Karl Przibram was the *Ordentlicher Assistent* at the institute from 1919 to 1938. He worked on the phenomenon of radiophotoluminescence, which he had discovered in 1921. He also extensively studied the coloring of crystals and the fluorescence of fluorites. His studies attracted a number of researchers, many women among them. His main collaborator was Maria Belar, but Elisabeth Kara-Michailova, Luisa Gröger, Marie Hoschtalek, Berta Karlik, and Berta Zekert also worked with him.[79]

Furthermore, at the Radium Institute there was no confusion over responsibility for keeping space and equipment neat for the next user or over the ownership of instruments and laboratory benches. From several of the institute's publications, it becomes obvious that possessiveness concerning instruments was clearly acknowledged and respected by personnel. In 1927, for example, Kara-Michailova thanked Georg Stetter for his generosity in lending her his apparatus for her experiments on the brightness of scintillations.[80] In a letter to Pettersson in 1933, Karlik mentioned that "Ortner had actually borrowed the Cenco-Hyrac-pump (we had agreed on that) and that he and Frau Sehork had just replaced it."[81]

Another facility of the institute—the library—reflects ways in which women made space their own. By doing so, they reinforced their role as members of their scientific community. The institute's library situated in the mezzanine was planned to serve as a bibliographical source, a space where researchers could consult recent publications related to radioactivity. Facing Boltzmanngasse at the right-hand corner of the building, the library had two external sides where windows allowed in the maximum amount of light. The literature stocked specialized exclusively in radioactivity. "For the remaining literature," as Meyer explained, "one was dependent on the big neighboring library of the Physics Institute (as well as the institutes of mathematics and chemistry)."[82]

The library served to reinforce the message. By moving to the Radium Institute, researchers working in the boundary field of radioactivity marked their own distinct community at the intersection of the physics and chemical institutes. They preserved their individuality, yet seriously depended on colleagues in the neighboring institutes. The move from the old Physics Institute at Türkenstrasse to the new natural-science quarter was followed by a corresponding move from the neighboring coffee houses where physicists used to spend their time to the more secluded institute library for social gatherings. It was a move from the public to the semi-private domain of the Radium Institute. Scientific discourse was moved out of cafés to a new institution that legitimized researchers in radioactivity and gave them a unique institutional persona. From the *Kaffehaus* to the institute's library, the physicists carried with them the Viennese cultural identity which permeated their scientific ethos. They shaped their library into a space where people went to meet friends and colleagues, to relax, and to consult the latest journals. It became the space where the Viennese "radioactivists" reinforced their feelings of belonging to a new discipline.

The social activities that took place in the library and surrounded the practice of radioactivity research allow us to identify more precisely the nature of the radioactivity community in Vienna. The small library was transformed into the heart of the institute by serving as the setting for exciting discussions on ongoing experiments, scientific developments, political upheavals, and the cultural life of Vienna. Every evening from the institute's inception until 1938 researchers met informally in the library to drink coffee and converse with their colleagues. "As I remember," Pettersson's daughter Agnes Rodhe recalled, "every day at four o'clock they had the so called 'jauze,' afternoon tea, and all the institute met, at least the ones who were free to drink coffee, chocolate, or tea and have desserts."[83] Strong friendships among the researchers started in that period and were developed through the collegiality of those coffee hours. "We all ended up close together," as Hess later put it, "and the collegial relations were extremely cordial. I thank here my friends Kohlrausch, Schrödinger, Przibram, Paneth, Hevesy, and Thirring, all in the old Exner-circle joined together, so to speak, we built a family."[84]

Those who tied the library to their social activities were not only the men of the institute. Archival sources document that women not only used the library for research, but that they participated in the social gatherings as well. As Karlik reported in a letter to Pettersson in 1938:

> The tea standard has been very high lately; no political remarks. Ortner getting more and more interested in theory and taking part regularly; a theoretician called Ludloff who has now got his *Dozentur* here (he comes from Breslau), Frau Dr. Seidl, occasionally even Sexl coming. Mattauch belongs already quite to the family.[85]

Franziska Seidl was one of the few women who, as early as 1923, worked as *Wissenschaftlicher Hilfskraft* (scientific assistant) at the first Physics Institute and in 1924, she was promoted to the position of *Assistentin* (assistant).[86] After her habilitation in experimental physics in 1932, Seidl became *Privatdozentin* (private lecturer) at the University of Vienna, teaching a course every semester from 1933 onwards. She focused on several topics in experimental physics including x-rays and their use.[87] Josef Mattauch and Theodor Sexl worked as *Assistenten* at the Physics Institute next door and Hans Ludloff had just arrived in Vienna from Breslau to teach a course on quantum theory.[88]

In addition to the regular "jauze," the women of the institute were visible during the formal visits of internationally known physicists. In 1933, when J. J. Thomson and his wife paid a formal visit to the Vienna Institute, Karlik entertained them socially with the rest of her colleagues. "At the Przibrams, we had the most delightful afternoon," as she reported to Pettersson on July 26, 1933.[89] As the episode suggests, women had established a lasting presence in the field. As in the case of Karlik, this was carried over even after the Second War World when she became the institute's director.

READING THE HISTORY OF RADIOACTIVITY THROUGH THE ARCHITECTURE OF ITS BUILDINGS

The concept of Vienna's Radium Institute and its architecture grew in response to the needs and interests of its designers and the concerns of the network of academics, their financial supporters, and the state. In 1910, the institute's planners created a place that had not existed before, both physically by building the institute and institutionally by creating a separate space for the new discipline. It was conceived as more than a research institution, since those in charge sought to provide both male and female researchers with a sense of scientific community. Yet over the years, the institute functioned as a contingent space, being constantly shaped by those who inhabited it. It grew in response to their demands, their scientific practices, the material culture that surrounded them and, most interestingly, their gender.

Architecture is more than the trace of these practices. I have argued that by designing for a new discipline, scientists and architects were collaboratively designing the new discipline in a strong sense. It is necessary but not sufficient to observe that the architecture and urban location of the institute were shaped by the subject matter of the new discipline and radium's properties as a trafficking material, allowing different scientific disciplines to coexist and to contribute to the making of radioactivity as a field, while protecting their distinct identities.

Most important is that the institute's planners saw the design of the new discipline as their main goal, in the sense of reinforcing specific disciplinary exchanges, forming a distinct community, and even encouraging the entrance of women into the field.

The institute proved to be an embodiment of the disciplinary, cultural and scientific status, and of the gender ideology of its researchers. The new aesthetics of the modern era, as they were expressed in the institute's architectural form, provided the physicists who specialized in radioactivity research with a new disciplinary identity. The building helped give tangible existence to the new field by making the researchers visible and institutionalizing their practices. It further added to their prestige and offered them unique opportunities for research. At the same time, the building's internal architectural arrangements made it possible to increase scientific collaboration and social interactions among scientists with a variety of expertise. Specific architectural features, such as the existence of women's toilets, point to the gendering of the laboratory space and to the fact that women's existence in the institute was both acknowledged and facilitated. Women working on radioactivity succeeded in acquiring "a laboratory bench of their own," indicating a shift in the political importance of the role of women in science. The reconfiguration of their space in laboratory work was in fact emblematic of their presence in the field and of their membership of the radioactivity community.

Radioactivity researchers considered themselves members of a distinct discipline, but at the same time ensured their connections to traditional ones as a way to empower their newly acquired identities. Indeed, the institute cannot be seen solely in terms of the forging of a singular community, that of "radioactivists." This would draw too tight a circle around those who lived and worked in and around it. Although it emerged in the city as a place apart, the Radium Institute was rather an inhabitant of the city, more specifically of the Mediziner-Viertel, a space which created a fluid relation between professional and private spheres. Its location manifested new modes of exchange between physics, chemistry, and medicine, which were enforced by the field of radioactivity. Both its urban location and architectural form were driven by the kind of interdisciplinary research done in its laboratories, the instruments in use, the hazardous properties of radium, its modes of trafficking, and, to a surprising extent, the gender identities of its personnel. Equally, the urban sitting of the institute in the Mediziner-Viertel reinforced exchanges of apparatus, ideas, and personnel among the institute and Vienna's medical institutions, as well as the physics and chemical institutes. In brief, as a space and as a concept, the Radium Institute took advantage of a transitional moment in the development of radioactivity as a discrete discipline; it was bound to traditional fields through a complex of architectural features.

NOTES

1. See, for example, Lang and Suess to the Austrian Ministry of Agriculture, January 15, 1904; Lang and Suess to Ritter von Hartel, January 15, 1904; Lang and Suess to the Ministry of Culture and Education, March 14, 1908, all in AÖAW.

2. Meitner, "Looking Back" (1964); H. Benndorf, "Worte der Erinnerung an Franz Exner" (1926), Nachlass Exner, AÖAW.

3. Benndorf, "Gedenkrede auf Franz Serafin Exner aus Anlass der Enthullung seines Denkmals in der Universitat Wien am 23 Janner 1937," Nachlass Exner, AÖAW.

4. Lang and Suess to the Ministry of Culture and Education, March 14, 1908, AÖAW; Lang and Suess to the Ministry of Work, December 21, 1908, AÖAW.

5. Badash, "The Suicidal Success of Radiochemistry" (1979), 12.

6. Hughes, " Modernists With a Vengeance" (1998), 342.

7. Hessenbruch, "Rutherford's 1901 Experiment" (2000), 419.

8. Boudia, "The Curie Laboratory, Radioactivity and Metrology" (1997); Boudia, *Marie Curie et son laboratoire: Sciences et industrie de la radioactivitè en France*, Paris, (2001); Roque, "Marie Curie and the Radium Industry" (1997); Roque, "Displacing Radioactivity" (2001).

9. Rutherford's similar references to "radioactive world" go back as early as 1905. In 1919, he informed Boltwood that it seems "to be an epidemic of marriages among the Radioactive people" as both Hevesy and Hönigschmid were getting ready to marry. Badash, *Rutherford and Boltwood: Letters on Radiaoctivity*, (1969), 322. See also Hughes, *The Radioactivists* (1993), 6. In his presentation speech for the Nobel Prize in chemistry in 1908, Hasselberg recognized that researchers such as Rutherford, Curie, and Becquerel "are closely allied to, and form a worthy continuation of work," referring to radioactivity research. See K. B. Hasselberg, *Presentation Speech, The Nobel Prize in Chemistry*, December 10, 1908. http://nobelprize.org/nobel_prizes/chemistry/laureates/1908/press.html.

10. Badash, *Rutherford and Boltwood* (1969), 205.

11. Hughes, "Radioactivity and Nuclear Physics" (2003).

12. See the special section on radioactivity research by Rentetzi, (coordinator), "Women Pioneers in Radioactivity Research" (2004).

13. Historian of Victorian science Roy MacLeod suggested long ago that medals "codified schools of thought and legitimized scientific paradigms." Macleod, "Of Medals and Men" (1971).

14. Lawrence, "The Origins of Big Science—Rutherford at McGill" (1979). See also Burcham, "Rutherford at Manchester, 1907–19" (1964).

15. Mould, "The Discovery of Radium" (1998); Peh, "The Discovery of Radioactivity and Radium" (1996).

16. Quinn, *Marie Curie* (1995), 203.

17. Badash, *Rutherford and Boltwood*, (1969), 158–59.

18. For example, embarrassed by his provisional laboratory, Franz Exner, the director of the Second Physics Institute, left Vienna before his international colleagues arrived for the meeting of German scientists and physicians in 1894. Moore, *A Life of Erwin Schrödinger* (1994), 57.

19. Wettstein, "Karl Kupelwieser" (1928). See also Löffler, "Limnology in Austria" (2001), 15; Reiter, "Vienna: a Random Walk in Science" (2001), 476. Janik and Veigl, *Wittgenstein in Vienna* (1998), 38–39.

20. Occasionally, Kupelwieser also made donations to the medical society of Vienna. Fischer, *Geschichte der Gesellschaft der Ärtze in Wien 1837–1937* (1938), 115. His was such a powerful figure in fin de siècle Vienna that when the state decided to construct a huge hydroelectric station close to the Biological Station, Kupelwieser was able to maneuver politically and cancel the decision. Wettstein, "Karl Kupelwieser" (1928), 14.

21. Karl Kupelwieser to the Academy of Sciences, August 2, 1908, AÖAW.

22. For example, Franz Serafin Exner characterized Kupelwieser as "the patriotic friend of science." *Almanac of the Austrian Academy of Sciences* (1912), 327, AÖAW.

23. Badash, *Rutherford and Boltwood* (1969), 193.

24. Suess and Lang to the Ministry of Culture and Education, March 14, 1908, AÖAW.

25. Pawkowicz, "Die Österreichische Akademie der Wissenschaften und die Bauten für ihre Forschungsinstitute" (1978), 72. Unfortunately, I was unable to locate any specific information on the architects of the project.

26. Ipser and Schönfeld, *Die Geschichte der Chemie an der Universität Wien und des Instituts für Anorganische Chemie* (1996), 22. As Wolfgang Reiter argues, the Vienna Radium Institute became the model for the Institut du Radium in Paris which architecturally resembled it. Reiter, "Stefan Meyer: Pioneer of Radioactivity" (2001), 114–16.

27. Suess and Lang to Auer von Welsbach, July 10, 1901, AÖAW.

28. As Karlik and Schmid put it, "Exner had also carried out essential organizational work for the University of Vienna. This concerned the planning, the preparation, the construction, and the establishment of the new Physics Institute at the corner of Boltzmanngasse and Strudlhofgasse." So besides working at the Radium Institute, Exner played a fundamental role in building the whole natural-science quarter. In collaboration with his friend Ernst Lecher, director of the neighboring institute, and his assistant Edward Haschek, he provided the plans for the Physics Institute. Karlik and Schmid, *Franz Serafin Exner und sein Kreis* (1982), 88. On Exner, see also Przibram, "Errinerungen an ein altes physikalisches Institut" (1959), 3; Schweidler, "Bericht des Generalsekretärs" (1927); Stöltzner, "Franz Serafin Exner's Indeterminist Theory of Culture" (2002).

29. Exner, "Le nouvel Institut pour le radium à Vienne" (1910).

30. Reiter, "Stefan Meyer: Pioneer of Radioactivity" (2001). The letters concerning the construction and furnishing of the institute were all addressed to Meyer. See, for example, the files "Bau des Institut, Erstausstattung," AÖAW.

31. Hoffmann, "The Design of Disturbance" (2002), 9.

32. Cahan, *An Institute for an Empiire* (1989), 168–75.

33. Benndorf, "Gedenkrede auf Franz Serafin Exner aus Anlass der Enthullung seines Denkmals in der Universitat Wien am 23 Janner 1937" Nachlass Exner, AÖAW.

34. Philosophische Dekanat 1895/6, Akt 117, Archive of the University of Vienna, quoted in Coen, *A Scientific Dynasty* (2004), 228.

35. Golitschek and Elbwart, *Der Neubau des Physikalischen Institutes der Wiener k.k. Universität* (1915), 1.

36. Hittner, *Geschichte des Studienfaches Physik* (1949), 73–75.

37. Hittner, *Geschichte des Studienfaches Physik* (1949), 73.

38. Schönfeld and Ipser, "Die Geschichte der Chemie" (2000), 22.

39. Pawkowicz, "Die Österreichische Akademie der Wissenschaften und die Bauten für ihre Forschungsinstitute" (1978), 73.

40. Meyer, "Das erste Jahrzehnt des Wiener Instituts für Radiumforschung" (1920), 3.

41. Hess, "Persönliche Erinnerungen aus dem ersten Jahrzehnt des Instituts für Radiumforschung" (1950), 44.

42. Hans Pettersson's report on the investigations regarding artificial disintegration, first half of 1926, GUB (in English).

43. Rona, *How it Came About* (1978), 24–5.

44. Rona, "Laboratory Contamination" (1979), 724.

45. Rona, *How it Came About* (1978), 29.

46. Rona, "Laboratory Contamination" (1979), 724.

47. Rona, "Laboratory Contamination" (1979), 724, 727.

48. Karlik to Pettersson, January 16, 1966, GUB.

49. Meyer, "Das erste Jahrzehnt des Wiener Instituts für Radiumforschung" (1920), 3.

50. Brinda-Konopik, "Robert Wilhelm Bunsen und die chemische Schule an der Universität Wien (Alma Mater Rudolphina)" (1992), 3; Rentetzi, "The City as a Context of Scientific Activity" (2004).

51. After its discovery, it became clear that radium affected human tissues, a finding that provoked medical practitioners to use it in cancer therapy. It is a surprise that although Viennese physicists as well as physicians were among the pioneers in the new field, when the Radium Institute was established medical research was explicitly excluded. Kupelwieser did not state the reasons for such exclusion. See also Paneth, "Über die Arbeit des Instituts für Radiumforschung" (1915).

52. In 1910, for example, Ernst Lecher taught experimental physics for medical students and a year later, he published one of the first textbooks on physics for physicians and biologists (Schweidler, Schweidler, "Bericht des Generalsekretärs" (1927), 178). Egon Schweidler offered a course on experimental physics for pharmaceutical students and Felix Ehrenhaft lectured on the theory of the microscope. The same year, Skraup Zdenko taught inorganic chemistry for pharmaceutical students and Herzig Josef offered pharmaceutical chemistry (Vorlesungen an der k.k. Universität zu Wien, 1910–1911, AÖAW).

53. Marietta Blau, curriculum vitae, 1941, GDSCA.

54. Meyer to Riehl, June 7, 1912, AÖAW.

55. Fernau to Meyer, October 17, 1912; Fernau to Meyer, February 8, 1913; Meyer to Fernau, May 10, 1912; Fernau to Meyer, June 10, 1912, all in AÖAW.

56. Meyer, "Die Vorgeschichte der Gründung und das erste Jahrzehnt des Institutes für Radiumforschung" (1950), 20. I was unable to identify the first name of Goldschmidt, the woman who collaborated with Rona.

57. Meyer to Rutherford, December 5, 1913; Meyer to Rutherford, December 16, 1913, Rutherford Papers, CUL.

58. Kupelwieser to the Academy of Sciences, August 2, 1908, AÖAW. Probably Kupelwieser himself decided on the project architects as these were the only ones he paid directly. For the rest of the expenses, he transferred money to the account of the academy. Suess to Exner, October 28, 1910, AÖAW.

59. Borngässer, "Architecture From the Turn of the Nineteenth Century to the Present Day" (1999), 273.

60. Schorske, *Fin-de-Siècle Vienna* (1980).

61. Pintaric, *Vienna 1900* (1989).

62. Schorske, *Fin-de-Siècle Vienna* (1980), 73.

63. Pawkowicz, "Die Österreichische Akademie der Wissenschaften und die Bauten für ihre Forschungsinstitute" (1978), 81.

64. O. Wagner, *Die Baukunst unserer Zeit*, Wien, 1979, 45.

65. American Trading Company to Meyer, May 20, 1910, AÖAW.

66. Johnson, *The Kaiser's Chemists* (1990).

67. Meyer, "Das erste Jahrzehnt des Wiener Instituts für Radiumforschung" (1920), 6.

68. Hans Pettersson's report to the International Education Board, Part III, April 1928, GUB (in English).

69. Rayner-Canham, M. and G. (eds.), *A Devotion to Their Science* (1997), 12.

70. Galison, *Image and Logic* (1997), 150.

71. Forgan, "Eine angemessen Häuslichkeit? Frauen und die Architecture der Wissenschaft im 19. Jahrhundert" (2003), 145.

72. Auer, *Das Physikalische Institut der Universität Heidelberg an der Albert-Überle-Strasse* 7 (1983), 15. I owe thanks to Christoph Hoffmann for pointing out this article to me.

73. Bischof, *Frauen am Wiener Institut für Radiumforschung* (2000), 23.

74. Meyer, "Das erste Jahrzehnt des Wiener Instituts für Radiumforschung" (1920), 4. From the early days of the institute to 1938, Karl Kornher was the mechanic and until 1922, Stanislaus Kijowski served as maid. In 1924, Kijowski was promoted to laboratory assistant (*Laborant*) until his retirement in 1936, when his son, Julian Kijowski, took over the position. In 1930, Josefine Schörg was hired as an additional cleaning lady. *Almanac of the Austrian Academy of Sciences*, 1920–38, AÖAW.

75. Vogt, "'The Timofeeff-Ressovsky's—a Couple in Science" (2000), 4.

76. Rona, *How it Came About* (1978), 15.

77. Karlik to Pettersson, 1 August 1933, GUB.

78. Karlik to Pettersson, 9 March, 1934, GUB. Added emphasis.

79. Karlik, "Karl Przibram Nachruf"(1974); Przibram, *Irradiation Colors* (1956).

80. Kara-Michailova, "Helligkeit und Zällbarkeit" (1927).

81. Karlik to Pettersson, August 1, 1933, GUB.

82. Meyer, "Das erste Jahrzehnt des Wiener Instituts für Radiumforschung" (1920), 4.

83. Agnes Rodhe, interview by the author, September 22, 2001, Göteborg, Sweden.

84. Hess, "Persönliche Erinnerungen aus dem ersten Jahrzehnt des Instituts für Radiumforschung" (1950), 45.

85. Karlik to Pettersson, February 3, 1938, GUB.

86. Bischof, *Physikerinnen* (1998), 13; Hittner, *Geschichte des Studienfaches Physik* (1949), 243.

87. *Vorlesungen an der k. k. Universität zu Wien*, especially for the summer semester of 1935, AÖAW.

88. At the end of 1938, Mattauch moved to Berlin as director of the Kaiser Wilhelm Chemistry Institute. Hittner, *Geschichte des Studienfaches Physik* (1949), 241–4. See also Vorlesungen an der k. k. Universität zu Wien, winter semester 1937/38, AÖAW.

89. Karlik to Pettersson, July 26, 1933, GUB.

3. GENDER, SCIENCE, AND THE CITY

In 1886, when the physicist Ernest Mach asked, "How is humanity to progress safely if not even half of it is walking on an enlightened path?" women had not been accepted yet to the University of Vienna. As he argued in his popular scientific lectures, "The level of education and choice of profession for women should in no way be restricted."[1] In the mid-1890s, and while the acceptance of women at the University of Vienna was still under discussion, Mach, with his colleagues and the directors of the first and second Physics Institutes, Victor von Lang and Franz Exner, founded a committee for the support of women's admission to university studies.[2] Surprisingly, leading figures in Vienna's physics community, such as Exner, Mach, and Lang, all members of the intelligentsia and bourgeoisie, contributed immensely to creating a friendly environment for the young women of their social strata to create a space for themselves within the field of physics.

Cradled in fin de siècle Vienna, such physicists as the young Stefan Meyer inherited their cultural meanings about sexual difference and a collegial ethos of working in physics from Vienna's cultural and scientific milieu. At the turn of the century, Vienna was an extraordinary cultural setting where psychoanalysts, architects, artists, writers, musicians, politicians, philosophers, and scientists transformed their fields of knowledge, established new ones, and defined new ways of being. The city's intelligentsia produced distinct schools of thought in architecture, music, painting, philosophy, and psychology. Figures such as architects Adolf Loos and Otto Wagner, social critic Karl Kraus, composers Gustav Mahler and Arnold Schönberg, painters Gustav Klimt and Egon Schiele, philosopher and author Robert Musil, and the founder of psychoanalysis, Sigmund Freud, are only a few of those who marked the turn of the century, and they are among those who contributed to the myth of Vienna as one of the most creative milieus in Europe.

Starting in the early 1960s, the work of historian Carl Schorske became the dominant paradigm in understanding 1900 Vienna as the birthplace of a major part of modern culture and thought. The defeat of liberalism and the political crisis of the 1890s provided the context for elucidating the cultural success of the city. As Schorske argues, "the writers of the nineties were children of a threatened liberal culture."[3] Establishing their power in a struggle against aristocracy, liberals had their heyday between 1860 and 1880. By 1900, new political forces such as feminists, Social Democrats, and the anti-Semitic Christian Socials had gained enormous political influence. Although defeated by the new political forces, the liberals had already provided the space where the new social and political powers could evolve. The feverish urban reconstruction of the late 1880s and 1890s had created potential spaces of cultural and scientific activity in the margins of Ringstrasse. Viennese modernism was produced through the cultural elite's reaction to this "failure of liberalism."

As Steven Beller explains, the Schorskean paradigm turned Vienna into a fashionable topic of research, and until the 1980s Schorske's work was accepted as the canonical view of fin de siècle Vienna. Posing the question of the Viennese Jewish contribution to the modern culture, historians such as George Steiner and Ivar Oxaal seemed to threaten the Schorskean model by the mid-1980s. The acknowledgment by Schorske of the Jewish dimension to the production of Viennese culture was not enough. The new argument fully acknowledged the Jewish tradition and experience in Vienna's modern culture. More recent studies have further undermined the theses of the failure of Austrian liberalism, showing that the liberal bourgeoisie retained a great influence in the Austrian state through their economic and political power.[4] What has emerged from challenging Schorske's model is a more complex picture of the relationship between politics and culture in fin de siècle Vienna.

Part of the intellectual and cultural cartography of the era has been the strong women's movement that developed at the turn of the century. Overshadowed by the Schorskean paradigm, women's history in 1900 Vienna has been mostly neglected. According to Harriet Anderson's *Utopian Feminism*, the most frequently cited work on fin de siècle feminism, progressive Viennese women did not limit their interests to practical goals but aimed to a new social order and deeper sexual and moral reforms. Yet, they did campaign for more vocational opportunities such as the right to vote and a state-supported secondary and higher education.[5] They certainly succeeded in the latter cause. Before the end of the century, women were finally accepted to the University of Vienna thanks to the persistent petitions and campaigns by feminists, supported by the Social Democrats, and despite the counteractions of the conservative Christian Socialists.

However, the city of Vienna was not just the stage, a mere social and political setting for women's emancipation and high cultural reforms. It was also a sociospatial setting where the practices and characteristics of the urban environment molded meanings of sexual difference and progressive politics.[6] If the oldest historical parts of the inner city were identified with the nobility of the aristocrats, the area around the new university, the Mediziner-Viertel, nurtured the reformists. Science and culture, politics and philosophy encountered one another in a small quarter, all within a 15-minute walk. Instead of speaking about women's segregation from this extraordinary milieu, I choose to focus on the ways they were integrated into the culture of Viennese science, taking advantage of the locally specific articulations of class and gender in a space such as the Mediziner-Viertel.

DESIGNING THE UNIVERSITY OF VIENNA

After the failure of the liberal revolution in 1848, in which the university faculty and students played an essential role, the imperial army of the Habsburg Empire occupied the old university in the inner city. As a result, institutes were scattered in the outer districts presenting the university with a disciplinary diaspora. The chemistry and physics institutes were moved into the third district of the city. The medical faculty was housed in the Josephinum at Wäringerstrasse in the ninth district and later moved to the old artillery industry in the corner of Schwarzspanierstrasse and Wäringerstrasse, behind the Votivkirche. Only the faculty of theology had moved to Stadkonvikt, in Dr. Ignaz-Seipel Platz next to the Academy of Sciences in the inner city, where lectures on philosophy and law were also held.[7] Clearly, the emperor aimed to dispel the intellectuals, enfeebling their political influence. At the same time, expelling most of them from the inner city was a strategy for ensuring the aristocrats' safety from any revolutionary threat.

During the 1850s, Vienna experienced an immense economic growth. Economic enterprises doubled, as did the population of Vienna. The textile industry became the main financial source for the empire and attracted a considerable number of workers from the countryside. Consequently, the Emperor Franz Joseph I decided in 1857 to extend the inner city of Vienna, decreeing that the military area around it be converted to civilian use. After assuming power in 1860, the liberals meanwhile saw Vienna as the center of their intellectual life and as a symbol of their ideological commitments. One of their first concerns was the architectural transformation of the city and soon the core of Vienna's urban reconstruction was the Ringstrasse.

A 60-foot-wide tree-lined boulevard in the shape of a ring replaced the walls around the inner city of Vienna and offered to the newly emerging bourgeoisie the space to fulfill its expectations. For example, textile manufacturers belonged to the group with the second highest ration of home owners in the Ringstrasse area after the nobility. As Schorske argues, "the art of building, used in the old city to express aristocratic grandeur and ecclesiastical pomp, now became the communal property of the citizenry, expressing the various aspects of bourgeois cultural ideal in a series of so-called Prachtbauten (buildings of splendor)."[8]

The Opernhaus (1869), the Parliament (1883/4), the Rathaus (1883), and other monumental buildings along the Ring indicated their purpose by their impressive architecture: parliamentary government, municipal autonomy, dramatic art, and liberal ideas. Since Vienna's university, however, signified the revolutionary and oppositional forces that played a central role in the revolution of 1848, attempts to include its new building on the Ringstrasse generated longstanding conflicts within the government and delayed its construction.

Already in 1849, Count Leo Thun-Hohenstein, the minister of education and religion, took a special interest in restoring the scientific and administrative autonomy of the university. The reform program included the architectural grouping of all university faculties and administrative offices. Thun's plan was to group several buildings around the main university site and create a *civitas universitatis* in Gothic style. In 1853, along the same lines, two of the most important architects of the time, Eduard van der Nüll and August von Sicardsburg, proposed a neo-Gothic design for the university. It was a complex of four buildings arranged around two inner courtyards. The first and main building hosted the philosophical faculty and the law faculty, lecture halls, and administration offices. The library was located in a separate building as were the chemistry and the physics institutes. A sacred edifice functioned as the university's church.[9] They proposed to locate it behind the Votivkirche, a Gothic basilica designed by Heinrich von Ferstel. Nonetheless, the plans met with the immediate objection of Ferstel, who disliked the proximity of two Gothic monumental buildings. For the next 15 years, given the obstructionism of the city, the lack of money, and Ferstel's opposition, Thun and his collaborators worked in vain to create the new university. It was only thanks to the liberals' initiatives that the university finally took its place on Ringstrasse in front of the Votivkirche.[10]

In 1868, the mayor of Vienna, Kajetan Felder, assigned Ferstel to provide new site plans for the university. Changing the architect in charge meant also modifying Thun's original ideal of the medieval *civitas universitatis*. Increasing needs related to Vienna's economic growth and industry as well as to scientific developments required different structures and architectural forms. Besides adding research laboratories, seminar rooms, and clinics to the traditional lecture halls, Ferstel tried to create a Universitätsviertel, a quarter of science buildings open to the city but sufficiently close together to facilitate the work of the faculty and students. Intentionally, he planned to accommodate Vienna's Allgemeines Krankenhaus (general hospital) in the quarter and locate all different buildings within a ten-minute walk. As his model, Ferstel chose the Italian universities and their Renaissance designs. In 1871, he traveled to Italy, visiting the "cradle of modern humanistic learning," the traditional universities of Bologna, Genoa, Padua, and Rome.[11] Eventually, when the university opened on October 11, 1884, the modern humanistic ideals of education that both the liberals and its architect endorsed were clearly indicated in its Renaissance façade.[12]

FROM THE UNIVERSITÄTSVIERTEL TO THE MEDIZINER-VIERTEL

The Mediziner-Viertel, a triangular piece of land defined by Währingerstrasse, Alserstrasse, and Spitalgasse acquired its existence in fin de siècle Vienna not only by bringing several scientific institutes together in that specific location but by transforming the area into a vital cultural and epistemic centre in the city. Originally, it was dominated by the *Zinhausen*, several-story buildings designed as apartments and offered as a solution to the significant housing shortage in the city. Built to house the working class, they usually contained 16 units and were absolutely inferior to the *Adelspalais*, the aristocratic palaces they were trying to mimic architecturally.[13] As Emil Kläger and Max Winter, the Viennese journalists of *Arbeiterzeitung*, described the housing on the margins of Ringstrasse, where apartments averaged around 43 square meters, "There are four beds, in which some seven adults and one child sleep. The apartment is damp to a height of 1.1 meters. Rent is 20.40 crowns per month. There is no water in the house and the roof is badly in need of repair."[14] By the turn of the century, the face of the city's ninth district was changed radically by the number of scientific institutes that were founded and the new culture they established.

As early as 1783 to 1784, impressed by the Central Hospital in Paris, the Emperor Joseph II commissioned his architect Isidor Canevale and the court physician Joseph Quarin to design Vienna's Allgemeines Krankenhaus. The idea was to close the smaller hospitals that were dispersed around the city and group all medical facilities in one extensive complex. In the final plan and in addition to the main hospital building, there was a maternity wing, an asylum known as the *Narrenturm*, an infirmary, and a foundling hospital, all in separate edifices around several courtyards. Considered to be the last Baroque building in Vienna, the Josephinum, located at 25 Währingerstrasse in front of the Narrenturm, housed the surgical and medical faculty and part of the hospital. It was originally designed to accommodate army surgeons who did not have medical degrees and were organized in craftsmen's guilds. The building hosted a spectacular collection of 1,192 wax models of normal human anatomy, designed and prepared at the famous La Specola, the Royal Museum of Physics and Natural History in Florence. The models were used for the teaching of the Viennese surgeons and constituted the subject of an ongoing controversy between medical professionals during the late eighteenth and early nineteenth centuries, bringing the building to the public's constant attention.[15] The whole complex was situated between Währingerstrasse and Alserstrasse.

During the second half of the nineteenth century, the hospital was repeatedly expanded while new medical specialties were created. The Sanatorium Löw, later known as Sanatorium Wiener, the anatomical, physiological, and pharmacological institutes, the Röntgen Laboratory of the second Medical University Clinic of Vienna, the new university clinics, and the Röntgen Department of the Viennese Polyclinic were all built in the area around the end of the century. The architectural grouping of medical institutes in the university quarter contributed to the strengthening of the community of medical practitioners and university professors whose reputation had no precedent. The medical faculty of the University of Vienna was the only one that had a considerable number of foreign students who came to study under leading physicians such as the surgeon Theodor Billroth, the anatomist Joseph Hyrtl, and the physiologist Ernst Brücke. In addition, from the early 1870s through the turn of the century, the medical faculties were overcrowded with students, most of whom were coming from previously underrepresented ethnic, social, and geographic backgrounds. The Viennese physicians and their institutions had such a dominant presence in the city that they quickly dominated the Viertel and turned it to a Mediziner-Viertel—the physician's quarter—a designation that slowly replaced the Universitätsviertel—the university quarter. Through a piecemeal but consciously and politically chosen process, the Viertel clustered not only medical institutions and the main university building but soon included the natural sciences as well.

THE CHEMISTRY INSTITUTE

The main university building, however impressive and monumental it was, offered space only for administration offices and teaching facilities. Everyone else had to occupy separate buildings around the same area, realizing the plan for a university quarter. The institute of greatest priority was chemistry (1869–72).[16] The Theresianum, the building where the Viennese Chemistry Institute had been housed since the end of 1860s, proved insufficient for both research and teaching purposes. Its small and cheaply built laboratories could not meet the demands of well-trained chemists, working in the empire's increasing industry, nor chemistry teachers in the *Realschulen*, high schools with a strong focus on technical education. The nature of chemistry was changing drastically. The shift from inorganic to organic chemistry gave way to fruitful experimentation and created the demand for bigger research laboratories. In addition, by being situated in the third district, the institute was far away from Vienna's medical institutions. As the teaching of chemistry was mainly targeted to medical students, the common theme of the Viennese press was the student's complaint of being unable to attend classes given the long distance between the hospital and the Chemistry Institute.[17]

Not surprisingly, Joseph Redtenbacher, professor of chemistry, member of the Viennese medical faculty, and director of the Chemistry Institute together with his colleagues voiced their demands for a modern edifice located in the Viertel. The idea of a separate experimental laboratory for chemistry was drawn mainly from the well-established institute (1828) in Giessen, Germany.[18] According to the German tradition, institutes included three spatially diverse facilities under the same roof: lecture halls, laboratories, and an apartment, usually on the ground floor, for the building's caretaker. Often, the director of the institute was offered a free apartment on the upper floor of the building as part of his appointment agreement.

Besides supplementing the low salaries of the time, the symbiosis of teaching, research, and residence offered the directors the ability to have absolute control over their institutes and their assistants. Using this as his main argument and paying attention to Redtenbacher's suggestions, Ferstel, the architect of the Chemistry Institute, included in his plans not only the director's residence but professorial apartments as well. During the realization of the project, cost overruns raised questions about the usefulness of this peculiarity. When the institute was eventually opened in 1872, its impressive Renaissance style reflected not only Ferstel's flexibility to accommodate changes in political power in his architectural style, but also both his and Redtenbacher's ability to incorporate and support the new scientific culture. Large lecture halls, spacious laboratories, and professorial apartments were grouped under the same roof. In their architecture, they embodied the complex task of bringing together different disciplines and therefore, several professors as well.

Apparently, the location of the Chemistry Institute in the Mediziner-Viertel reflected not only the political changes that made possible the design of the university quarter in the first place but also changes in the culture of science. The shift was underway from teaching-oriented institutes with limited working space to institutes that combined research facilities with modestly advanced educational laboratories and lecture halls designed to accommodate a large number of students.

Redtenbacher was the first who took advantage of the political changes, using the enthusiastic support of the liberals for humanistic education as a lever to alter the culture of chemistry. Eventually, the impressive Renaissance-style Chemistry Institute was located on Währingerstrasse, a minute walk from the university, and it reflected not only Ferstel's flexibility to accommodate changes in political power in his architectural style but also both his and Redtenbacher's ability to incorporate and support the new scientific culture. Large lecture halls, spacious laboratories, and professorial apartments were grouped under the same roof. In their architecture, they embodied the complex task of bringing together different disciplines and therefore, several professors as well.

By locating the institute in the Mediziner-Viertel, within walking distance from the Viennese medical institutions, students were able to attend one lecture after the other with no delay. At the same time, the interdisciplinary nature of chemistry was reflected in the education and research of the institute's chemists. Starting with Redtenbacher, all the directors of the laboratories housed in the Chemistry Institute combined pharmaceutical, medical, and chemical expertise. Franz Cölestin Schneider, a pharmacist and medical chemist, and, later, Redtenbacher's successor, clearly stated that the institute should be designed to cover the purposes and needs of pharmacists.[19] The interrelation of chemistry and medicine was also reflected in the journals of the time. For example, one of the most prestigious periodicals, the *Annalen der Chemie und Pharmacie*, combined both pharmacy and chemistry in its title and content. Chemistry was obviously in the crossroads of pharmacy and medicine. Most of the chemists' teaching was indeed done for medical students. This was evident in the fact that after the institute's establishment, besides the 300 to 400 students of chemistry that attended lectures, 100 more pharmaceutical and medical students crowded into the lecture halls. Although originally designed with the capacity for 140 students in its laboratories and 400 students in the big lecture hall, it turned out to be too small. Chemistry was rapidly expanding and in 1870, a second professorship was already established.[20] The physicists, nonetheless, had to wait a long time to realize their requests for a significant site for their own institute.

THE PHYSICS INSTITUTE

After the revolution of 1848, the Physics Institute (Physikalisches Institut) had been housed in Erdbergstrasse in Vienna's third district close to the Theresianum. In 1855, it was moved to a nearby building that was formerly used as a prison. Both edifices were totally insufficient for the number of students attending physics courses. Furthermore, the Physikalisches Kabinett, the private collection of the Habsburgs' scientific instruments, machinery, and curiosities, was more a museum-like collection that did not satisfy the research needs of the physicists either. As Ludwig Boltzmann later recalled, "We always had plenty of ideas and were only preoccupied with the lack of apparatus."[21]

In 1865, Victor von Lang, a Viennese physicist who had worked with Michael Faraday in London, became the director of the Physikalisches Kabinett and acquired new apparatus, but he only slightly improved the institute's extremely modest conditions.[22] In 1872, the medical faculty introduced experimental physics as a new, two-semester course requirement for its students and Lang taught it. His lectures were soon overcrowded. Lang finally arranged to lecture in the Josephinum and moved his apparatus to the big lecture hall, saving his students from commuting.[23] Obviously, there was a high necessity for physics lectures and laboratory work for the medical students. While Lang's course evolved into one of the more well-attended lectures, it consequently provoked the urgent demand for apparatus and the restructuring of the course to meet the needs of the medical students. Shortly after, Lang received an endowment of 4,500 florins, a considerable amount for the time, and an assistant position.[24]

Facing similar issues as the chemists, the physicists expected similar treatment. They were after a new institute, preferably located in the Mediziner-Viertel. Amid Redtenbacher's negotiations, Josef Stefan, the director of the Physics Institute, and Ernst Brück, director of the Institute of Physiology, agreed to establish their own laboratories at the corner of Hörlgasse and Währingerstrasse. Ferstel was once again their architect. In 1872, he submitted his plans of the two institutes as a continuation of the Universitätsviertel project. The longer side of the building was attached to the Chemistry Institute and several halls permitted communication between them. The plan was to facilitate teaching and laboratory work and to encourage research and exchanges among the different institutes. The interconnection of both physics and chemistry to medicine and pharmacy and the dependence of the prestigious medical students on chemists and physicists for their education offered the latter a similarly prestigious location within the university quarter.

In the meantime, some natural scientists vigorously reacted to the massive, monumental, and impressive style of the new university. As they argued, a simpler architectural design could better fit the plain and austere nature of their sciences. In a faculty petition on August 4, 1871, they claimed that the new buildings, the Chemistry Institute and the university, by resembling the architecture of the Italian universities did not serve the purpose of furthering the natural sciences. "These flowered elsewhere in the universities of Berlin and Munich, in the Polytechnic of Zurich, in the Collège de France in Paris, and in the Kings College in London, where the exact science can feel comfortable—fewer floors of the same height, several but larger courtyards, and all straight lines fitting to austere requests."[25]

One of the most politically flexible architects of his time, Ferstel for once faced resistance from conservative scientists and not politicians. Caught between obstinate natural scientists and the crash of Austria's stock market in 1873, Ferstel's plans for the Physics Institute were never realized. Instead, in 1875, the university purchased a block of flats and converted it to an institute. The Physikalisches Kabinett and Stefan's Physics Institute moved into the four-story building located in Türkenstrasse 3, a short side street crossing Währingerstrasse. Under the same roof on the third floor was located the Institute for Physical Chemistry. In 1867, Redtenbacher, Lang, and Stefan proposed the establishment of a professorship of physical chemistry. At the time, Josef Loschmidt was working on gas theory, combining chemistry and physics in his research. He collaborated closely with Stefan and used the facilities of the Physics Institute when it was still housed in Erdbergstrasse. Thus, it is not surprising that the physicists suggested Loschmidt as the most appropriate person for the position in physical chemistry.[26] He soon became director of the Institute of Physical Chemistry and his lectures attracted 100 to 120 students regularly. In his laboratory courses, Loschmidt had around twenty students as trainees and his physics course, designed for pharmacists, had the highest attendance.

The building in Türkenstrasse lacked basic features of a modern science institute. "The inner space was not designed as a laboratory," Karl Przibram later recalled. The ceiling beams were so rotten that they seemed as if they had been chewed by termites. As the daily *Arbeiterzeitung*, a Social Democratic paper, satirized the shabbiness of the building, "Once again, a student has registered in the Physics Institute on the Türkenstrasse; unhappiness in love is said to be the motive for the deed." Later, when Lise Meitner entered the University of Vienna and attended a physics course at the institute, she expressed fear that if a fire broke out, very few could actually survive.[27] Although insufficient for the research that modern physics demanded, the collegial atmosphere at the Physics Institute balanced the inadequacy of the space. The Viennese physicists were fortunate to empower their relations and acquire a sense of community through their regular gatherings at the house of Franz Serafin Exner, director of the II. Physikalisches Institut.[28]

FRANZ EXNER'S CIRCLE AND HIS ETHOS OF WORKING IN PHYSICS

At the turn of the century, the physics community was affected by the reorganization of its institutes and big cuts in their finances.[29] In 1902, the Physikalisches Kabinett was renamed I. Physikalisches Institut, and Lang remained its director. The Physikalisches Institut that Stefan directed until 1894 was renamed Institut für Theoretische Physik and Ludwig Boltzmann became the director.[30] Loschmidt's Institute of Physical Chemistry was renamed II. Physikalisches Institut and Franz Serafin Exner took over the directorship. Given the reorganization of the physics institutes, Boltzmann's apartment in the building at Türkenstrasse was ceded to Exner's institute as working rooms.[31] The collection of scientific instruments formerly belonging to the Physikalisches Kabinett was transferred to Exner's supervision.[32]

At the time, Exner was *Ordinarius Professor* at the University of Vienna and one of the most prominent experimental physicists, in charge of a considerable number of students. Research on atmospheric electricity, color theory, spectral analysis, and radioactivity occupied his interest. Open to scientific challenges, he was the first to report Konrad Röntgen's discovery of x-rays on January 7, 1895, to the Institute for Physical Chemistry in Vienna.[33]

Exner's earlier career started in Strasburg. Disappointed by the monolithic culture of the physics community there, he returned to Vienna. As he complained, "Day after day from 8:00 a.m. to 10:00 p.m., physics and again physics, this no decent human could stand."[34] What Exner missed, as Berta Karlik and Erich Schmid explained, was Viennese culture and art, especially music, and it was this culture that Exner maintained in the physics community of Vienna upon his return. As best described by his assistant Hans Benndorf, Exner established a new ethos for working in physics:

> A circle of like-minded friends, we surrounded our admirable and beloved teacher as a big family. The most wonderful and also cheerful hours of the day were the tea at the institute during late afternoon, where the "little father," this is how we called him among us, with his pipe in his hand, was regularly presiding. There we were talking about God and the world, often passionately discussed, and about controversial

> scientific questions. And we had good arguments with him, and he could bear any contradiction as long as it was not personal. Also, we never felt his superiority and he always seemed to be another youth among the young. We are indebted to those hours for the important stimulation of our own work.[35]

Przibram recalled the same picture of Exner in his white lab coat and the inevitable pipe in his hand: "The picture reflects something of the coziness that ruled the circle in those days."[36] Passionate debates on science and politics and Exner's narratives of his long trips to Greece and Asia often accompanied by music took place at his home on Währingerstrasse 29 every Saturday evening.[37] In his welcoming house right in the heart of the Mediziner-Viertel next to the Josephinum, Exner cultivated the Viennese ethos of collaborative work in physics and mentored nearly all the experimental physicists in Austria. His circle included Hans Benndorf, Egon von Schweidler, Stefan Meyer, Maria Smoluchowski, Friedrich Hasenöhrl, Karl Przibram, Felix Ehrenhaft, Erwin Schrödinger, and Hans Thirring among others.

Exner's circle[38] or Exner's school,[39] as it was known, promoted not only the integration of physics with Viennese culture but also the work of women in the field. Exner was among those professors who in 1893 formed a committee to support women's admission to the University of Vienna.[40] Due to his strong beliefs in the liberal, bourgeois ideal of education, Exner's support of women's education does not come as a surprise. To Austrians, the figure of his father, Franz Exner, embodied the liberal pedagogical reforms proposed in 1840. Deborah Coen argues that "While Exner's reform plans of 1848–9 were to be no more than guidelines, they remained the standard, the ideal, against which Austrians judged the success or failure of the empire's schools for over half a century."[41] His sons inherited their father's authority.

As Franz Serafin Exner acknowledged in 1908 participating in the debate for the reform of the middle-school education, "The traditions of the author of the organizational proposal of 1849 are still alive in me." His brother, the physiologist Sigmund Exner, tried very hard to have an impact on the medical reforms in education when in 1899 his friend von Hartel became minister of culture and education. To reconcile the ideals of liberal education with the concerns of socially progressive critics, he took a positive stance on the issue of women's training as physicians. Although believing that women tended to focus more on the exams than on utilizing their education, he had never had a problem with teaching men and women together. His wife, Emilie von Winiwarter, was also an activist for academic education for women. From 1901 to 1906, she was president of the Viennese Women's Employment Association (Frauen-Erwerbverein), an association which was fighting for the entrance of women into appropriate professions.[42] Among those professions that Emilie considered suitable for women was that of a pharmacist as it allowed women to work at home and preserve their ideals of maintaining a good and happy family. Preserving their liberal ideas, Emilie and the Exner family, were actually giving a right to education and a voice in the political domain to their progressive critics.

WOMEN'S ADMISSION TO THE UNIVERSITY OF VIENNA

During the 1860s, the economic distress in imperial Austria forced many women to search for employment. For example, the majority of workers in the textile industry were women.[43] Proliferating in these new professions, lower-middle-class women needed special training in new technologies. In 1867, the Viennese Women's Employment Association (Wiener Frauen-Erwerbverein) emerged to cover such needs. Its leader, Iduna Laube, having the absolute support of the association, soon established a school for sewing linen and a commercial school with courses in embroidery, lace-making, dressmaking, and housekeeping. In 1871, the association founded an academic school in the form of a *höhere Bildungsschule*, providing four years of practical training. It was primarily established by upper-class women—Emilie Exner among them—concerned with facilitating women's "entry into female white-collar employment," who ran the school.

Although the association vehemently stressed women's right to education, it did not have any further political aims. The suggestion to establish a gymnasium for girls (a school equivalent to that for boys preparing for university studies) was considered so revolutionary that none of the association's women was willing to pursue it. Without challenging the conventional female identity of a housewife, their goal was to professionalize traditional female activities. A number of other vocational schools followed that of the association's, but not one of them was "suited to raise the level of education of girls from the middle-class intelligentsia and therefore to open profitable sources of income to them," argued Marianne Hainisch, one of the prominent feminists of the time.[44]

It was not until 1888 that the Association for Extending Women's Education (Verein für Erweiterte Frauenbildung) was founded as an attempt to support the establishment of a gymnasium for girls. The goal was not modest since the school would be the first one in Vienna to raise girls up to the academic standards necessary for university entrance. Thus, along with a petition for the establishment of the school, the association handed to the imperial government an appeal for the admittance of women at the arts and medical faculties of the

University of Vienna. This was signed by 3,644 women, members of a number of different Viennese groups, such as the Association of Housewives (Hausfrauen Verein) with 2,601 members, and the Association of Teachers and Nursery School Teachers (Verein der Lehrerinnen und Erzieherinnen) with 600 members. Women scientists from Germany joined the campaign and offered their experiences as invited speakers at the association's activities. The ophthalmologist Rosa Kerschbaumer and the surgeon Agnes Bluhm were among the first to support Viennese women's right to education. The right to enter the university as full students was the pivotal feminist demand, so central that it was presented as self-evident in the feminist discourse of the time. "We believe," read the petition of 1890, "that women's right to scientific education needs no further evidence but women's self functionality and opportunity."

In a certain sense, the appeal of the Association for Extending Women's Education was limited to a few thousand middle-class Viennese women and probably directly affected only those who had the means to pursue private education. In another sense, however, their education reform campaign shaped the ideals of a number of young women who held different personal expectations concerning their university admission.

The association and its demands provided a crucial impetus for the transformation of Viennese culture and society. A new, collective female identity emerged and traditional meanings of sexual difference were challenged. Since the Austrian government did not respond to the feminists' requests for educational changes, the association opened up a private gymnasiale Mädchenschule in 1892 based on the curriculum of the boy's gymnasium. Prominent intellectuals and officials such as the director of Vienna's training institute for gymnasium teachers, Emmanuel Hannak, supported women's educational initiative. Setting high standards from its establishment, the school was equivalent to the akademisches Gymansium, the school preparing boys for higher education.[45]

The school was not established in a theoretical vacuum and apart from the feminist discourse of fin de siècle Vienna. The investigation of early Viennese feminist movements by Harriet Anderson changes the conventional picture of Vienna as merely the city of Sigmund Freud and Gustav Klimt, of music, art, and smoky coffeehouses crowded by the Viennese intelligentsia. "There was in fact a flourishing culture of political opposition in which men and women worked together for a vision of a society which would not dehumanize its members but permit them to go 'in purity' through life, without concealment and without regret."[46] The Viennese feminists played a crucial role in the formation of this culture of political opposition. In the last decade of the nineteenth century, an organized political feminist movement emerged in Vienna, interwoven with the broader cultural changes. The interplay of different cultural and philosophical critiques led to an increasing political awareness. Women became part of the political game, bringing into focus ethics and sexual morality as well as women's rights and equal opportunities to education. Two autonomous feminists groups, the General Austrian Women's Association (Allgemeiner Oesterreichischer Frauenverein) and the League of Austrian Women's Associations (Bund Oesterreichischer Frauenverein) marked the Viennese feminist scene.

Auguste Fickert, a primary school teacher with a strong personality and radical ideas, was the leading figure of the women's association. Her views, extreme for the time, included the argument that changing external conditions alone would not be enough in changing women's lives. Better wages and equal rights for women at work and at home were necessary but insufficient without the "merging of intelligence with morality."[47] Both the women's association and Fickert emphasized women's education, their intellectual awakening, and the raising of their consciousness.

On the other hand, Marianne Hainisch and the League of Austrian Women's Associations (Bund Oestereichischer Frauenverein) that she established in 1902 had less ambitious and radical goals. The wife of a cotton factory owner, Hainisch founded the league as part of the international feminist network based on the principles of the International Council of Women. It functioned as a public relations body of the Austrian women's movement and united a number of associations, attempting to support and not compete with them. Adopting the discourse of equality and difference, Hainisch fought for the equality of women and men, stressing their differences without challenging the status quo of power relations. One of the strongest advocates for a women's gymnasium, she envisioned education as women's means to enter better professions. Hainisch and the league strongly emphasized their apolitical character, and they argued less in terms of women's rights and emancipation than in terms of morality. Despite their differences, Hainisch and Fickert had one common goal, that of women's *Bildung*. Along with the Association for Extended Women's Education, the league organized a number of talks inviting male intellectuals to speak about women's rights from a philosophical, historical, or juristic point of view.

Men played a significant role in the middle-class women's movement. A number of university professors, artists, and left-wing middle-class intellectuals supported the women's movement, spoke at their meetings, joined their associations, and offered their expertise for the improvement of women's education. In 1900, the historian Ludo Moritz Hartmann and the Union of Austrian University Teachers initiated an association known as Athenäum: Association for the Holding of Academic Courses for Women and Girls. Julius Tandler, an anatomist at the University of Vienna and a main figure in the Social Democratic party later in the 1920s, offered free lectures. The zoologist Carl Brühl lectured on natural sciences in a seminar room flooded with women. The philosopher Friedrich Jodl was one of the most passionate supporters of women's education along with his wife Margarete Jodl, president of Vienna Women's Club (Wiener Frauenclub), a cultural network of upper-class Viennese women.[48]

Surprisingly, of all the physics communities in Austria, only Vienna's played an essential role in promoting women's education at the academic level.[49] Victor von Lang, firmly convinced that women should not be held back from studying at the university, welcomed them to his lectures before they were officially accepted as formal students.[50]

Ludwig Boltzmann supported the Association for Extended Women's Education with his full membership. In the mid-1870s, when Henriette von Aigentler, the woman who later became his wife, was refused permission to unofficially audit lectures at the University of Graz, Boltzmann encouraged her to appeal.[51] Although Aigentler's appeal was successful, she was able to audit classes only for one semester. The following semester, the philosophical faculty approved a rigid rule to exclude women from their lectures at the University of Graz.[52] Their daughter, Henriette Boltzmann, later became one of the first women to take the *Matura*—exams for university entrance—in 1901.[53]

The above examples of the support by academics of women's right to be admitted to the University of Vienna and for attempts to reform their everyday lives indicate that in the late nineteenth century, the forging of a new identity for women was underway. This identity was generated by feminist discourse and was documented in journals such as the *Dokumente* and the *Neues Frauenleben*, both publications of autonomous feminist groups. The feminists attempted to alter women's lives and envisioned them as socially active intellectuals. Thus, in 1896, when women were finally granted the permission to sit for the *Matura*, they embraced the chance. Since the only women's gymnasium in existence did not have the license to set up its own exam, women had to sit as *Externisten* (outside students) in one of the gymnasiums for boys under strenuous and unpleasant conditions. "We were fourteen girls in all," Lise Meitner recalled, "and took a not altogether easy exam (only four of us got through) at a boys' school, the akademisches Gymnasium in Vienna."[54] Despite the frustrations, women were finally accepted to the philosophical faculty of the University of Vienna in 1897 and three years later, the doors of the medical and pharmaceutical faculty of the university opened for them.[55]

Soon after their admission to the University of Vienna, the number of female students increased beyond expectations. In the first academic year, three women were registered as matriculated (*ordentliche*) and 34 as non-matriculated (*au;szerordentliche*) students.[56] Among the first three was Elise Richter, later the first female doctoral student (*Dissertantin*) in the philosophical faculty.[57] Coming from a wealthy upper-class family of the Viennese Jewish intelligentsia, Richter grew up in a disciplined but cultured environment. Her father was a medical doctor who assured her private education. Elise took the *Matura* as

Externistin and studied romance languages and literature. In 1907, she became the first *Dozentin* (lecturer) at the University of Vienna.

Richter's background is not exceptional among the women who chose to enroll in the university. The women who entered the University of Vienna and its surrounding institutes were mainly from the prosperous middle and upper classes. The only school with official state recognition to prepare women for academic studies was the private gymansiale Mädchenschule that the Association for Extending Women's Education founded in 1892. It was not until 1906 that the school was able to set up its own *Matura* exams and changed its name to Mädchen Obergymansium (Upper Gymnasium for Girls).[58] Receiving no subsidy from the Austrian government, the school depended on its wealthy pupils, most of them coming from the families of businessmen, civil servants, and the Viennese upper class. Since Jews were among the leading economic force in Vienna, 35 to 40 percent of the students were of Jewish descent. In the early days of the school, classes took place in the Natural History Museum. Despite the criticisms of radical feminists for its elitist character and despite the hostility of the educational authorities, the school soon flourished, partly due to the contributions of patrons. One of the most important patrons of women's education was Marie von Najmajer, who in 1898, just a year after women's admission to the university, set up an annual grant for full-time university students of 150 florins. In 1901, she donated an additional 40,000 kronen to support the school.[59]

Even though the academic elite viewed female students as intruders, women won admission to the medical faculty in 1900. Thus, during the academic year 1900–1901, the total number of female students substantially increased. There were 31 matriculated students on the philosophical faculty and 10 on the medical faculty, while the number of the non-matriculated students for the two faculties was 87 and 25 respectively.[60] (See Table 01/3.)

During the academic year 1897–98, the ratio of women to men was 1:183; during the academic year just before World War I, it had fallen to 1:12. Within the first 17 years of women's admission to the University of Vienna, their number had multiplied 21.5 times. (See Chart 01/3.)[61]

Despite the steady increase of female students, their number in relation to the inhabitants of the city of Vienna was about 1:36,000 in 1913–14.[62] Considering the fact that the university was accessible only to the women of the Viennese elite who had the finances to pursue their secondary education at the Mädchen Obergymnasium, the above ratio is not surprising. In 1900, the Ministry of Culture and Education, after conducting an inquiry about the need for a state gymnasium for girls, rejected the feminists' request and established instead a *lycée*, a school preparing women for more "feminine" professional posts.[63]

What the above numbers indicate, nonetheless, is that the feminist movement and its members were not intellectual or political outsiders. They had an influence on women's choices, decisions, and lives. For instance, despite the modest beginnings of some feminist groups, the total membership of the league reached 40,000 women in 1914.[64] Whatever expressions were used and whatever goals were anticipated, feminists and their supporters had forced a rearrangement of Viennese social institutions. Admission of women to the university was not the result of a natural process, but was engendered by the persistent efforts of the feminists. It clearly emerged out of a discourse that constructed a collective identity for upper- and middle-class Viennese women.

WOMEN ENTER THE FIELD OF PHYSICS

After their admission to the University of Vienna, women were welcomed to physics lectures and laboratory courses. Out of 32 Austrian women registered in the University of Vienna for the academic year 1897–98, 10 took classes in physics.[65] In 1903, Olga Steindler (1879–1933) was the first to complete a dissertation in physics. As a daughter of an attorney, Steindler studied at the private Mädchengymnasium of the Association for Extended Women's Education and took her exams as *Externistin* in Prague. In the fall semester of 1899–1900, she registered at the philosophical faculty and studied physics and mathematics. In her *Rigorosen,* the oral examination required for the degree, Exner and Boltzmann were her examiners.[66] Aware of women's need for emancipation, Steindler lectured at the Athenäum while still a student. It was at the University of Vienna that Steindler met Felix Ehrenhaft, Exner's *Assistent* and member of his circle, whom she married in 1908. They were the same age and shared a strong interest in physics, culture, and politics. Steindler devoted her career to women's *Bildung,* serving as director of Handelsakademie (Business Academy) and becoming one of the first female school directors.

The same enthusiasm for physics brought a second woman, Lise Meitner, to the Physics Institute at Türkenstrasse. She graduated in 1906, the same year as Selma Freud; the latter, however, did not pursue any further work in the field. Meitner and Freud shared a work room in the institute at Türkenstrasse.[67] After her graduation, Meitner stayed one more year at the institute and worked with Stefan Meyer on radioactivity research in a room next to Przibram's. In a well-documented biography, Ruth Sime highlights the meaning of the ethos and collegiality of Exner's circle in Meitner's career:

> One of Lise's fellow students, Karl Przibram, remembered Exner for his contagious enthusiasm and for the community spirit that went far beyond the usual relationship

> between teacher and students. This sense of community was essential for Lise in finding her way. She had come to the university on her own, very conscious of how few women there were and how visible she was, how some of the men went out of their way to be pleasant and others, just as conspicuously, did the opposite.[68]

Slowly and persistently, women were engaging in the physics community. Apparently, they were not only accepted in lectures, but they were also assigned work space within physics laboratories, as Freud's and Meitner's trajectory illustrates. Friedrich Hasenöhrl with two female students at the Physics Institute c 1909.Marking the transition from exclusion to integration in fin de siècle Vienna, women gained access to the University of Vienna and to almost all of the scientific institutes in the Mediziner-Viertel. As Table 02/3 illustrates, from the fall semester of 1897–98 to the spring semester of 1913–14, women were actively interested in traditionally male-dominated fields.[69]

During the first academic year, 31.3 percent of the total number of the enrolled female students chose the field of physics. In chemistry the percentage was 28.1 and in mathematics 37.5 for the same year. These percentages were very close to 40.6 percent for pedagogy and not far off from 56.3 percent for philology, fields conventionally characterized as female ones.

By being able to attend university studies, women were also able to enter the cultural and physical space that surrounded the university building. A further argument to be made is that the Mediziner-Viertel serves as the midpoint for understanding how women assimilated themselves into the academic world, especially the physics community, and became part of it instead of invisible outsiders. It was in that physical and intellectual space that women associated with their male colleagues, exchanged ideas, and acquired the *Bildung* that both feminists and Social Democrats valued and aspired to so strongly. The city stood not merely as the social and political setting for the development of scientific institutions, but it also molded both scientific knowledge and a hierarchy on those that possessed it.

THE MEDIZINER-VIERTEL AS A LABORATORY IN THE FIELD: THE COFFEEHOUSE CULTURE

One could read the history of the sciences in early-twentieth-century Vienna as a chapter in the urban reconstruction of the city, the creation of the Mediziner-Viertel, and the architectural grouping of diverse laboratories. Scientists of fin de siècle Vienna were strongly attached to the locality of their institutes and the surrounding culture. The face-to-face interaction of researchers with various experts created a scientific network enmeshed in the cultural life of the city, a laboratory in the field. Apparatus for educational demonstrations, instruments for experimental use, and theoretical ideas of different scientific disciplines crossed Währingerstrasse the same way that scientists crossed their institutional boundaries. Studying the importance of the Mediziner-Viertel in the cultural and scientific scene of fin de siècle Vienna, one encounters both the broader spectrum where knowledge was produced and circulated in the city and also those involved in the production.

The coffeehouses in the area particularly provided them with a space for social interaction where endless discussions on science enforced feelings of collegiality. The role of coffeehouses as locations of scientific exchanges, as sites of learning and even, in the case of eighteenth-century England, as spaces where scientific principles and experiments were turned into commodities to be sold by philosophers, has recently drawn the interest of historians of science.[70] Conversations on chemistry and the phlogiston, for example, took place in the eighteenth-century London coffee shop where the Coffee House Philosophical Society held its open conversations. As Ian Golinski argues, the coffeehouses were known as sites of "free flowing and open discourse" and served both the goals of the group of holding egalitarian type of conversations and the ideals of science as public culture.[71] With each coffeehouse attracting a particular clientele, in the latter half of the eighteenth century, Benjamin Franklin spent his Thursdays as a colonial agent in London at St. Paul's Coffeehouse, the Club of Honest Whigs, a political club with overtones of libertarian politics and a house for many of the members of the Royal Society and the Society of Arts.[72]

In Vienna, coffeehouses have been associated with the city's cultural preeminence at the turn of the nineteenth century and have been portrayed as clubs of philosophical, political, and artistic circles. They were the spaces that maintained the cohesiveness of the Viennese elite and functioned as institutions where the Viennese intellectuals mingled with the professional elite and shared ideas and values with each other. Since 1880, painters had their *Stammtisch* in Café Sperl; Victor Adler, the leading figure in the social democratic party and his fellows met at Café Griensteidl; the political critic and radical intellectual Karl Kraus was a habitué of Café Imperial; and the architects met at Café Museum.

Between 1907 and 1912, Philip Frank, Hans Hahn, and Otto Neurath, the "first Vienna Circle," gathered at Café Central, usually on Thursdays. Ulla Heise argues that without these coffee shops, most of which were located in the inner city, the Viennese Sezession probably could not have existed nor made their significant contribution to the literature of 1900 Vienna.[73] This tradition continued in the 1920s and 1930s. In exile, Bertolt Brecht singled out the Viennese cafés among all its highlights: "As every newspaper reader knows, the city is built around a few coffeehouses where the population sits together and reads papers." The Viennese writer Stefan Zweig called the coffeehouse "the best school of everything new."[74]

Symbolic of a carefree existence, the Viennese coffeehouse also carried a paradox: it reflected the hard realities of life in Vienna at the same time that it seemed to embody a relaxed way of spending a day, by reading the papers and enjoying the delicious pastries. Due to the housing shortage, the working-class apartments were small, inadequate, and unbearable, especially during the cold Viennese winters. Coffeehouses offered a pleasant and warm environment throughout the day.[75] The same architectural absurdity led the physicists, working at Türkenstrasse, to the charming coffeehouse at the corner of their laboratory, even though it was less impressive than the coffeehouses of the first district. It was there they contributed to the confluence of scientific disciplines and discussions. Of that "idyllic time," Przibram later recalled with nostalgia that "the young generation of physicists can hardly imagine the passion of the debates, echoed in those days particularly in the above mentioned coffeehouse."[76]

WOMEN ENTER THE FIELD OF RADIOACTIVITY

In the crossroad of physics, chemistry, and medicine, women found it convenient to enter the field of radioactivity, taking advantage of its interdisciplinarity. From 1910 to 1920, eight women out of 48 authors published work in the *Mitteilungen*, the annual bulletin of the institute. Taking into account the canonical stories of women's invisibility in the physical sciences, the percentage of women authors (16.7 percent) is surprising.[77] Each of those women came into radioactivity by a different route. Each of them had a different life pattern, however, which embodied the local circumstances and was affected by the spatial relations developed in the Mediziner-Viertel. Women's involvement to radioactivity research during the 1910s had to be seen as an urban practice, as a flow of trafficking materials and practices, which linked the confined space of the laboratory to other sites of knowledge production and sites of consumption. The women's everyday life in the local context of the Viertel and the city in general influenced the science they were able and most importantly, allowed to produce.

For example, to enter the institute, Friederike Friedmann used her experimental skills in physics. Her research topic was testing the theoretical hypothesis concerning the variations in the range of alpha particles. In the early days of radioactivity research, the main problem was to determine the chemical identity of the new elements, define their properties, and explain the decay-series transitions. One way to determine atomic weights was by calculating the number of alpha transitions from the known elements such as uranium, radium, and thorium. Thus, the properties of alpha radiation were essential in the identification of elements. Friedmann made an experimental investigation of the variations in the range of individual alpha particles emitted from a source. The fact that all alpha particles did not penetrate the exact same distance in air was tested statistically by Karl Herzfeld in 1912.[78] In 1913, it was Friedmann who offered an experimental confirmation of Herzfeld's theoretical results for the case of polonium.[79]

In order to gain greater technical expertise for her experimental work, Friedmann sought further education at the Technische Hochschule Wien. To Friedmann, the city could afford an important opportunity to access the technicians and their knowledge, which was needed for the background work on her scientific research. The Technische Hochschule was located walking distance from the Radium Institute at Karlsplatz, on the exact opposite side of Ringstrasse from the university. Although women had already been accepted at the university for more than a decade, the Technische Hochschule was still a space of tension and transition as it remained closed to female students.

Women's technical education definitely challenged the dominant discourse of womanhood in Vienna in the 1910s. In 1913, the Ministry of Education issued a restrictive decree, allowing the Technische Hochscule to accept women as non- matriculated students and only for some courses that were required for teacher's exams. The same year, Friedmann applied as a non-matriculated student for a course on electrotechnics. Despite the ministry's decree, her application was rejected. Friedmann's resistance to the space restrictions imposed upon her life, both personal and professional, yielded results only in the academic year 1918–19, when women were finally accepted as matriculated students at Vienna's Technische Hochschule. For example, Elisabeth Kara-Michailova, one of the women who entered the institute in the early 1920s, was able to follow courses at the Polytechnic from 1921 to 1923 in electrical and radio engineering and was also admitted to practical training in these subjects.[80]

Crossing their space boundaries enabled women to map themselves differently within new spaces. They tried to use the city as a resource for gaining access to and maintaining their presence in public spaces such as the laboratory or elsewhere in the university. For some, such as Stefanie Horovitz and Marietta Blau, the city provided ready access to the Radium Institute and to the network of science-related institutions in the Viertel. City life and social relations offered women visibility and opportunities to develop their professional careers.

STEFANIE HOROVITZ: DISPELLING THE MYTH OF THE INVISIBLE ASSISTANT

For Stefanie Horovitz, it was her knowledge of chemistry, her connection to Lise Meitner, and the topography of the Mediziner-Viertel that opened the door to the Radium Institute. By the end of the 1910s, a considerable number of chemically non-separable pairs or groups of radioelements had begun to accumulate very rapidly. As Frederic Soddy put it colloquially, "Their atoms have identical outsides but different insides." These elements, identical in their whole chemical character and not separable by any method of chemical analysis, were later called isotopes.

In 1913, Soddy succeeded in placing all the known radioelements in the periodic table, despite the fact that there were more of them than places available. He did so by locating more than one radioelement in the same box based on the elements' atomic numbers. Even though these elements belonged to different decay series they were chemically inseparable. In order to confirm the existence of these identical and at the same time different substances scientists had to determine and compare the atomic weights of the isotopic elements.[81]

The experimental work of determining atomic weights was painstaking and time consuming. The substances to be weighed had to be capable of isolation in a pure state—a challenging task—and the experimenter needed to be in a position to determine even the minutest quantity of the substance that might get lost during the quantitative experiment. The world's leading expert in the field was the chemist Theodore Richards, professor at Harvard University and Nobel laureate in 1914 for the accurate determinations of the atomic weight of a large number of chemical elements. Otto Hönigschmid, another atomic weight expert from Vienna, was asked by both Soddy and Kasimir Fajans to undertake atomic weight experiments using lead from radioactive sources. He repeated the experiments, employing the same Harvard method with the advantage of quartz apparatus and using lead from Joachimsthal pitchblende. The transfer of knowledge from Harvard to Vienna does not come as a surprise. Hönigschmid had spent 1909 working with Richards in Boston, hoping to learn his techniques. In the fall of 1910, he returned to Vienna to undertake work at the Radium Institute. While carrying out numerous fractionations and crystallizations, he determined the weight of radium and prepared a radium standard, which became the official substitute to the original in Paris.

Performing atomic weight experiments was not a task Hönigschmid could conduct alone given that, in 1911, he had already accepted the directorship of the laboratory for inorganic and analytic chemistry in the Deutschen Technischen Hochschule in Prague, where he had also become an *Ordinarius* professor.[82] In

January 1914, he asked Lise Meitner, who was already in Berlin, whether she knew of someone in Vienna qualified to assist him in his atomic weight determination project.[83] The advantage of acting within a space such as the Mediziner-Viertel, where scientific institutes were within walking distance of each other, was that the few women in science were very visible. Meitner had worked at the Physics Institute at Türkenstrasse, and she remembered Stephanie Horovitz, who was studying under Guido Goldschmiedt at the second Chemistry Institute nearby. Thanks to Meitner's recommendation, Horovitz was offered the job. As Hönigschmid wrote to Meitner a few months later, "I am sending you greetings from Miss Horovitz, who does not believe that you still remember her. I have just argued with her about that."[84]

Like Meitner, Horovitz was a Jew. She was born in Warsaw in 1887 and moved to Vienna with her family in 1890. In 1913, Hönigschmid approached her when she was a young chemist who had just graduated from the University of Vienna. Wet chemical techniques and the experimental identification of atomic weights were definitely within her capabilities and, by June 1914, Horovitz and Hönigschmid were working closely together. "Miss Horovitz and I," Hönigschmid informed Meitner, "worked like coolies. On this beautiful Sunday, we are still sitting in the laboratory at 6 o'clock."[85]

Horovitz and Hönigschmid purified lead out of 100 kg of lead sulfate from the Joachimsthal pitchblende, a time-consuming and meticulous assignment. The atomic weight of radioactive lead was found to be 206.73, lighter than ordinary lead (207.21). On May 23, 1914, Hönigschmid presented their results at a congress of the Bunsen Geselschaft in Leipzig. Conscious of the importance of their work, they immediately sent their article first to the *Monatshefte für Chemie* instead of the institute's *Mitteilungen* and, shortly afterwards, they published a version in the French *Comptes rendus*. As Badash points out, Hönigschmid and Horovitz offered the most convincing confirmation of the Harvard work. For the next two years, they continued to copublish on the atomic weights of uranium, thorium, and ionium.[86] In addition, their research showed that ionium was not a separate element but just an isotope of thorium.

Accounts of Hönigschmid's and Horovitz's collaboations always present her as his "protégé," "research student,"[87] or simply as his "student"[88] who assisted him in determining the atomic weight of radioactive lead. Attempts to address the imbalance reach the other end of the spectrum by references to the results as "hers" even though copublished papers are cited.[89] To disentangle the politics of collaboration between men and women who work in partnership proves to be a difficult undertaking. Common publications do not reveal who actually took the lead in each project, who was the assistant, and who was assisted. In the case of the Hönigschmid-Horovitz cooperation, there is no doubt that he was the mature partner and project leader. He introduced Horovitz to experiments of atomic weight determination and welcomed her both to the Radium Institute and, it seems, to his laboratory in Prague. In a letter to his friend Max Lembert, Hönigschmid forwarded Horovitz's greetings: "With best wishes from Fräulein Doctor Horovitz, the beautiful graduate." Without a doubt, Horovitz was more than an able assistant who followed instructions by her mentor.

The way Hönigschmid described to Meitner his research project in 1914 is revealing for its emphasis on Horovitz's input in the work. "*We* now isolate lead from pure Joachimsthal pitchblende . . . *We* hope that in the next two weeks before the holidays *we* will analyze these preparations of lead . . ."[90] As a chemist, Horovitz brought her expertise from the Chemistry Institute to the neighboring Institute for Radium Research and entered the field of radioactivity as a young researcher instead of a student. In his Nobel lecture in 1922, Frederick Soddy acknowledged Horovitz's presence as Hönigschmid's partner in quite the reverse to the account of Richard's team in Harvard. "Simultaneously, work on lead from uranium minerals by T. W. Richards and his students at Harvard, and by Hönigschmid and Mlle. Horovitz, gave values all below the international figure."[91]

Nevertheless, the end of the First World War also brought an end to the Hönigschmid-Horovitz collaboration. He accepted a position at the University of Munich, and she left the institute to return to her hometown in Poland. Long afterwards, Kasimir Fajans informed Elisabeth Rona that, "Stefanie moved there [to Warzawa] to join her married sister after the First World War, after her parents had died in Vienna. She was not active in chemistry, and she and her sister were liquidated by the Nazis in 1940." Unable to bear the burdens imposed by the Nazis, Hönigschmid and his wife committed suicide on October 14, 1945.[92]

MARIETTA BLAU: TAKING ADVANTAGE OF A TRAFFICKING MATERIAL

If Horovitz's involvement in radioactivity research serves as a clear example of women's entrance into the field because of the essential interplay of chemistry and physics, the case of Marietta Blau points out the significance of medicine and clinical radiotherapy to radiophysics. Once again, the Mediziner-Viertel functioned as a locus of scientific practice, shaping Blau's opportunities for interdisciplinary research. Her expertise on radium was her vehicle to an advanced career.

Blau was born in fin de siècle Vienna as the third child of an upper-class Jewish family. Her father, Markus Blau, was a lawyer in the Kaiser's courts and an important music publisher. Her mother, Florentine Goldenzweig, was the sister of Josef Weinberger, the main publisher of Gustav Mahler's works in Europe. During her childhood, she attended some of the best Viennese schools and, in 1905, was sent to the private Mädchen Obergymansium, the only school with official state recognition to prepare women for academic studies. In 1914, when the young men were drafted and women had more access to education, Blau enrolled at the University of Vienna to study physics and mathematics. For two semesters, she conducted her *Praktikum* (practical training) at Exner's Second Physics Institute and eventually moved to the Institute for Radium Research.[93] In 1918, Blau submitted her dissertation on the absorption of diverging gamma rays—"a radiological subject"—and her first paper appeared both in the annual bulletin of the Radium Institute and in the *Sitzungsberichte* of the Austrian Academy of Sciences. [94]

Blau's research topic turned out to be important in the clinical treatments of cancer. Discovered by a French physicist, Paul Villard, gamma rays had occupied the interest of the radioactivity community since 1900. In 1904, Marie Curie performed the first gamma radiography and in 1914, Rutherford showed that gamma rays were a form of electromagnetic light.[95] Eventually, it was Blau's radiological research topic that opened the door of Holzknecht's radiological institute for her. As Blau put it in a later curriculum vitae, after defending her thesis in 1919, "I conducted theoretical studies and at the same time was a research assistant at the Laboratory for Medical Radiology at the Holzknecht Clinic, where I studied medical physics."[96]

Holzknecht's clinic was located at Vienna's Allgemeine Krankenhaus, not far from the institute and within the limits of the Viertel. Their physical proximity played a significant role in facilitating the path of radium and the transferring of personnel from Boltzmanngasse to the neighboring medical facilities. For instance, Meyer helped the physician Gustav Riehl set up the Radium Station of the Allgemeine Krankenhaus and provided the expertise of his colleagues for the preparation of radium for therapeutic use.[97] Available to all the clinics and departments of the hospital, the Radium Station, along with Holzknecht's Röntgen Laboratory, provided a network of medical practitioners to the Institute for Radium Research. Blau simply took advantage of these connections in order to maintain her work on radioactivity and straddled the boundary of medicine and physics for the rest of her career.

In 1921, Blau left Vienna to accept a position as a physicist in Eppens and Co., which manufactured x-ray tubes in Fürstenau, Germany. A year later, she moved to the Institute of Medical Physics in Frankfurt am Main where she worked as a research assistant.[98] For more than a year, she instructed doctors in radiobiology while she conceived and elaborated a theory on the effect of x-rays on biological objects. At that time, the Frankfurt Institute was Germany's epicenter of target theory, which described the impinging of radiation on living tissues as particles hitting a target. The theory's aim was to use radiation to probe the structure of the organic world.

Blau played an instrumental role in developing the statistical analysis of biophysical processes that constituted "hits," which was the main research project of Friedrich Dessauer, the director of the institute.[99] Working with Kamillo Altenburger, she studied the number of "hits" that were required for a biological process to occur. Her research career in Germany was interrupted abruptly in the fall of 1923 when her mother became ill and she had to return to Vienna. Although she left Dessauer's laboratory, Blau retained her ties. Later, she contributed a piece on photographic investigations of radioactive rays in a multiauthored volume edited by Dessauer on the boundary between physics and medicine.[100]

GETTING OUT OF THE LABORATORY

I propose that the plan of early-twentieth-century Vienna represented the map of radium's trade and exchange. Dealing with a trafficking material such as radium, physicists were in need of interdisciplinary collaborations that became embodied in the city's roots. During this period, work on radioactivity could not be confined in the walls of the laboratory. Although research on radium's physical and chemical properties was based on the institute's laboratories, biological investigations of radium's impact on humans and other living organisms moved to the physiology laboratory in the Viertel. Moving along with them were the scientists and techniques, and always the material itself.

Similarly, the clinical usage and medical studies of radium were hospital-based. It was the city that provided the large numbers of cancer patients in need of radium treatments as well as the physicians who practiced radium therapy. Vienna was indeed the magnet of Europe's medical elite. Standardization of radium and the production of Austria's official radium standard by Otto Hönigschmid further placed the Radium Institute in a privileged position among the medical sites of the city and increased its esteem to the civilian authorities. Radium preparations for medical use crossed Währingerstrasse the same way that radium researchers moved over to the hospital as radium experts.

Respected professors such as Exner and Meyer taught the new methods of radium extraction, preparation, and handling at university amphitheaters and institutes' lecture halls. When the institute's building became inappropriate for accurate measurements, the neighboring laboratories served as alternatives whereas their libraries served as invaluable resources. Negotiations about the trade of radioactive materials and price regulations were taking place among the institute, the Academy of Sciences, and the ministries of agriculture, culture, and education. Sites of radium consumption, such as stores selling emanators, were soon established around the institute. The interconnectedness of Vienna's medical, scientific, and intellectual elites, including the physicists, was taking place in the salons and coffee shops of the city. To the "radioactivists," the basic unit for community life had a spatial definition: the Viertel. Their sociability was centered at Exner's house and at the coffee shops of the area while their students intermingled in the university aula and on their way to various institutes on Währingerstrasse. By retaining the scientific discourse outside the laboratory, the Viennese physicists constructed a strong sense of collegiality, taking advantage of the stimulating coffee shop culture.

Despite the large number of diverse pursuits of radium researchers, Vienna's scientific center in which they located their practices was quite expedient: starting from the ring in front of the university and walking up Währingerstrasse for five minutes, one passed the Physiology Institute on the left, then the Josephinum, and across from that was the natural-science quarter. From there, it was about a ten-minute walk to the general hospital. From the ring and walking toward the center on the opposite side of the Viertel, it was about a quarter-hour's walk to reach the academy, unless one stopped at the major coffee shops of the inner city, such as the Central.

Thus, the city itself played a unique and prominent part in shaping radioactivity research, filling it with both obstacles and opportunities. For the men of the field, their professional life was as motivating as their social one. Working at the Radium Institute gave them opportunities to teach at the university and to advance their careers as academics. Women, however, had a longer way to go. Although accepted as students at the city's university, they were not supposed to claim technical education until the early 1920s. Vienna's Technische Hochschule remained an obstacle for women's technical education until 1919. Moreover, their career opportunities reached a zenith with positions such as laboratory assistants but not as academic professors. The partial isolation—spatial and cultural—from the university did not necessarily mean women's exclusion from city life or other vocational institutions. For instance, the visibility that the city's confined scientific center provided to women facilitated Horovitz's assimilation into the radioactivity community. As in the case of Blau, the general hospital's radium station and Holzknecht's Röntgen Laboratory provided an opportunity to get into medical physics and stumble on related jobs.

Rediscovering the area around the academic and medical institutes and looking out of the laboratory, reveals a great deal about the city's impact on the production of scientific knowledge. City disturbances dictated new architectural forms for scientific institutions. The grouping of physics institutes with medical establishments presented a chance for material and intellectual exchanges. Yet focusing on the city places science in the context of the local conditions and gender politics beyond the walls of the laboratory. "Places of knowledge" now became sites such as the coffeehouse, the street, the university lecture halls, and the hospital where gender relations were more likely inscribed differently from the laboratory and where knowledge emerged as a practice more obviously subjected to the local culture.

Although in 1920s Vienna both the local culture and the gender politics were still inscribed in its streets and buildings, they underwent major transformations. Unique political circumstances at the academic and state level fostered equally unique gender roles in the Viennese political scene and, by extension, at the Radium Institute. Red Vienna, the Viennese socialist experiment from 1919 to 1934 was not only reflected, but was to a great extent constructed within the Mediziner-Viertel.

NOTES

1. Mach, *Populär-wissenschaftliche Vorlesungen* (1910), 355.

2. Bischof, *Frauen am Wiener Institute* (2000), 23.

3. Schorske, *Fin-de-siècle Vienna* (1980), 6.

4. For a detailed presentation of the relevant literature, see Beller, *Rethinking Vienna 1900* (2001).

5. Anderson, *Utopian Feminism* (1992).

6. For a recent collection on the urban history of science that raises similar issues and reveals the interrelation of the history of science to urban history, see Dierig, Lachmund, and Mendelsohn, *Science and the City* (2003).

7. Wibiral, and Mikula, *Die Wiener Ringstrasse* (1974).

8. Schorske, *Fin-de-siècle Vienna* (1980), 31.

9. Wibiral and Mikula, "Heinrich von Ferstel" (1974), 45.

10. On the architecture of Vienna's university see Wibiral and Mikula, *Die Wiener Ringstrasse* (1974); Plassmeyer, "Nineteenth Century Architecture" (1999).

11. Schorske, *Fin-de-siècle Vienna* (1981), 40; see also Flevaris, *Universität Wien* (2001), 2; Wibiral and Mikula, "Heinrich von Ferstel" (1974), 57.

12. See Plaschka, "Universität 1884" (1986).

13. Schorske, *Fin-de-siècle Vienna* (1980), 47.

14. Janik and Veigl *Wittgenstein in Vienna* (1998), 20.

15. Maerker, *Model Experts* (2005).

16. Wibiral and Mikula, "Heinrich von Ferstel" (1974), 49. As Jeffrey Johnson argues in the case of Germany between 1866 and 1914, chemistry was the first among all

disciplines in the philosophical faculties to acquire its own institute. The German chemical industry, particularly coal-tar dye, and the demands from fields like pharmacy and medicine for workers with chemical training, justified the government's investment. See Johnson, "Academic Chemistry," (1985), 501. See also Meyer-Thurow, "The Industrialization of Invention" (1982). In Vienna, the situation was similar. The textile industry, the most prosperous industry in Austria, required chemists as well. See Schorske, *Fin-de-siècle Vienna* (1980), 58–59; Okey, *The Habsburg Monarchy* (2001), 229–233.

17. Michl, *Geschichte des Studienfaches Chemie and der Universität Wien* (1950).

18. Redtenbacher had been a student and close collaborator of Justus Liebig, director of Giessen's chemical laboratory and the central figure in the development of German organic chemistry in the early nineteenth century. For a recent account of the Giesssen model of chemistry, see Rocke, "Origins and Spread of the 'Giessen Model' in University Science" (2003). On the architecture of the Chemistry Institute, see Wibiral and Mikula, *Die Wiener Ringstrasse* (1974).

19. Before Schneider accepted his position as director of the second Chemistry Laboratory, he was the director of the Chemistry Laboratory in the Josephinum, the medical school for military physicians. Schönfeld and Ipser, *Die Geschichte der Chemie an der Universität Wien und des Instituts für Anorganische Chemie* (1996).

20. Brinda-Konopik, "Robert Willhelm Bunsen" (1992), 5; Michl, *Des Geschichte des Studienfaches Chemie* (1950), 61.

21. Cercignani, *Ludwig Boltzmann* (1998), 6.

22. Reiter, "Vienna: A Random Walk in Science" (2001), 466. In 1872, Lang was able to obtain the considerable amount of 1,210 florins and an annual donation of 1,000 florins, according to Hittner, *Geshichte des studienfaches Physik* (1949), 58.

23. Karlik and Schmid, *Franz Serafin Exner* (1982), 38; Hittner, *Geshichte des studienfaches Physik* (1949), 58.

24. Hittner, *Geshichte des studienfaches Physik* (1949), 58–59.

25. Wibiral and Mikula, "Heinrich von Ferstel" (1974), 61.

26. Karlik and Schmid, *Franz Serafin Exner* (1982), 33; For the establishment of the Institute for Physical Chemistry, the government offered the generous endowment of 6,400 florins and an annual donation of 800 florins in Hittner, *Geshichte des studienfaches Physik* (1949), 60.

27. The *Arbeiterzeitung* was among the newspapers that openly supported the Social Democrats; see Rabinbach, *The Crisis of Austrian Socialism* (1983), 27. Karl Przibram was the one who recalled the incident; see Przibram, "Errinerungen" (1959), 1. See also Benndorf, "Gedenkrede auf Franz Serafin Exner" (1937), Nachlass Exner, AÖAW; Sime, *Lise Meitner* (1996), 11.

28. For the history of Vienna's Physics Institutes, see Hittner, *Geschichte des Studienfaches Physik* (1949); Lintner and Schmid, "Das II. Physikalisches Institut der Universität Wien" (1965); Reiter, "Vienna: A Random Walk in Science" (2001).

29. Cercignani, *Ludwig Boltzmann,* (1998), 30.

30. In 1894, Boltzmann succeeded Stefan as director of the *Physikalisches Institut.* In 1900, dissatisfied with political and professional conditions in Austria, he accepted

an appointment as Professor of Theoretical Physics in Leipzig. He returned to Vienna in 1902 (Hittner, *Geshichte des studienfaches Physik* (1949), 64–65; Cercignani, *Ludwig Boltzmann* (1998), 28–29).

31. Reiter, "Vienna: A Random Walk in Science" (2001), 471. Between 1894 and 1900, Boltzmann, as director of the *Physikalisches Institut*, was also responsible for conducting laboratory courses. Because of the load of his own research, he asked to be excused from the above duty. Exner was the one who had to take over the responsibility. In addition to his own laboratory courses in physics, Exner was obliged to direct Boltzmann's as well. The odd situation ended when Boltzmann left Vienna, accepting the position in Leipzig. On his return in 1902, the rearrangement of the physics laboratories revealed the old tension between Exner and Boltzmann, creating the only exceptional case of a conflict between colleagues at Türkenstrasse; see Cercignani, *Ludwig Boltzmann* (1998), 28–30.

32. Reiter, "Vienna: A Random Walk in Science" (2001), 471; Lintner and Schmid, "Das II. Physikalisches Institut" (1965), 379; Karlik and Schmid, *Franz Serafin Exner* (1982), 65.

33. Karlik and Schmid, *Franz Serafin Exner* (1982), 63. Exner's close friendship with Röntgen went back to the time that he spent the third year of his studies in Zurich under the physicist August Kundt and later in Strasburg where both worked as *Assistentent* at the university.

34. Karlik and Schmid, *Franz Serafin Exner* (1982), 63.

35. Benndorf, "Worte der Erinnerung an Franz Exner" (1926), Nachlass Exner, AÖAW.

36. Przibram, "Errinerungen" (1959), 3.

37. Benndorf, "Worte der Erinnerung an Franz Exner" (1926), Nachlass Exner, AÖAW.

38. Karlik and Schmid, *Franz Serafin Exner* (1982).

39. Benndorf, "Worte der Erinnerung an Franz Exner" (1926), Nachlass Exner, AÖAW.

40. Bischof, *Frauen am Wiener Institut* (2000), 23.

41. Coen, *A Scientific Dynasty* (2004), 52.

42. Karlik and Schmid, *Franz Serafin Exner* (1982), 52. See also Coen, *A Scientific Dynasty* (2004).

43. Barea, *Vienna* (1966), 254.

44. Anderson, *Utopian Feminism* (1992), 26, 29.

45. Bandhauer-Schöffman, "Frauenbewegung und Studentinnen" (1990), 50–54.

46. Anderson, *Utopian Feminism* (1992), 253.

47. Anderson, *Utopian Feminism* (1992), 11.

48. Anderson, *Utopian Feminism* (1992), 113; on Friedrich Jodl see also Korotin, "Auf Eisigen Firnen" (1997), 292.

49. Based on her unpublished data and research, Brigitte Bischof argues that the participation of women in physics at the University of Vienna was much higher than, for example, that at the University of Graz. The support of the physics community made the case of Vienna unique within Austria. (personal communication with Bischof).

50. Bischof, *Frauen am Wiener Institut* (2000), 23.

51. Sime, *Lise Meitner* (1996), 14.

52. Cercignani, *Ludwig Boltzmann* (1998), 11.

53. Sime, *Lise Meitner* (1996), 9.

54. Meitner, *Looking Back* (1964), 2.

55. Heindl, "Zur Entwicklung des Frauenstudiums in Österreich" (1990), 17.

56. Tuma, "Die Österreichischen Studentinnen der Universität Wien" (1990), 87.

57. Andraschko, "Elise Richter" (1990).

58. Anderson, *Utopian Feminism* (1992), 31. Many of the women who later worked at the Institute for Radium Research attended the *Obergymnasium*.

59. Bandhauer-Schöffman, "Frauenbewegung und Studentinnen" (1990), 66. The amount of 150 florins was equivalent to the annual tuition to the *Gymansiale Mädchenschule*. See also Anderson, *Utopian Feminism* (1992), 31.

60. The figures in this table are based on Renate Tuma's work on Austrian women students in the University of Vienna; Tuma, "Die Österreichischen Studentinnen der Universität Wien" (1990), 80. See also Plech, "Das frauenstudium an den philosochischen Fakultäten" (1968).

61. This chart includes both Austrian and international students.

62. Tuma, "Die Österreichischen Studentinnen der Universität Wien" (1990), 81.

63. Anderson, *Utopian Feminism* (1992), 31–32.

64. Anderson, *Utopian Feminism* (1992), 90–91.

65. Tuma, "Die Oessterreichischen Studentinnen der Universitaet Wien" (1990), 87.

66. Bischof, "Olga Steindler (1879–1933)" (2001).

67. Przibram, "Errinerungen" (1959), 3.

68. Sime, *Lise Meitner* (1996), 12.

69. Tuma, "Die Osterreichischen Studentinnen" (1990), 87. The total in the table refers to the actual number of women enrolled at the philosophical faculty of the University of Vienna. Those were able to attend several courses in different fields. This is why the total does not match the numbers given by field.

70. Stewart, "Other Centers of Calculation" (1999); Trevor, "Public and Private Science" (2002).

71. Golinksi, "Conversations on Chemistry" (2002).

72. Crane, *The Club of Honest Whigs* (1966).

73. Heise, *Kaffe und Kaffeehaus* (2002); Schorske, *Fin-de-siècle Vienna* (1980), xxvii.

74. Robinson and Keen, "Café kultur: the Coffeehouses of Vienna" (1996).

75. Janik and Toulmin, *Wittgenstein's Vienna* (1973), 34.

76. Przibram, "Errinerungen an ein altes physikalisches Institut" (1959), 1.

77. The percentage of 16.7 refers to the number of female authors from 1910 to 1919. There were actually eight women: Eleonore Albrecht, Marietta Blau, Friderike Friedmann, Hilda Fonovits, Bertha Heiman, Stephanie Horovitz, Grete Richter, and Helene Souczek. The proportion of articles that were published or coauthored by women is up to 8.6 percent. The first volume of the *Mitteilungen* includes articles of Ernest Rutherford, Bertram Boltwood, and William Ramsay that I did not take into account for calculating the above percentages.

78. Herzfeld, "Über die Schwankungen der Reichweite" (1912).

79. Friedmann, "Experimentelle Bestimmung" (1913).

80. Mikoletzky, Georgeacopol-Winischhofer, and Pohl, ""Dem Zuge der Zeit entsprechend . . .'" (1997). Friedmann remained at the Radium Institute until 1919. On Kara-Michailova see Sretenova, "Elizabeta Karamichailova" (2003).

81. Badash, "The Suicidal Success of Radiochemistry" (1979), 250.

82. Zintl, "Otto Hönigschmid zum 60 Geburtstag" (1938).

83. Sabine, "Lise Meitner an Otto Hahn" (1992), 149. The original letter from Meitner to Hönigschmid, January 26, 1914, is deposited at the Churchill College Archives, Cambridge University, MTNR 5/78.

84. Rayner-Canham, M. and G. "Stefanie Horovitz" (1997), 194.

85. Rayner-Canham, M. and G. "Stefanie Horovitz" (1997), 194.

86. Badash, "The Suicidal Success of Radiochemistry" (1979), 252. Hönigschmid and Horovitz, "Über das Atomgewicht des "Uranbleis'" (1914); Hönigschmid and Horovitz, "Sur le poids atomique du plomb de la pechplende" (1914); Hönigschmid and Horovitz, "Über des Atomgewichtes des "Uranbleis'" (1914); Hönigschmid and Horovitz, "Über des Atomgewichtes des "Uranbleis II" (1915); Hönigschmid and Horovitz, "Zur Kenntnis des Atomgewichtes" (1915); Hönigschmid and Horovitz, "Revision des Atomgewichtes des Thoriums," (1915); Hönigschmid and Horovitz, "Zur kenntnis des Atomgewichtes des Ioniums" (1916).

87. Rayner-Canham, M. and G. "Stefanie Horovitz, Ellen Gleditsch" (2000), 104. Rayner-Canham, M. and G. "Stefanie Horovitz: Crucial Role in the Discovery of Isotopes" (1997), 192.

88. Badash, "The Suicidal Success of Radiochemistry" (1979), 252.

89. Rayner-Canham, M., and G. Rayner-Canham, "Stefanie Horovitz" (1997), 194.

90. Hönigschmidt to Meitner, June 24, 1914, Churchill College Archives, Cambridge University, MTNR 5/78 (my emphasis).

91. Soddy, Frederick. 1922. "The Origins of the Conceptions of Isotopes" Nobel Lecture, December 12, 1922, http://nobelprize.org/nobel_prizes/chemistry/laureates/1921/soddy-lecture.html.

92. Fajans to Rona, August 31, 1963, BHL; Rayner-Canham, M. and G., "Stefanie Horovitz" (1997), 195.

93. Marietta Blau, Curriculum Vitae, July 4, 1918, Rigorosenakt 4557, AUW.

94. Blau, "Absorption divergenter Gamma-Strahlung" (1918).

95. Gerward and Rassat, "Le decouverte des rayons gamma" (2000).

96. Marietta Blau, Curriculum Vitae, 1941, GDSCA.

97. Meyer to Riehl, June 7, 1912, AÖAW. Meyer, "Das erste Jahrzehnt" (1920), 12; Meyer, "Die Vorgeschichte" (1950), 20; Protokoll Z 8167, November 30, 1911, AÖAW.

98. Marietta Blau, Curriculum Vitae, 1941, GDSCA; Galison, "Marietta Blau" (1997), 42.

99. Beyler, "Imagine a Cube" (1997), 39–46.

100. Blau, "Über photographische Untersuchungen" (1931).

4. THE INSTITUTE FOR RADIUM RESEARCH IN RED VIENNA

> As this work has now been organized after several years of tentative efforts each collaborator has *his* or *her* [emphasis mine] particular share to take in making the practical preparations necessary for an experiment. Besides each has *his* or *her* particular theme for research which he pursues and where he can count on the help from one or more of his fellow workers. Such help is freely given certain workers having spent months preparing the means required for another workers theme.[1]

When Hans Pettersson submitted this description of the work at the Radium Institute in a report to the International Education Board in April 1928, several women physicists were already part of his research team on artificial disintegration. A number of other women explored radiophysics and radiochemistry as collaborators of the institute, formed their own research groups, and worked alongside some of the best-known male physicists in the field. More specifically, between 1919 and 1934, more than one-third of the institute's personnel were women. They were not technicians or members of the laboratory support stuff but experienced researchers or practicum students who published at the same rate as their male counterparts.

Marelene Rayner-Canham and Geoffrey Rayner-Canham have already drawn our attention to the fact that women clustered in radioactivity research in the early twentieth century. Identifying three different European research schools on radioactivity—the French, English, and Austro-German—the Rayner-Canhams argue that women "seemed to play a disproportionately large share in the research work in radioactivity compared to many other fields of physical science."[2] Through prosopographical studies of important women in these three locations, the authors address the puzzle of why so many women were attracted to this particular field. The explanations they propose focus mainly on the character of mentors, directors, and chief collaborators. The more supportive these figures were, the easier it was for women to be accepted into physics laboratories specializing in radioactivity. Once they gained professional access to the field, women developed strong networks and maintained close contacts with one another, thus fortifying and sustaining their positions.

A second set of explanations touch on research practices—the fact that radioactivity involved meticulous, routine, and repetitive work. The Rayner-Canhams, like many others, hold that women were more willing than men to perform monotonous tasks and endure the hardships of dealing with radioactive materials.[3] Finally, they point to the structure and the subject matter of the discipline as a reason for the unusually high concentration of women in it. In the early twentieth century, radioactivity was a comparatively new field that was thought to be on the edge of mainstream science.[4] Because it lacked the strong male hierarchies of more established disciplines, it was easier for women to gain access; moreover, it had an aura of excitement for anyone seeking a meaningful career.[5] In addition to such explanatory approaches, scholars have explored the role of women in early radioactivity research through a variety of biographical studies that aim to provide missing pieces of the history of the discipline, recover forgotten women scientists, or do justice to the role of Marie Curie as a pioneer in the field.[6]

While these approaches reveal the importance of women's scientific contributions, they also create a master narrative that presents radioactivity research in general terms, hiding differences among the various geographic localities and historical contexts. The lack of comparative analyses that take into account the particular institutional, cultural, and political circumstances of the different research schools strengthens the assumption that women clustered around radioactivity research in all European institutes. It also encourages a monolithic understanding of disciplinary practices and laboratory cultures by assuming that work in radioactivity involved tedious tasks that women performed more willingly than their male colleagues throughout the different institutional settings.

It is a fact that during the early twentieth century, women could be found in the most prestigious laboratories of the day, probing the chemical, physical, and biological phenomena of radioactive substances. Ellen Gleditsch in Oslo, Lise Meitner in Berlin, and Jarmilla Petrova in Prague are just a few of those who worked extensively in radioactivity.[7] Nonetheless, studies indicate that the only institutes with a significant percentage of women researchers were the Radium Institute in Vienna and Marie Curie's laboratory in Paris. From 1919 to 1934 in the Vienna Institute, women accounted for 38 percent of the total number of researchers at the institute. At Curie's Institut du radium, women made up between 25 and 30 percent from 1906 to 1934.[8] In contrast, although Ernest Rutherford welcomed and supported women's participation in science, there were no women among his students or collaborators in Cambridge during the 1920s. Gleditsch seems to have had only a couple of female students in her laboratory, and it is still unclear whether there were any women experimenters in Meitner's group.[9]

Focusing on Vienna, I would like to complicate historians' general assumptions that there were a disproportionately large number of women engaged in radioactivity research in the early years. Largely unaddressed in earlier studies is the significance of the institutional context, particularly with regard to work culture and gender politics of collaboration in the major radioactivity research centers. I argue that a unique constellation of progressive politics and supportive, politically aware personalities created the local conditions that led women to play an extraordinarily significant role in the Radium Institute.

Universal suffrage in 1919 brought the Social Democratic Party (SDAP) to power in Vienna and gave them absolute control of the municipal government. In what has come to be called "Red Vienna," Social Democrats retained their power from 1919 to 1934 and carried out extensive social reforms. From the housing projects and educational reforms to the admission of women to those academic institutions that had tenaciously kept their doors closed to them even after the end of the First World War, the basic motive of the Social Democrats was to reconstruct the working and middle class according to socialist standards. Among the ongoing reconstructions was that of the meaning of sexual difference in Viennese society. The construction of gender was put forward through the party's discourse and its reformative social projects.

Having established their intellectual and political power through the control of the municipal government and some of the academic institutes, Social Democrats were able to provide a new conceptual framework for understanding the relationship between men and women. They challenged the traditional and dominant images that described social gender roles, reattributed meaning to sexual difference according to socialist ideology, and offered practical venues in which both men and women could personify these novel discursive constructions. It was within this framework that women conceived themselves and were able to become active participants in the new social and political status.

The impact of the First World War on the gendering of occupational positions played an additional role. As the Viennese mathematician Olga Taussky-Todd recalls:

> Careers for women before World War I were, as far as I remember, primarily as teachers in girl's schools, secretaries, shop assistants, domestic service, nurses, dressmakers, and things of that sort. All this was changed greatly during the war and it never went back the way it had been, though some of the positions acquired during the war years went back to men afterwards. I remember very well in the buses and trams, the fares were collected by women and even the drivers were occasionally women. All secretaries were now women. Nurses had been given very intense training, including university courses, and their profession became highly respected. Women teachers had to have far greater training than was required before—even a Ph.D. in the better high schools.[10]

In that ongoing and busy construction of gender, the academic community occupied a central role. Although the socialist reforms were aimed mainly at the proletariat, the reformists, many of them members of the academic community, were nurtured in the Mediziner-Viertel. From the party leader, Otto Bauer, to the councilor for health and social welfare, Julius Tandler, to the key figure in the Vienna Circle, Otto Neurath, the socialists gathered in the Viertel, making politics through intellectual discussion. The Mediziner-Viertel was transformed into one of the main experimental laboratories where the Social Democrats instigated their social reforms. Benefiting from this ongoing political process, women entered several academic institutions. It is indicative that before the war, only one woman entered the university for every 12 men, but at the end of the Red Vienna era, the ratio had increased to 1:2.8.[11]

The Institute for Radium Research and the field of radioactivity was just one example of areas where women were gaining entrance to a scientific field and succeeding with notable achievements. The political reforms merged well with the peculiarities of radioactivity research. As an interdisciplinary field, radioactivity offered scientists unique opportunities to cross disciplinary boundaries. This often proved a successful strategy for women in sustaining their role as active experimenters. However, only a few scientific institutes in Vienna were "Red." In general, well-established institutions such as the University of Vienna and the Austrian Academy of Sciences retained their conservative politics, proving the limitations of the socialist power.

RED VIENNA: POLITICS AND GENDER

Peace treaties at the end of the First World War led to the establishment of the First Austrian Republic, with Vienna as its capital, in November 1918. In the elections of February 1919, the Social Democrats emerged as a strong political power. Supported by the working class, especially in Vienna, the party was run by a group of progressive Viennese intellectuals, many of them Jewish, including Otto Bauer and Karl Renner. Realizing that they could not rule alone, they were forced to form a coalition government with the conservative Christian Social party.[12] A second coalition, formed at the end of 1919, with slight differences in parliamentary composition, gave the socialists the time to put forward reforms in education and the military and to secure the full legal equality of women under the constitution. However, the coalition was dissolved by June 1920, marking the end of the socialists' power at the national level.

Having achieved the two practical goals of the prewar feminist movement—the right to vote and admission to higher education—during these years, women seemed to have lost any specific motive for further political engagement.[13] The lack of an independent feminist political forum, in combination with the support of women's rights by the Social Democrats, prompted many women to participate on the party's side. Apparently, feminists' support was decisive in the municipal elections of May 1919 when the Social Democrats won an absolute majority on the city council, holding 100 out of 165 seats. According to Anson Rabinbach, "The shift to Socialist hegemony" was in part "a result of the extension of the franchise to women and young adults."[14] It is not a coincidence that by 1921, women accounted for 26 percent of Social Democrat party members in Vienna.[15]

While in power, the aim of the Social Democrats was to transform working-class culture and education, to alter the behavior of the workers, and to use the city as a laboratory for the party's social experiment. The architecture of Red Vienna is indicative of this transformation. The severe housing shortage forced the socialists to relieve the crisis by building new dwellings and improving the living conditions of the Viennese working class. The design and layout of the new buildings embodied communitarian values that the socialists wished to foster. Apartment buildings and complexes were located around a central courtyard. Their interior was characterized by small apartments with two rooms and minimal kitchen and bathroom facilities. Specific architectural structures imposed the party's culture and introduced unique gender relations and roles. The communal laundry and cooking facilities, a central housing office and meeting rooms for the tenants, and enforced building rules for access to space were intended to provide quasi-private environments. Playgrounds and kindergartens were designed to facilitate women's everyday lives between work and family. Although they were designed to reduce the burden on women, the new apartment buildings kept intact the underlying assumption that children's care was a gendered task. The building planners and socialist designers sought to protect a few more traditional values despite their overall attempts to alter women's social position. For example, gender roles within the family became tangible in the managerial structure of communal facilities. In each building, there was a laundry supervisor who, besides scheduling the wash days for each family, "kept all but the women out of the washing facility (on the prudish grounds of protecting female modesty)."[16] The gender division of labor within the household was apparent by the assignment of the traditional role of housekeepers to women.

Besides using architecture as a way to impose socialist ideology, the Social Democrats challenged the traditional gender roles and attempted to construct gender according to their views. The popular magazines of Red Vienna, the party's newspapers, the trade unions literature, and the mass media such as cinema and radio shaped the gender discourse of the 1920s, while further reforms made it possible for Viennese men and women to endorse the party's efforts.

As Helmut Gruber emphasizes, the "new woman" was presented as intelligent, educated, engaged in politics and social life, dressed in unrestrictive garments, and professionally successful. "The working class woman of yesterday—careworn in appearance, imprisoned by her clothes, unapproachable by those who needed her—was to be abolished by waving a magic wand."[17]

The strategies for these transformations varied. By equalizing male and female wages, the socialists made it possible for women to gain status through their work. The city's social support services, such as nurseries and kindergartens, recognized women's double role as workers and mothers, while access to birth control and abortion became part of the socialists' program. New domestic technologies such as electric irons, sewing machines, and vacuum cleaners were supposed to reduce the burden of housework. Most important to women's emancipation were the educational reforms and the increasing role of public education in character formation (*Erziehung*) and intellectual training (*Bildung*).[18] Especially for the Jewish women, as Edith Prost argues, "*Bildung* was a possibility for emancipation—emancipation as a Jew and as a woman."[19]

The Social Democrats' tradition of supporting women's active social personae goes back even before the 1920s. The Verein Jugendlicher Arbeiter, the party's major youth organization, admitted women for the first time in 1914. In 1908, the party had established the Kinderfreunde, a parents association through which they aimed to reduce the burdens of childcare that fell disproportionately on mothers.[20] Despite these efforts, some critics argue that the party's intended cultural, economic, and legal changes did not fundamentally alter women's subordinate social role. As Johanna Gehmacher claims, "The Social Democratic model of comradeship in gender relations merely covered the increase of women's reproductive tasks within the party's reform program."[21]

It is true that various municipal programs lacked sensitivity to gender issues in many ways. Nonetheless, it would be shortsighted not to acknowledge that, in fact, the Social Democrats provided the framework for a potentially unique political culture that would enable women to shape spaces for themselves in the factory, the laboratory, and the household. These advantages came at a high price, however. Women working in academia, especially, were often obliged to sacrifice motherhood, to accept an overly burdensome workload, and to settle for a diminished personal and intimate life. Yet the Social Democrats made it possible for them to enter the academic scene. Despite constant feminist petitions and political pressure before the First World War, women had been excluded from several academic institutions. The socialist political agenda in education forced these institutions to accept women as students.[22]

In the academic year 1918–19, the Technische Hochschule, Vienna's Polytechnic, opened its doors to women. The next year, 20 women registered as matriculated (*ordentliche*) students and 17 as non-matriculated (*ausserordentliche*), most of whom chose the field of technical chemistry.[23] In 1919, the faculty of law did likewise, opening up the opportunity to women for the higher civil service, one of the higher status professions.[24] Additional reforms secured girls' entrance to boys' secondary schools and the state undertook girls' secondary education.[25] At the same time, the labor and feminist movements of Red Vienna gave a boost to the number of women enrolled at the city's university.[26] From 1919 to 1934, the number of women attending university studies was exceptionally high compared to the number of male students. (See Table 01/4.)[27]

That means that immediately after the First World War, only 13 percent of students at the university were female; by the end of 1933–34, their numbers had reached 25 percent. (See Chart 01/4.)

As the next table indicates, the study of physics and chemistry was among the most favored fields at the University of Vienna. (See Table 02/4.)[28]

The substantial increase in women's admission to university studies during the 1920s, as Edith Probst argues, came as a result of the reforms of the workers and feminist movements of Red Vienna.[29] What is also striking is the role of the academic community of the Mediziner-Viertel in that ongoing construction of gender.

SCIENCE IN THE "RED" CITY

Science in Red Vienna was interwoven into the social, political, and gendered fabric of the city, with the Mediziner-Viertel as its central cocoon. The urban space in and around the university functioned as a complex material and symbolic environment where the socialists implemented their educational reforms. Believing that real educational opportunities were vital to working-class progress, school reforms became the central piece of socialist politics. To transform secondary school education, however, required the establishment of a network of communal scientific institutions, which could support the socialist program. Otto Glöckel, president of the Vienna School Council in interwar years and undersecretary of education in the coalition government of 1919, was the leading figure of the educational reform movement. To realize his reform program, Glöckel first needed a new generation of teachers with academic training in recent child development and psychology methods.

In 1922, Glöckel offered the eminent German psychologist Karl Bühler one of the philosophy chairs at the University of Vienna. He was eager that the university appoint a philosophy professor with expertise in experimental psychology and Bühler fit his expectations. Bühler would be accompanied by his wife, Charlotte Bühler (née Malachowski), a Berliner Jew who was also a child psychologist. To the eyes of Glöckel, the two of them could turn Vienna into the center for psychology research. He was proved right.

The University of Vienna lacked a specialized department on psychology and thus negotiations with Bühler were threatened to fail. Negotiating for the city of Vienna, Glöckel guaranteed the establishment of a municipal institution that would be devoted to advancing teacher training and be used as a university institute on psychology. The agreement led to the establishment of the Institute of Psychology under the auspices of the city of Vienna in 1923.[30] Bühler was obliged to teach at the university while the city agreed to provide him the rooms and the funds for one assistant and one mechanic at the Vienna School Supervisory Council, located at the Ring close to the university. In the following years, because of its prominence, the institute was the only European institution to receive funds from the Rockefeller Foundation to develop a biologically grounded theory of development.

The structure of the institute reflected the political tensions between the Social Democratic party and the Ministry of Education, controlled by the Christian Socialists. It was known both as the *Institute of Psychology of the City of Vienna* and the *Institute of Psychology of the University of Vienna*, revealing its double status. The one who most benefited from it was Charlotte Bühler, who was hired by the city as *Ordentlichen Assistentin* at the institute and de facto as a university assistant. It was not until 1929 that the university faculty accepted her as an associate professor. Charlotte also directed the Reception Center for Children (Kinderübernahmestelle), a modern observation center which decided the fate of abused or neglected children from problematic families. This was an initiative of the Public Welfare Office concerned with the lives of Viennese working-class families, a terrain where socialists tried to impose their norms of the orderly. The center provided for Charlotte the institutional space to advance her practical studies on children's psychology.[31]

The socialists' attempts to document and resolve urban problems such as child abuse or disorderly behavior resulted in new ways of organizing scientific research—such as the around-the-clock observation of children's behavior at the Reception Center. The topography of the socialists' reforms pointed once again to the Viertel. The Reception Center for Children was intentionally located in the ninth district, close to the university and the Rathaus, the center of the social power. Karl Bühler held a Wednesday night psychology colloquium at the institute. Participants frequently moved the meeting at late evening for dinner to the Regina Hotel, across from the Chemistry Institute on Währingerstrasse.[32]

As Sheldon Gardner and Gwendolyn Stevens argue, "Viennese psychology revolved around the university; even the private mental health practitioners sought the prestige that accompanied academic affiliation."[33] Thus, the city was shaping science in two ways: through the Social Democratic party—its political representative—and through its urban space which provided the subjects and the sources for psychological research as well as the cultural environment for social intermingling.

Besides sponsoring innovative projects, Charlotte Bühler was opening doors for women. Many young and committed educationalists entered the institute, among them prominent female psychologists such as Marie Jahoda, Edith Weisskopf, Else Frenkel-Brunswik, Else Köhler, and Editha Sterba.[34] Lote Danziger was Charlotte's assistant and Käthe Wolf offered courses in child psychology. The boldness of Jahoda's words is indicative of women's new self-conception during Red Vienna: "At the time, I was completely convinced that I would once become socialist minister of education in Austria. There was no question about it!"[35]

Psychoanalysis was another attractive scientific enterprise for women during Red Vienna. As Allan Janik and Hans Veigl remind us, "Freud's clientele consisted principally of the sons and, above all, the daughters of Ringstrasse participants."[36] The city, and especially the area around the Viertel, provided an essential cultural milieu to be used as a source for Freud's practice as well as ready access to the desired academic respectability that the title of a professor at the University of Vienna carried. From 1873 to 1885, Freud studied at the University of Vienna, lived his passion for physiology and neurology, and after graduation, took a lower post at the general hospital.[37]

In late summer of 1891, Freud chose to move into a spacious middle-class apartment at Berggasse 19, a street running parallel to Türkenstrasse in the Mediziner-Viertel.[38] His works on hysteria, the interpretation of dreams, and the psychopathology of everyday life found a wide audience among the upper-class, especially Viennese women. After he was appointed as an associate professor at the University of Vienna in 1902, his apartment on Berggasse became the meeting place for four physicians. Every Wednesday, the group, known as the Wednesday Psychological Society, discussed psychoanalytic issues. Including more than a dozen participants by 1908, the group adopted the designation of the Vienna Psychoanalytical Society and was registered under this name as a local group of the newly formed International Psychoanalytical Association in Vienna. Freud moved his office to Berggasse 35, an ideal location for his work.

During the First World War, practically all activities of the society ceased and only resumed operations after its end. A second-generation of analysts, most of whom had already undergone formalized training, such as Freud's daughter Anna, Helene and Felix Deutsch, Edward and Grete Bibring, Wilhelm Hoffer, Wilhelm Reich, Richard Sterba, Otto Fenichel, and Siegfried Bernfeld took over the society. Women played an unusually prominent role in the group.

In 1922, an "enthusiastic group of young rebels," as Dora Hartmann, the wife of the analyst Heinz Hartmann, characterized the psychoanalysts, founded a psychoanalytic outpatient clinic and three years later, the Psychoanalytic Training Institute, a center for training analysts.[39] Being instrumental in establishing it, Helene Deutsch became its director, a position which she maintained until 1933 when she immigrated to the United States.[40] The same year, Anna, Freud's daughter, became the second deputy of the Vienna Psychoanalytic Society and, since 1927, she had been general secretary of the International Psychoanalytical Association.[41] By 1931, the membership of the society listed 39 men and 19 women, meaning one-third of the participants were women.[42]

The socialists, however, never felt comfortable with the work of Freud. In 1927, in an effort to support the psychoanalytic society, the municipal principals offered Freud a plot of land at Berggasse to build an institute and house the society.[43] Despite the fact that the city was supportive, the offer was turned down due to the lack of additional financial sources. In 1936, the society finally acquired a place of its own at Berggasse 7, where the training institute, the outpatient clinic, and the publishing house of the society were also accommodated. By then, however, most of the society's members who were Jewish had fled Austria to avoid Nazi persecution.[44]

A research center for experimental biology, the Biologische Versuchsanstalt, known also as the Vivarium, is also worth noting for the number of women who worked there during Red Vienna. The institute was founded on the basis of a private initiative and not as a direct municipal plan. It was devoted exclusively to laboratory research in zoology, botany, biochemistry, and physiology.

In 1903, the zoologist Hans Przibram, in collaboration to the botanists Wilhelm Figdor and Leopold von Portheim, purchased a bankrupt show/aquarium in the Prater, Vienna's entertainment park close to the center, and they turned it into an institute for experimental biology.[45] In 1914, the institute was donated to the Imperial Academy of Sciences and its founders remained honorary directors whereas a board of curators appointed by the academy ensured its smooth function.

Przibram, the brother of Karl Przibram, belonged to the Jewish bourgeoisie and did not hide his progressive ideology. He was acquainted with the architect Adolf Loss and the Sezession, where he presented some of his graphics during the winter exhibits of 1899–1900 and 1900–1901.[46] The other two directors of the institute, Leopold Portheim and Wilhelm Fidgor, also came from prominent Jewish families with close connections to the industrialists and culturally progressive circles of Vienna.[47]

The picture of the Vivarium as one of the city's most liberal institutes remains incomplete without mentioning Paul Kammerer, who started his work at the institute as Przibram's assistant in the planning of the aquaria and terraria.[48] The talented Kammerer studied music at the University of Vienna and was acquainted with the artistic circles of the city, an interest he shared with Hans Przibram. In 1926, Kammerer took his own life, offering an odd resolution to an ongoing controversy with William Bateson from Cambridge.[49] Bringing the institute to the center of scientific attention, the Viennese and international presses covered the episode extensively.

Controversially or not, under the auspices of Przibram and Portheim, the work of a number of other prominent researchers such as Kammerer boosted the Vivarium to the forefront of research in experimental biology. Through its connections to the Social Democrats, the Vivarium also emerged as an important site of attempted social reforms. Supporting the municipal projects of educating the public (*Volksbildung*) during the Red Vienna era, Przibram played an instrumental role in reviving an exhibit in the old aquarium of the Vivarium.[50] Within the first half year, in 1932–33, the institute hosted around 6,500 visitors and the profits from the tickets reached 2,600 schillings.[51] The exhibit was also open to the public schools of the city.

Przibram's role was decisive in also supporting women's participation at his institute. The zoologist Leonore Brecher, one of his students, completed her habilitation in 1923 and during the last two years of her study, she was appointed as Przibram's private *Assistentin*.[52] In 1925, the opportunity arose to appoint an *Assistent* as a state employ at the institute, but Brecher had already received a fellowship from the *Deutsches Notgemeinschaft* to spend research time at the Rhoda Erdmann Institute in Berlin. Her colleague, Paul Weiss, was finally appointed instead.[53] Brecher's list of publications is notable during the five years she remained at the Biological Institute. Other women who did research at the Vivarium were the botanists Rosi Jahoda and Helene Jacobi, the zoologist Auguste Jellinek, and the botanist Irma Pisk-Felber.[54] From 1920 to 1934, there were 39 women out of the 109 scientists, or 36 percent, who listed either as research students or personnel. This high percentage does not come as a surprise. "Many Jewish, liberal, and social scientists," as Ute Deichmann mentions, "worked at the Vivarium."[55] Only few of them had positions at the University of Vienna.

The intellectual circles that flourished on the margins of the academy and the traditional university were definitely more open to women. It is not by chance that in both of the scientific institutes created through the city's initiatives and in private endeavors that benefited from municipal support, women accounted for one-third of the research personnel and often held high positions in the organizations' hierarchies. Besides the personal ideology of those who founded or directed the institutions, the political agenda of the socialists seems to have also played a crucial role in accepting a large number of women in science. As a means to understand the status quo of postwar Vienna and a way to change it, socialists such as Otto Glöckel, the party leader Otto Bauer, and the philosopher Otto Neurath regarded science and education as a crucial component of their socialist cultural experiment. Thus, they provided the epistemic framework and the political power for social transformations, including the altering of traditional gender roles in society and academia.

Neurath emphasized the importance of economics and natural sciences to the socialist program through his involvement with the Vienna Circle, the philosophical group of logical positivists, the socialist housing project, and the creation of a museum of economics in Vienna.[56] His strong political activity in the circles of Social Democrats and his connections to the scientists and philosophers of the Vienna Circle shed light on the engagement of philosophical circles with the politics and culture of Red Vienna. With his fellow students Philip Frank, Hans Hahn, and Olga Hahn, Neurath shared an interest in new trends in theoretical physics, mathematics, logic, and philosophy. In 1922, the mathematician Moritz Schlick was appointed to Ernst Mach's former chair at the University of Vienna.[57] Beginning in 1924, with Schlick's initiative, the Vienna Circle started to meet regularly and discuss their philosophical interests. Regular participants included Neurath, Hahn, Rudolf Carnap, Kurt Gödel, Karl Menger, Herbert Feigl, Friedrich Waismann, Rose Rand, and Olga Hahn, later Neurath's wife.

"The circle was open to women" as Malachi Hacohen notes. "The presence of women was insignificant, however, in comparison to other liberal circles, such as those around Bühler and Ludwig von Mises. Philosophy was not, in interwar years, an expanding profession likely to attract women."[58] Justifying Hacohen's claim, it is not by chance that both Rose Rand and Else Frenkel, a woman who irregularly attended the meetings, were practicing psychology. In 1930, Frenkel was employed at the psychological institute of the University of Vienna, and she led the department of biographical psychology, working together with

Karl and Charlotte Bühler. From 1930 to 1937, Rand worked at the Viennese Psychiatric Clinic and was the only woman whose name was in the protocol list of the Vienna Circle meetings.[59] She was admitted to the group when still a graduate student, something that Schlick did not do often.[60] The core group remained quite exclusive anyway as a personal invitation by Schlick was required to attend. In 1938, Carnap highly recommended Rand for an exchange fellowship to the United States, praising her knowledge in logic and philosophy of science. Later, Popper pronounced her "a philosopher of distinction."[61] She became known, however, as the one transcribing the meetings of the circle.

The gathering point for every Thursday evening meeting was a small lecture room at the Physics Institute at Boltzmangasse 5, next to the Radium Institute.[62] Loyal to the Viennese coffeehouse culture, the group continued the meetings in Café Herrenhof in the inner city.[63] The mathematician Walter Mayer, "one of the little mathematics crowd that met in the Café Herrenhof," introduced one more woman, the mathematician Olga Taussky-Todd to the circle.

It is important to mention that the Viennese coffeehouse, one of the most important public institutions in the city, was in general not gender exclusive. Among other places, it was in the same coffee shops of the Viertel that women associated with their male colleagues, exchanged ideas, and made themselves visible. In 1900 Vienna, the bourgeoisie Viennese women such as the Jewish mother and daughter Josephine and Franzi von Wertheimstein formed salons. In the interwar years, women of the middle class met in the city's coffeehouses.[64] By the 1920s, coffee shops around the university were crowded with female students along with male colleagues.

When it opened in 1918, Café Herrenhof stood as the meeting place of those who were "politically and erotically revolutionary." These included artists, Freud's students, writers, and many representatives of the exact sciences. Taussky later recalled that after getting acquainted with the group at this coffee shop, she took a course by Schlick, attended one of his seminars, "and even later [attended] the meetings of one of his private circles. He ran an even more esoteric circle to which I could later have been admitted, but by that time, I had withdrawn from these studies in order to spend more time on my mathematical pursuit."[65]

Drawing on the above instances, it is clear that the Radium Institute was not unique as the only Viennese scientific organization with a concern for gender politics. While they held municipal power, the socialists set the stage for a radical program of social reforms, some of which had a direct impact on the Radium Institute. As a counterbalance to their lack of national political influence, the socialists in Vienna tried to create a state within a state, paying particular attention to city services such as public housing, education, and the welfare system. A key figure in transforming the latter was Julius Tandler. His efforts to reshape the public health and welfare system of Vienna put his ties to the physics and medical community of the city to use and gave Stefan Meyer and his institute the chance to play an important role in the socialists' reforms. At the same time, Meyer's connections to Tandler offered practical opportunities for financing some of his personnel, including women.

Tandler, a prominent anatomist, was one of the few Jews with a chair at the medical faculty of Vienna's university.[66] His inside knowledge of Vienna's medical system and his strong socialist ideology made him an ideal guide for the Social Democratic reforms in public health. Appointed as the city's councilor of welfare in 1920, Tandler was soon able to reshape the medical clinics and the general hospital of the city. An increased budget was essential in improving the quality of services and making them accessible to all citizens.[67]

Among other initiatives, Tandler's socialist program promoted new scientific methods in medicine, including the use of radium. In 1929, he asked Meyer to provide the municipal hospital in Lainz with 5 grams of radium. Eventually, he envisioned a more ambitious project such as the establishment of a radium station and a pavilion for cancer therapy.[68] Hoping to profit from the long experience of French physicians, Tandler visited Paris twice in 1919. During the summer of 1930, he went to the Radiumhemmet in Stockholm, one of the leading centers for radium therapy in Europe.[69] A few months later, at a city meeting at the *Rathaus* (city hall) on December 20, 1930, Tandler was ready to promote his plan.[70] He depended heavily on Meyer's help to realize this ambitious and costly endeavor. As Emil Maier, a physician at the hospital in Lainz, informed his colleagues in Stockholm, Meyer and his personnel at the Radium Institute offered not only to provide the radium but also to build a "radium gun," a device with a strong radium preparation used in cancer therapy.[71] Furthermore, as Maier added, "The consultant of the municipality of Vienna for the radium purchase is Herr Professor Stefan Meyer."[72]

The pavilion for radium therapy opened in 1931 and a year later, the radium station was established.[73] Known as the Physikalische Laboratorium am Strahlen-Institut, the station functioned as the point of entry to the field of radium therapy for two of the Radium Institute's collaborators. Franz Urbach directed the Physikalische Laboratorium and worked on radium dosimetry and instrumentation from 1932 to 1934 when Hilda Fonovits-Smereker succeeded him.[74] Through his connections to Tandler and the physicians at the hospital in Lainz, Meyer offered both of them the chance to cross the scientific boundary between physics and medicine, bringing their expertise in instrumentation and experimentation from the Radium Institute to the municipal hospital. Radium became the vehicle for such exchanges. As we shall see, these kinds of career paths were common in the institute's laboratory throughout the 1920s and early 1930s.

THE TRANSITIONAL PERIOD: THE RADIUM INSTITUTE DURING THE EARLY 1920s

Inside the walls of the Radium Institute, postwar politics took on interesting dimensions. Franz Exner's academic politics in the period before the First World War and his affiliation to the Austrian Academy of Sciences enabled physicists to acquire their own quarters, expanding the scope and effectiveness of their research. In the Habsburg Empire, the institute became a source of power for physicists and their emblem in the international scientific world. However, by the end of the war, the transformation of national politics from a monarchy to a republic and the financial deterioration of Austria posed new challenges to the physicists of Vienna. Austria did not include Bohemia anymore; that part of the old Habsburg Empire now belonged to Czechoslovakia. Redrawing the political map of the area meant more than demobilizing soldiers and the rearrangement of populations.

The cost to the institute was the loss of their main source of radioactive materials, the St. Joachimsthal mines in Bohemia. "Dr. Ulrich was expelled last summer from Joachimsthal," Meyer complained to Rutherford. "He does not know the Czech language, and as nobody in Joachimsthal speaks it, it was quite unnecessary, but this crime was sufficient to dismiss him."[75] Ulrich, who served as the mines' director before the war, found shelter as a guest at the institute and continued his work, but the main advantage of the Viennese over the rest of the radioactivity community was now gone.[76]

Meyer was kept busy seeking financial resources to ensure the survival of his institute, supporting the experimental work of his colleagues, and engaging in academic politics for the sake of his research personnel. Although he was not yet the official director, Meyer had been acting as the main administrator and the leader of the research carried out in it since the establishment of the institute. With Franz Exner's retirement imminent in 1920, he seemed about to become his undisputed successor. Surprisingly, in a letter to Rutherford on February 8, 1920, Meyer expressed his concerns. The shift in directorship was not an easy issue; it was connected to the relations to the neighboring institutes, the political situation in Austria, and the culture of physics that Meyer wanted to establish in Vienna.

> Professor Exner is now 71 years old and will leave his institute soon. We want his successor to give lectures on experimental physics based on theoretical foundations in courses of 2 years. The best men, I think, we can get for the post are Jäger or Schweidler or Mache or Benndorf. Personally I myself do not want to be in the combination as long as one of these gentlemen, to whom I am in terms of heartily friendship are

> to get the charge. Now there is also Ehrenhaft, who pushes himself forward in a very intrusive way and who has some acquaintance in the momentary reigning government sustain his candidature. I would be very much obliged to you, if you would write me as soon as possible your opinion on the works of Ehrenhaft and if there is anybody in England who believes in his "sub-electrons." I cannot deny that I do not sympathize with his scientific way or with his personage but as I am not willing to be guided by my own prejudices, I would be very pleased to have your objective judgment.[77]

Rutherford's response came a few days later. He did not find Ehrenhaft's scientific work convincing but could not comment on his personality. As he mentioned, he had no acquaintance with him. He was definitely surprised, as he "always supposed he [Ehrenhaft] was one of the band connected with the Radium Institute."[78]

In fact, Felix Ehrenhaft was one of Exner's students. The son of a physician, he was born in 1879 in Vienna. He studied physics with Exner and in 1904, he became an *Assistent* at the first Physics Institute under Victor von Lang.[79] His main focus was on theoretical physics and, with Exner's retirement, Ehrenhaft envisioned himself as his successor, utilizing some political connections. Meyer's main concern, on the other hand, was to sustain the experimental research at the Radium Institute. Besides ensuring that Exner's successor would be able to teach and so favor experimental physics, Meyer needed the connections to the Physics Institute for practical reasons.

The use of instruments and library resources was essential for the research at his institute. As it turned out the issue was finally resolved by political intervention, since Exner's retirement resulted in the reorganization of the physics institutes. On October 1, 1920, instead of becoming the director of the second Physics Institute, Ehrenhaft was appointed head of the newly established third Physics Institute which focused on general physics.[80] The same year, Gustav Jäger, one of Meyer's close collaborators, succeeded Exner in the directorship of the second Physics Institute while Meyer was officially named the director of the Radium Institute.[81]

The end of the First World War proved to be a difficult time for both the country and the institute. As Meyer wrote to Rutherford, "The so-called peace has aggravated the difficulties enormously and I fear, we will not be able to continue scientific work, if at all we continue our life."[82]

Inflation was out of control—the Austrian currency had only 2 percent of its prewar value—and food and energy sources were in short supply. Meyer's institute could not afford to purchase even the most prestigious science journals.[83] Without knowledge of the foreign literature, research became problematic. Although the academy was supposed to cover the running costs of the institute, it was able to appropriate only 64 dollars a year. The upkeep of the building—servants, mechanical needs, light, heat, and water—as well as Meyer's and his assistants' salaries were actually paid by the University of Vienna. Although the value of radium owned by the institute was up to 300,000 dollars before the war, it had now dropped to 200,000 dollars, as some of it had been sold to defray certain costs of maintenance since the war.[84] As a temporary solution and in response to Meyer's gentle request, Rutherford kindly arranged to purchase the radium that had been lent to him before the war by the Austrian Academy of Sciences. By the end of the year, Meyer received a check for over 500 pounds which contributed to the ongoing research at the institute and temporarily relieved its financial problems.[85] Additionally, some of the institute's international friends provided the institute's library with subscriptions to *Nature*, *Philosophical Magazine*, and a number of the other prominent scientific journals necessary for keeping up research at the institute.[86]

One significant problem was that the institute's former staff and key figures who had worked there during the 1910s were now scattered around Europe. Hönigschmid had moved to the University of Munich, Horovitz left for Warsaw, and Hevesy went to Budapest. Paneth had accepted the position of an *Extraordinarius Professor* at the University of Hamburg in 1919 and Karl Herzfeld took over his tasks as *Ausserordentlicher Assistent* for a short time.[87]

In December 1919, the duties of an assistant were turned over to Hilda Fonovits, but without a salary.[88] Born in Vienna in 1893, Fonovits was one of the women who received her education at the University of Vienna during the First World War. A student of Exner and of the theoretical physicist Gustav Jäger, Fonovits graduated in June 1919 with a doctoral degree in physics. Her dissertation focused on the factors necessary to obtain alpha ray saturation.[89] Her practicum was done at the Radium Institute and thus her first article was published directly from her dissertation and appeared in the *Mitteilungen* in 1919.[90] Just a year later, in December 1920, Fonovits was formally accepted as *ausserordentlicher Assistentin* in the institute with a monthly salary of 1,000 Kronen.[91]

In the meantime, Fonovits married and as Hilda Fonovits-Smereker held a joint appointment as *Assistentin* to the Second Physics Institute and the Radium Institute during 1920–21.[92] When her son was born in 1922, she found it difficult to combine motherhood with a scientific career. In a letter to Meyer, she confessed that "Unfortunately, I have not been successful despite all my searching in finding a reliable employee to substitute for me during the day for my child's care and so it is impossible for me to keep my position as an assistant."[93] Fonovits was ready to give up her career in order to fulfill the obligations of motherhood. "I am very sorry," she admitted, that "I have to quit the job I loved, but I have not found a way to combine my professional and domestic duties."[94] The position was then given to Sebastian Geiger, a Swiss engineer, who retained it for two years.[95] In June 1924, Gustav Ortner, who had received a doctoral degree in physics from the University of Vienna at the end of 1923, succeeded him.[96]

The position of the *ordendtlicher Assistent* also became vacant just after the end of the First World War as Victor Hess accepted the job of *ausserordentlicher Professor* at the University of Graz. The physicist Karl Przibram succeeded him in 1919.[97] The son of Gustav Przibram, a Hungarian Jewish industrialist, Karl belonged to a dynasty. His mother, Baroness Charlotte Schey, came from one of Vienna's richest families. As Przibram described it, "In my parents' house the prevailing spirit was that of the cultivated Jewish, middle-class liberal era with its unconditional belief in progress and its open-mindedness to all the achievements of the arts and sciences."[98] Raised in this stimulating environment, Przibram studied physics at the University of Vienna under Exner and Boltzmann and then moved to the University of Graz to work with the physicist Leopold Pfaundler.

During the academic year 1902–3, Przibram visited the Cavendish Laboratory in Cambridge to work with J. J. Thomson. In 1905, he completed his habilitation at the University of Vienna and worked as *Privatdozent*. In 1912, with Meyer's encouragement, he entered the Radium Institute with a research project on coloring and luminescence caused by radioactive rays. In 1916, he was honored with the title of *ausserordentlicher Professor* at the University of Vienna. Przibram was the obvious choice to succeed Hess at the institute. In 1921, he discovered the phenomenon of radiophotoluminescence. He also worked extensively on the coloring of crystals and the fluorescence of fluorites. His investigations led to one of the two main research projects of the institute throughout the 1920s and 1930s and attracted a number of women researchers. His main collaborator was Maria Belar, but he also worked with Elisabeth Kara-Michailova, Luisa Gröger, Marie Hoschtalek, and Berta Zekert.

Information on the professional staff members who held the main administrative positions at the institute—the director and the two assistants—is summarized in Table 03/4. It also contains information on the affiliation of these scientists with the University of Vienna up to 1938–39.

HANS PETTERSSON: BRINGING A PROPITIOUS WIND TO THE RADIUM INSTITUTE

In November 1921 in the midst of the institute's postwar reorganization, Meyer received a letter from the Swedish physicist Hans Pettersson asking permission to use the facilities for his own research.[99] Hans was the son of Otto Pettersson, the founder of oceanography in Sweden and a professor of chemistry at the Stockholm Högskola. He studied physics at Uppsala University under Knut Angström, who encouraged him to work on radioactivity. His first published paper on the heating effects of radium appeared in 1910.[100] From October 1911 to August 1912, he worked at the University College in London under William Ramsay, a close friend of his father.[101] It was the dominant Otto who had arranged his son's early involvement in radioactivity research in Ramsay's laboratory, giving him "a year of freedom to go around the world."[102]

Arriving in London, Pettersson had little idea of what he was supposed to work on and Ramsay suggested the construction of a microbalance for his dissertation topic. In this early work, Pettersson addressed the problem of accurate physical measurements and aimed to construct a balance with a sensitivity of more than 1/250,000 mg.[103] Working with Ramsay, Pettersson acquired skills invaluable for his later experimental research in physics. His daughter characterized him as "a devil at building apparatus." It is not by chance that in the following years after his return to Sweden, Pettersson showed a great interest in constructing instruments useful in oceanography.[104] In 1913, he was appointed to the staff of Svenska Hydrografiska Biologiska Kommissionen. A year later, he defended his dissertation and obtained a lectureship at Göteborg Högskola, torn between work in radioactivity and hydrography, a conflict his daughter Agnes Rodhe has described as a struggle between his father's wishes and his own interests.[105] His position as a lecturer paid so poorly that he had to work as an assistant hydrographer at his father's oceanographic station in Bornö to supplement his income.

In the summer of 1921, Pettersson approached Rutherford, suggesting some experiments with radium. Rutherford's response was not very encouraging for the young Pettersson for it implied that Pettersson would need his own radium sources. As Rutherford informed him, "I am not sure from your letter whether you have the use of 200 mg of radium for several years for your experiments." Rather than issuing the invitation Pettersson must have hoped for, Rutherford continued, "I am sorry that I will not be in Edinburgh this year but will be in Cambridge in the 4th week of September."[106]

In his response, Pettersson mentioned that some years earlier, Stefan Meyer had offered him the use of radium at the institute in Vienna. "I am unfortunately not able to get any large quantity of radium in this country," Pettersson explained. He went on to wonder "whether Meyer is able to keep his offer open under the present state of things." Rutherford did not seem willing to extend a warm invitation and in his last response, he claimed that the laboratory would be closed for the first three weeks of September: "I am afraid that this will make it rather difficult to see you unless you are able to stay in England over some time." Even Pettersson's strategic mention of his father did not help: "My father, late chairman of the Nobel committee for chemistry, sends you his best remembrances."[107]

Probably because his correspondence with Rutherford had not yielded results, Pettersson accepted an invitation from Prince Albert I of Monaco to work on the radium concentration of the deep sea sediments collected by the Challenger expedition.[108] Unfortunately, the Musèe Oceanographique—the institute in Monaco—lacked the important apparatus that Pettersson needed for his work. At the end of 1921, aware of the low cost of living in postwar Vienna and the remarkable instruments at the Vienna Institute, Pettersson turned to Meyer, his old contact.[109] His request was very modest: besides measuring the radioactivity of sea sediments, he hoped to work on the disintegration of radioactive elements if a small amount of radium bromide could be made available to him.

Meyer had been hospitable to foreign scientists on other occasions and Pettersson offered a further incentive to accede to his request. In a postscript, he added, "I bring with me a sensitive thread electrometer with a voltammeter and my institute in Göteborg Högskola, Sweden, will provide me with the necessary resources for my work."[110] Since the Radium Institute had a hard time supporting its own scientists, it would have been impossible to provide Pettersson with more than work space. By this point, Meyer had brought stability to the institute. The positions vacated by the old core workers had been redistributed to younger researchers and, most importantly, there were funds available and promising research projects to work on. In the following period, the personnel of the Vienna Institute produced an impressive body of work. Przibram's group contributed greatly to the understanding of radioluminescence and Pettersson's collaborators seriously challenged assumptions about the nature of the atomic nucleus put forward by Rutherford and his colleagues at the Cavendish laboratory in England.

Pettersson brought a new era of experimentation to the institute. The need for specialized personnel led a considerable number of women to engage in radioactivity research, elevating the Radium Institute to what Galison called a "Mecca" for women working on radioactivity research.[111] During the 1920s, researchers such as Ewald Schmidt, Max Kindinger, Blau, Rona, Karlik, and Kara-Michailova formed a closely connected group and boosted the institute to one of the most prestigious radioactivity centers in Europe. A number of female doctoral students such as Hertha Wambacher, Theodora Kautz, Erna Bussecker, Felicitas Weiss-Tessbach, Selma Schneidt, and Elsa Holesch oriented their research projects around the themes of Pettersson's group.[112]

When Meyer received Pettersson's letter, he could not have envisioned the drastic changes that the young physicist was going to bring to the institute.

Friendly and hospitable as usual, he posted the note, "Willkomen," on the upper corner of Pettersson's letter, welcoming the chance of having one more international colleague on his staff. After Meyer's positive response, Pettersson and his wife, the chemist Dagmar Pettersson, finally settled in Vienna.[113] Impressed by the "friendly and stimulating atmosphere" at the institute, Pettersson threw himself into intensive research on artificial disintegration, establishing a strong research team and enlisting a number of patrons to support the work.[114] They both blended very nicely into Viennese culture and felt at home with the ethos of studying physics at the institute.

For the next three years, Pettersson divided his time between radioactivity research in Vienna and his lectures in Göteborg, feverishly searching for financial support. Amazingly energetic and ingenious, Pettersson approached Albert Einstein, securing his prestigious recommendation to the Swedish patrons. The connection seems to have been made through Otto Pettersson, thanks to his high status in the Swedish scientific community.[115] During Göteborg's jubilee exhibition in 1923, Einstein lectured to a large audience. As Hans wrote to him afterwards, "Unfortunately, I had no opportunity to thank you during the Naturforschertag in Göteborg for your very friendly recommendation for the upcoming research on atomic disintegration that I worked on with Dr. Kirsch and Dr. Geiger."[116] A long list of the needed instruments for his experiments followed his gracious opening and a detailed work program closed the letter.

Apparently, Pettersson had inherited father's gift for raising funds. Between 1923 and 1925, he secured the equivalent of 7,000 dollars from Swedish institutions, several societies, and patrons.[117] At the time, "The total grant of the Radium Institute in material and equipment was equivalent to a sum of about 110 dollars per year which was totally inadequate and eliminated all intentions to investigate problems for which new instruments had to be bought or constructed."[118] By the mid-1920s, the Austrian economy had stabilized by the flow of foreign loans. The institute was finally in the position to support its personnel.

STEFAN MEYER: A SUPPORTIVE AND POLITICALLY AWARE DIRECTOR

As the Rayner-Canhams argue, one of the reasons women chose to enter the field of radioactivity was "the presence of supportive supervisors who acted as mentors for them."[119] Elisabeth Rona's experience supports this idea. In early 1913, she chose to work with Kasimir Fajans on radioactivity instead of Georg Bredig on physical chemistry because the latter was an "autocratic German professor," a model that she hoped Fajans did not fit. Her second experience, an "exciting and pleasant" one, was with Georg Hevesy in Budapest. Rona was free to use her own imagination in her research and described Hevesy as someone who "did not feel the need to keep his students in their place." At the Kaiser Wilhelm Institute in Berlin-Dahlem, Rona felt also at home with Otto Hahn and Lise Meitner and said she was "fortunate to work in such a stimulating environment."[120] She spent almost two years at the radioactivity department starting in 1921. Similarly, during the interwar years, Meyer also played a decisive role in making opportunities in radioactivity research available to a number of women.

Of the 32 women who appeared as authors in the *Mitteilungen* between 1920 and 1934, 20 entered the institute as Meyer's students. He kept in touch with some of them even after their departure.[121] On several occasions, he used his connections in the international scientific community on behalf of the institute's women. In 1930 for instance, he arranged for Berta Karlik to work at William H. Bragg's laboratory in London.[122] Rona's later account of Meyer's personality and his role as a director comes as no surprise.

> The atmosphere at the institute was most pleasant. We were all members of one family. Each took an interest in the research of the others, offering help in the experiments and ready to exchange ideas. Friendships developed that have lasted to the present day. The personality of Meyer and that of the associate director, Karl Przibram, had much to do with creating that pleasant atmosphere.[123]

Blau especially appreciated Meyer's kindness after she experienced Robert Pohl's "authoritative style" of directorship at the physics institute in Göttigen where she spent the fall of 1932 on a fellowship from the Association of Austrian Academic Women.[124] The same collegiality is echoed in Otto Hans's recollection of Meyer: "As a special characteristic of genuine collegial loyalty toward his many coworkers, I always have felt that he gave every individual far-reaching freedom in his work and allowed him always to publish alone, although for many investigations, he nevertheless was the intellectual stimulus.[125]

Meyer's role was also crucial in supporting women's networks in the wider community of radioactivity researchers. The Norwegian radiochemist Ellen Gleditsch paid several visits to the Radium Institute over the years at his invitation.[126] Well- known in the field and active in feminist politics, Gleditsch functioned as a mentor of the younger women scientists, paying attention not only to their scientific work but to their personal lives as well. In 1934, with the perspective of one who was older and more experienced, Gleditsch warned Rona about the hazards of radioactivity. "My dear Elisabeth, pay attention before it is too late," she urged her when Rona was suffering from anemia.[127] Karlik greatly appreciated Gleditsch's warm friendship and took care of her when she spent time at the institute in late 1937.[128]

Given the close collaboration between Meyer and Gleditsch, exchanges of publications, instruments, materials, and even research students between the two institutes became a common practice.[129] In 1934, Gleditsch arranged for her assistant Ernst Föyn to spend some time in Vienna. The visit led to an important research project on the effects of bombarding radioactive isotopes with neutrons, conducted in collaboration with Rona, Kara-Michailova, and Hans Pettersson.[130] The exchange of researchers was reciprocal. Rona and Karlik paid frequent visits to Gleditsch's laboratory while working at Pettersson's oceanographic station in Bornö, Sweden, over the course of several summers.[131] Moreover, amid the political turmoil that characterized Vienna in the late 1930s, Rona and Blau found shelter at Gleditsch's laboratory.

Meyer was one of the best-loved and most respected persons in the international scientific community. The outstanding feature of his personality, as Paneth later recalled, was his "never-failing kindness."[132] In his obituary of Meyer, Robert Lawson, recalls his "personal charm and good nature, his warm friendship and his innate kindliness."[133] Lawson, an Englishmen, came to work at Meyer's institute in the end of 1913, but with the outbreak of the First World War, he was trapped in Vienna. When the police harassed him as an "enemy alien," Meyer came to his aid and kindly offered him research space in his laboratory until the end of the war. "[Meyer] supplied me with money on trust and free of interest, the amount being left to my discretion, and he established intermediate but adequate contact with my parents."[134]

There is no doubt that Meyer shaped the collegial ethos among his researchers and encouraged women to take active roles in the institute and in the international radioactivity community. Within the general political context of Red Vienna, however, Meyer's role takes on additional dimensions. A few liberal and well-placed men anywhere might have made occasional exceptions to incorporate a few women into their own scientific settings, but individual initiative is not enough to explain the more persistent phenomenon of women's participation in the Radium Institute. Changing attitudes toward women's fitness to work in science and creating opportunities to do so require more than kindness. It requires political intention.

Although Meyer was not a member of the party, it seems to have held the same progressive ideas about women's role in science and society as the Social Democrats. In his youth, following the lead of important figures in the Viennese physics community such as Exner, Ernst Mach, and Victor von Lang, Meyer contributed immensely to creating an environment friendly for women who wanted to pursue careers in physics. When the Radium Institute was established in 1910, Meyer had a remarkable degree of flexibility in administration and in setting the scope of the research, and he consistently supported women's participation there.[135] Rona writes in her autobiography of the particular effort he made in inviting her to join the institute's research team in the mid-1920s. In the summer of 1925, Meyer was spending his holidays in Bad Ischl, a summer resort in Austria, when he knocked unexpectedly on Rona's door and offered her a position in his institute. She was already a well-known physicist as she had previously worked with some of the main figures in the community of radioactivity.[136]

Moreover, in a strongly anti-Semitic city like Vienna, the fact that Meyer was Jewish gave him a distinct standpoint. As Helmut Gruber argues, anti-Semitism was deeply rooted in Austrian society even during the years of Red Vienna. "It was the Viennese Jews prominent in professions and arts, in journalism and the rising mass media, in industry and high finance but especially in SDAP who were the targets of the hate campaigns which were a permanent fixture of the First Republic."[137]

A Jewish woman seeking to be hired and then promoted to the University of Vienna faced very high odds. When Blau attempted to get a position as *Dozentin* at the university, she was told, "You are a woman and a Jew and together this is too much."[138] For those women who wished to remain in academia after their student years, it was clearly a disadvantage to be Jewish.[139] In Meyer's institute, however, that was not the case. Jews—both men and women—were welcomed and attained important positions as the cases of Przibram, Blau, and Rona attest. The gender profile of the Radium Institute's personnel between 1919 and 1934 indicates Meyer's influence on women's careers as well as the interrelation of socialist politics and women's participation in science.

THE GENDER PROFILE OF THE INSTITUTE'S PERSONNEL: 1919–34

The almanac of the Austrian Academy of Sciences from 1919 to 1934 serves as a first indicator of the gender profile of the institute's personnel. A survey of the director's annual reports published there reveals that over those 15 years, there 113 scientists, 43 women and 70 men, who conducted research at the institute. (See Chart 02/4.)

During this period, 83 of them—73 percent of the total—remained at the institute from one to three years. Because the institute occupied a prominent position within the international scientific community, a few of those researchers were visitors from abroad.

One such visitor was Frances Wick, an associate professor at Vassar College in the United States. Wick graduated from Cornell University in 1908. In 1918, she became the first woman to work on airplane radios and gun sights in the U.S. Army Signal Corps. She came to Vienna twice and spent more than two years at the institute, joining Przibram's research group on radioluminescence.[140] In the summer of 1925, another American woman, Marie Farneworth, wrote to Meyer asking his permission to spend six months at the Vienna Institute. Farneworth had graduated from the University of Chicago and worked as a research chemist at the U.S. Bureau of Mines. "It is my intention to apply for a Guggenheim fellowship for advanced study abroad for the academic year 1926," she informed Meyer. Her expertise was on separating protactinium from carnotite and she had previously worked as a consultant for physicians and hospitals.[141]

Apart from the visiting researchers, most of the scientists who stayed for a short period at the institute were practicum students, taking advantage of the special training—the "radioactivity practicum"—that was offered every spring semester for young physicists.[142] Most of the practicum students published one or two papers related to their dissertation topics in the *Mitteilungen* and then left the institute. Overall, nearly 39 percent of those short-term scientists were women. Table 04/4 shows a list of those researchers who remained at the institute for more than four years. One-third of these were women.

In 1927, in addition to the two assistants and the director, the Austrian ministry of education offered 5,000 schillings for the appointment of a *wissenschaftliche Hilfscraft* (scientific assistant) at the Radium Institute.[143] Ewald Schmidt was the first to obtain the position, which he held as a joint appointment with the Second Physics Institute. Attracted by the work on artificial disintegration, he entered the institute during the academic year 1924–25. Pettersson reported to his sister, "My third assistant, Dr Schmidt, is a jewel." He was surprised that Schmidt, who was married, could manage on a salary of just 125 sek a month.[144] A few months after his appointment to the Radium Institute, Schmidt quit the position for the prospect of becoming an *ausserordentlicher Assistent* at the Second Physics Institute.[145]

The next scientific assistant was Elisabeth Kara-Michailova. Born in 1897 to a prosperous bourgeoisie family, Kara-Michailova spent her childhood in Vienna and received a private education.[146] Her father, Ivan Kara-Michailoff, was a Bulgarian physician and her mother, Mary Slade, was an English musician. In 1907, her parents decided to move to Sofia where they played an influential role in the artistic and scientific life of the city. Ten years later, Kara-Michailova returned to Austria alone to enter the University of Vienna. Between 1917 and 1921, Kara-Michailova studied physics, mathematics, chemistry, mineralogy, and philosophy, eventually taking a major in physics and a minor in mathematics with Meyer and Jäger as the referees of her final exams. Even before she completed her thesis, Kara-Michailova entered the institute and collaborated closely with Przibram on determining the phenomenon of radioluminescence. During the autumn of 1923, she returned to Sofia as a guest fellow at the city's university for a very short period before she was soon attracted by Hans Pettersson's work on the artificial disintegration of light elements in Vienna.

In November 1928, Kara-Michailova was offered the position of *wissentschaftliche Hilfskraft* at the Radium Institute, "with the salary of an *ausserordentlicher Assistant.*"[147] By the end of March 1933, however, both Austria and the institute were deeply affected by the wider European political crisis. Financial problems prompted Meyer to address the dean of the Faculty of Philosophy on March 21, 1933. He sought to retain the position, even if at only half the former salary.[148] Kara-Michailova, meanwhile, had decided to apply for a Yarrow Scientific Research Fellowship, a grant that aimed to support female scientists. As she informed Meyer from her parents' home in Sofia, she could rely on the financial support of her father, who wanted her to continue her scientific research even if she was unable to extend her stay at the institute after March.[149]

Karlik was the next to obtain the position. She had entered the Radium Institute as a practicum student in 1927. Karlik was born in 1904 to an upper-class Viennese family. Her father, Carl Karlik, was director of the national mortgage institution for Lower Austria and Burgenland. She lived in a small castle in Mauer, a Viennese suburb, where she always returned for her summer holidays during her adulthood. Adopting the status of her class, she learned to play piano and speak several languages while also taking classes on painting. Although she entered the University of Vienna in 1923, Karlik mainly wanted to take the exams that would enable her to become a teacher and study physics and mathematics on the side. The course of her life proved differently. Attracted by physics, she combined research at the Radium Institute and teaching at a *Realgymnasium* in Vienna. In 1930, her name appeared as the institute's collaborator at the almanac of the academy.[150] When Karlik took up the job of the research assistant on April 1, 1933, her monthly salary was reduced to 150 schillings from the 289.5 schillings that Kara-Michailova received in 1932.[151]

In addition to Fonovits-Smereker, Kara-Michailva, and Karlik all of whom were paid directly by the Radium Institute, Rona was appointed as an additional *wissenschaftliche Hilfscraft* for the academic year 1928–29 only. Erwin Zach, the Austrian-Hungarian general consul in Singapore, and the industrialist Ignaz Kreidl, whose son Norbert was working as a research student in Pettersson's group, funded her position.[152] Rona, known as "the polonium woman,"[153] was probably the most experienced experimenter among the women of the institute.

By the end of the First World War Rona was already considered as an experienced scientist, working on chemistry, physics, and biology. Her earlier work on animal physiology in Berlin and her knowledge of chemistry qualified her for a job she was eventually offered by Francis Tangl as a biochemist and physiologist at the University of Budapest. Tangl needed a scientist to set up complementary courses in chemistry for his medical students and Rona fit the description. Nevertheless, her task did not last long. The communist revolution in 1919 and the political upheavals that followed, forced Rona to resign in 1921; she was unable to bear the overload of work caused by the depletion in staff.[154]

At the time, the radioactivity community was small enough for researchers to be noticeable and international enough to promote mobility and scientific exchanges among different institutes. Rona's work, first with Fajans and then with Hevesy, offered her such visibility in the community that Hahn provided her a grant and an opportunity to work with him and Meitner at their radioactivity department of the Kaiser Wilhelm Institute in Berlin-Dahlem. During the two years of her stay, Rona became proficient in separating thorium-230 from uranium ores and when the economic situation in Germany deteriorated, she was transferred to the Kaiser Wilhelm Textile Institute. As she explained, "Only institutions whose research was important to the nation's economy could receive grants."

When her grant in Berlin and the political turmoil in Hungary were over, Rona finally returned to Budapest in 1923. She had already gained experience not only in radioactivity but also in industrial chemistry and textile technology as well. This knowledge guaranteed her a position in one of the biggest textile industries in Hungary. Given the economic situation in Europe and the low status of women in academia, Rona was right in arguing that "industry offered some hope." The cruelty, though, of the industrialists and the lack of research opportunities forced her to resign in a year. Unemployed, Rona joined her family in the Austrian resort in Bad Ischl for summer holidays in 1924. It was there that Stefan Meyer offered her a position at the Radium Institute in Vienna.[155]

Given the shortage of archival sources, it is difficult to draw any systematic picture regarding the financial support of those who conducted research at the institute The *Institutsverrechnung*, a notebook briefly recording monthly revenues and expenses, shows that from 1925 to 1928, Kara-Michailova received monthly checks of 200 schillings. From 1928 to 1932 Rona was systematically paid 250 schillings per month, while Blau's monthly checks ranged from 100 to 200 schillings from 1929 to 1932.[156] A few *Bestätigungen* (receipts) signed by female scientists confirm that some of them were paid for chemical and photographical tasks they performed as well as for the preparation of radioactive sources. The same can be said for many of their male colleagues who appear either to have received monthly paychecks or have been paid for individual tasks.

A list of those who conducted research at and were paid by the institute indicates the exceptional gender politics of employment and opportunities for research that were offered to women. A survey of the *Mitteilungen* reveals the gender division of labor and the level of participation of men and women in the institute's ongoing scientific research. From 1920 to 1934, 98 individual authors appeared in the *Mitteilungen*; 32 were women and 66 were men. (See Chart 03/4.)

Only six of the women and 15 of the men published more than two papers in the *Mitteilungen*. This confirms the pattern found in the almanac of the Austrian Academy—namely, that most of the practicum students remained at the institute only for a short period of time. Table 05/4 shows the publications of these people, using the *Mitteilungen* as the chief source. Surprisingly, women accounted for the 29 percent of the total. Moreover, if we exclude Pettersson and Przibram from the calculation, the productive women averaged about one more paper each than their male colleagues.

These women can be divided by age into two groups. The older generation who were born before the turn of the century had to overcome many educational obstacles. Most of them had private education and belonged to the Viennese upper class. Blau, for example, studied at the private Mädchen Obergymnasium. While education for women became more socially accepted and widespread in the interwar years, women came from less affluent families.

One can safely conclude that women were among the most productive individuals working at the Radium Institute. Given the number of their publications and the time they stayed at the institute, it is obvious that they were not merely assistants and members of the laboratory support staff, setting up experiments and performing tedious preparatory tasks for their male colleagues. Instead, they made steady contributions to radioactivity research and were as scientifically productive as their male counterparts, publishing not only in the institute's journal but in other periodicals as well. Articles by Blau, Kara-Michailova, Karlik, and Rona often appeared in prestigious journals such as *Zeitschrift für Physik, Physikalische Zeitschrift, Naturwissenschaften, Journal of the Chemical Society,* and *Nature.*

Further analysis of the coauthored papers that appeared in the *Mitteilungen* reflects the politics of collaboration in the institute. In the early years, many of the young female students collaborated with mature male researchers of the institute. For example, during the academic year 1919–20, Maria Hornyak was working under Meyer's supervision on the ionization of alpha rays in different gasses.[157] She then teamed up with Victor Hess and continued her research on alpha particles emitted specifically from polonium. Their coauthored article appeared in the *Mitteilungen* in 1920.[158]

The stereotype of a male mentor and a female student broke down in the mid-1920s when women such as Kara-Michailova, Blau, Rona, and Karlik gained status in the institute and participated in its core research groups. Those women worked closely together, forming 19 percent of the total collaborations. Depending on the topic under investigation, they alternated research partners and they never abandoned their own research projects, publishing as sole authors at the same time. Karlik's account of her work to Pettersson is revealing: "I should have liked to use the old Em[emanation] for *my own experiments*."[159] Additionally, collaborations between men and women covered 29 percent of the total and were mainly formed within the core research group of the institute. These percentages map not only the ongoing research projects but, most importantly, they should be taken as indicators of the positions of men and women in the laboratory. Women led not only their own projects, but they had also a place at the workbench next to their male colleagues and became part of the team instead of interlopers. (See Table 06/4.)

POLITICS MATTER, BUT HOW MUCH?

As we leave behind the days of bemoaning the "poor and excluded women" from science, one can recognize that women in the Vienna Institute were researchers in their own right. They were not anomalies in the laboratory or figures with secondary roles but well integrated in the radioactivity community. They were keen researchers, close collaborators with their male colleagues, and experienced experimenters. Some of them published extensively not only in the *Mitteilungen* but also in prestigious scientific journals of their time. A few advanced in the institute's hierarchy, gaining paid positions as assistants, or carried their knowledge on radioactivity to other institutions.

The high density of scientific and science-related institutions in the Mediziner-Viertel allowed for a continual exchange of personnel, ideas, practices, objects, and trafficking materials. Furthermore, it encouraged social interaction, most often spontaneous, given the close proximity of so many scientific institutes. "I had a very nice surprise the other day," reported Karlik to Pettersson. "I ran across Prof. Ewald from Stuttgart in the Währingerstrasse when going out to have lunch." Because of this unexpected meeting, "He came to the institute on Saturday morning to see Prz.'s [Przibram's] crystals and then we had lunch together all three of us."[160] Paul Ewald, known for his dynamical theory of x-ray diffraction, was then Rector at Stuttgart's Technischen Hochschule. Karlik had met him at William Bragg's institute in England after a series of lectures he had given at Imperial College.

Intellectually and socially, women played a crucial role not only within the walls of the building at Boltzmanngasse but also in the Mediziner-Viertel. By the 1920s, they had resisted the space restrictions imposed upon their lives, claiming space of their own in the laboratory and in the city. Places such as the coffee shops, the city streets around the institute, and various other public sites involved in the daily routine of scientists and university students provided women with many occasions for social interaction and the exchange of information. Probably more socially engaged than any of her single female colleagues, Karlik was in the circle of some young Austrians with interests in music and democratic politics.

In a seminar for female grammar school teachers, Karlik became friends with Rosi Jahoda who was studying botany and zoology at the University of Vienna and worked at the Vivarium. Franz Urbach and Gustav Kürti, both physicists working at the Radium Institute, were also members of their group.[161] According to Wolfgang Reiter, Karlik had more than amicable feelings for Kürti. Kürti, however, married Jahoda. Karlik retained her cultural and social contacts while splitting her time between teaching and physics research. In an interview with the Austrian radio station ÖRF, Karlik admitted to Reinhard Schlögl that she started her studies in 1923 with the intention of becoming a teacher. During her studies, however, she developed an interest in physics and decided to do her dissertation on radioactivity.[162] Even though Karlik could have immediately sought a career in physics by remaining in the institute, she followed the safer option and accepted a position at a *Realgymnasium* in Vienna.

Men and women at the institute were friends, respected colleagues, close collaborators, and at times, as in the case of Felizitas Weiss-Tessbach and Gustav Ortner, intimate partners. Ortner was the successor of Sebastian Geiger in the position of the institute's second assistant and she was an extended member of Pettersson's research group. As a doctoral student, Weiss-Tessbach worked on the absorption of gamma rays of radium C.[163] Their marriage took place in 1931. Occasionally, men became the immediate reason for women's access to the Radium Institute. Anni Federn entered the field of radioactivity because of her husband Franz Urbach. The daughter of psychoanalyst Paul Federn, Anni was the first woman to work at a juvenile court. To the social group that Urbach was acquainted with, which included Karlik, Anni provided the link to Vienna's psychoanalysts. After marrying Urbach, she started to study physics, carried out her practicum at the Radium Institute, and remained there even after the birth of their first son in 1934.[164]

Yet women's personal lives and choices within the institutional and ideological boundaries of Red Vienna were still in many ways curtailed. It is important to note that although many of the female research students that stayed at the institute for short time were married or got married later, none of the women who worked at the core research group ever got married. While the male counterparts were married with families, most of these women remained single.

Blau, Rona, Karlik and Kara-Michailova, all of whom fostered important careers in science, were unable to combine marriage and motherhood with scientific research. Fonovits got married, but when her son was born, she had to quit her job as the institute's assistant despite the socialists' efforts to provide day-care facilities for working mothers.

To conclude, women's active participation in laboratory life can be credited in large part to the social and intellectual interactions of the Radium Institute scientists that made the city of Vienna a crucial setting for science open to women, to Meyer's progressive politics, and to the political context of Red Vienna. However, because the socialists' power was restricted to the municipal level, people like Meyer had only a limited influence. Thus, the employment system of the University of Vienna, controlled by the conservative national government, remained fairly gender segregated. It is indicative that none of the women who were affiliated with the Radium Institute was promoted to the position of *Privatdozentin* at the University of Vienna before 1934. This had a significant impact on women's careers when the controversy over artificial disintegration between Pettersson's and Rutherford's groups was resolved in favor of the British at the end of the 1920s.[165] Soon thereafter, Petterson returned to Sweden and the loss of the key figure of the group led to disarray in the study of atomic disintegration. They also suffered from the ongoing financial crisis, which was already severe and growing steadily worse. Because the women at the institute lacked stable university positions and monthly payments from the state, they were the ones most deeply affected by the dissolution of Pettersson's group. Eventually, the political upheavals of the 1930s dramatically affected the faith not only of the women who had worked in Pettersson's group but also of the entire Radium Institute.

NOTES

1. Hans Pettersson, Report to the International Education Board, April 1928, GUB (in English) (emphasis mine).

2. Rayner-Canham, M. and G., *A Devotion to Their Science* (1997), 12.

3. The Rayner-Canhams base their claim on Margaret Rossiter's work. See Rossiter, "Sexual Segregation in the Sciences" (1978); Rossiter, *Women Scientists in America: Struggles and Strategies to 1940* (1984); Rossiter, *Women Scientists in America: Before Affirmative Action 1940–1972* (1995). They also compare radioactivity to other subdisciplines of physics such as astronomy and crystallography. See for example,

Portugal and Cohen, *A Century of DNA* (1977); Julian, "Women in Crystallography" (1989); Lankford and Slavings, "Gender and Science: Women in American Astronomy, 1859–1940" (1990).

4. Rayner-Canham, M. and G., *Women in Chemistry* (1998), 94.

5. Rayner-Canham, M. and G. *A Devotion to their Science* (1997), 18.

6. See for example, Weidler Kubanek and Grzegorek, "Ellen Gleditsch: Professor and Humanist" (1997); Rayner-Canham, M. and G., "Stefanie Horovitz, Ellen Gleditsch, Ada Hitchins, and the Discovery of Isotopes" (2000); Sime, *Lise Meitner* (1996); Kurner, *Lise, Atomphysikerin: Die Lebensgeschichte der Lise Meitner* (1998); Rife, *Lise Meitner and the Dawn of the Nuclear Age* (1999); Crossfield, "Irene Joliot-Curie: Following Her Mother's Footsteps" (1997); Pflaum, *Grand Obsession* (1989); Pycior, "Marie Curie's 'Anti-Natural Path'" (1989); Pycior, "Reaping the Benefits of Collaboration While Avoiding its Pitfalls" (1993); Pycior, "Marie Curie: Time only for Science and Family" (1997); Rosner and Strohmaier (eds.), *Marietta Blau* (2003).

7. On Petrova see Tesinska, "Women in the Field of Radioactivity: The Case of the Czech Physical Chemist and Radiobiologist Jarmila Petrova" (2004).

8. Schürmann, "Promoting International Women's Research in Radioactivity" (2004).

9. According to Horst Kant, from 1912 to 1938, there was close to 14 percent women in the Meitner/Hahn department at the Kaiser Wilhelm Institute for Chemistry in Berlin. However, the role of those women in the laboratory has not been researched. Kant, "Von KWI für Chemie zum KWI für Radioaktivität" (2002).

10. Tausky-Todd, "Olga Tausky-Todd: An Autobiography" (1985), 314.

11. Tuma, "Die österreichischen Studentinnen" (1993), 81–82.

12. Rabinbach, *The Crisis of Austrian Socialism* (1983), 23.

13. In November 1918, women were granted the right to vote for the National Assembly and the provisional assemblies and to stand for office. They won full legal equality under the constitution in 1920.

14. Rabinbach, *The Crisis of Austrian Socialism* (1983), 26.

15. Gruber, *Red Vienna* (1991), 20.

16. Gruber, *Red Vienna* (1991), 63.

17. Gruber, *Red Vienna* (1991), 148.

18. Esden-Tempska, "Civic Education in Authoritarian Austria" (1990), 192.

19. Prost, "Emigration und Exil österreichischer Wissenschafterinnen" (1987), 450.

20. Rabinbach, "Politics and Pedagogy" (1978).

21. Gehmacher, "Men, Women, and the Community Borders" (1998), 209.

22. The *Technische Hochschule* opened its doors to women in 1918–1919. The following year, 20 women registered as matriculated (*ordentlicher*) students and 17 as non-matriculated (*ausserordentlicher*) and most of whom chose the field of technical chemistry; see Edith Lassmann, "Das Frauenstudium an den Technischen Hochschulen Wien und Graz", Lassmann file, IAWA, Virginia Tech Library. In 1919, the faculty of law did likewise, opening up the opportunity to women for the higher

civil service, one of the higher status professions. See Heindl, "Zur Entwicklung des Frauenstudiums in Österreich" (1993).

23. Lassmann, Edith. "Das Frauenstudium an den Technischen Hochschulen Wien und Graz", Lassmann file, IAWA. See also Seidler, "Das Frauenstudium" (1927), 18.

24. Heindl, "Zur Entwicklung" (1993), 17.

25. Anderson, *Utopian Feminism* (1992), 120.

26. Prost, "Emigration" (1987), 468.

27. The figures in this table are based on Renate Tuma's work on the female Austrian students in the University of Vienna; see Tuma, "Die österreichischen Studentinnen" (1993), 81.

28. The figures in this table are based on Renate Tuma's work on the female Austrian students in the University of Vienna; see Tuma, "Die österreichischen Studentinnen" (1993), 88.

29. Probst, "Emigration und Exil" (1987), 468. For example, during Red Vienna, the number of female physicians increased substantially because, starting in 1922, many were hired by the city of Vienna. Gamper, ". . . so kann ich nicht umhin zu wendern, dass nicht mehr Ärztinnen da sind . . ." (2000). See also Arias, "Die ersten Ärztinnen in Wien" (2000); Walzel, "Ärztinnen in Wien" (2000).

30. Benetka, "The Vienna Institute of Psychology" (1995), 128.

31. Benetka, "The Vienna Institute of Psychology" (1995), 130.

32. Hacohen, *Karl Popper-The Formative Years, 1902–1945* (2000), 136.

33. Gardner and Stevens, *Red Vienna and the Golden Age of Psychology* (1992), 2.

34. Gardner and Stevens, *Red Vienna and the Golden Age of Psychology* (1992).

35. Benetka, "The Vienna Institute of Psychology" (1995), 128.

36. Janik and Veigl, *Wittgenstein in Vienna* (1998), 19.

37. Gay, "Sigmund Freud: A Brief Life" (1989), xi–xxix.

38. Janik and Veigl, *Wittgenstein in Vienna* (1998), 19.

39. Mühlleitner and Reichmayr, "The Exodus of Psychoanalysts from Vienna" (1995), 98.

40. Gardner and Stevens, *Red Vienna and the Golden Age of Psychology* (1992), 172–74.

41. Gardner and Stevens, *Red Vienna and the Golden Age of Psychology* (1992), 10.

42. Mühlleitner and Reichmayr,"The Exodus of Psychoanalysts from Vienna" (1995), 98.

43. Gardner and Stevens, *Red Vienna and the Golden Age of Psychology* (1992), 126. For the SDAP's attitude toward psychoanalysis and Freud, see Gruber, *Red Vienna* (1991), 227.

44. Burian, "History of the Vienna Psychoanalytical Society" (1999).

45. Koestler, *The Case of the Midwife Toad* (1971), 22; Reiter, "Zerstört und vergessen" (1999), 586. On Vivarium see also Hofer, "The Beginnings of Biological System Theory," unpublished manuscript.

46. Reiter, "Zerstört und vergessen" (1999), 592.

47. Reiter, "Zerstört und vergessen" (1999), 595–96.

48. Koestler, *The Case of the Midwife Toad* (1971), 23.

49. The dispute between the two biologists concerned the existence of the nuptial pads of the midwife toad. For an account of the episode, see Koestler, *The Case of the Midwife Toad* (1971).

50. Hofer, "The Beginnings of Biological System Theory," unpublished manuscript. See also Reiter, "Zerstört und vergessen" (1999), 608.

51. Almanach der Akademie der Wissenschaften in Wien (1932–33), 263.

52. Leonore Brecher, Rigorosenblatt, AÖAW.

53. Hans Przibram to Exner, November 5, 1925, AÖAW.

54. Almanach, Akademie der Wissenschaften in Wien, for the years from 1920 to 1934, AÖAW.

55. Deichmann, *Biologists Under Hitler* (1996), 18.

56. For Neurath's political engagement and philosophy, see Cartwright et al. *Otto Neurath* (1996), 56–82. For the cultural links between Neurath's and Vienna Circle's philosophical positioning and the Bauhaus architecture of Red Vienna, see Galison, "Constructing Modernism" (1996), 17–44.

57. Cartwright et all., *Otto Neurath* (1996), 77.

58. Hacohen, *Karl Popper* (2000), 187.

59. Korotin, "Auf Eisigen Firnen" (1997), 301–3. See also Rose Rand's curriculum vitae, RR 2-5–21, ASP.

60. Felix Kaufman, Letter of Recommendation to Rose Rand, 26 January, 1938, RRASP.

61. Rudolf Carnap, Testimonial, February 6, 1938, RRASP. Karl Popper, Letter of Recommendation, 20 July 1951, RRASP.

62. Reiter, "Vienna: A Random Walk" (2001), 478.

63. Tausky-Todd, "Olga Tausky-Todd: An Autobiographical Essay" (1985), 317.

64. Coen, *A Scientific Dynasty* (2004), 170–76. See also Herrberg, and Wagner, *Wiener Melange* (2002).

65. Taussky-Todd, "Olga Taussky-Todd" (1985), 318.

66. Sablik, *Julius Tandler* (1983).

67. Gruber, *Red Vienna* (1991), 65–73.

68. Meyer, "Die Vorgeschichte" (1950), 20; Alth, *50 Jahre Strahlentherapie Lainz* (1981), 12.

69. Gard to Maier, August 8, 1930, AHL.

70. Tandler to Ahlboom, December 20, 1930, AHL.

71. Maier to the Radiumhemmet, December 30, 1930, AHL.

72. Maier to Ahlboom, January 3, 1931, AHL.

73. Alth, *50 Jahre* (1981), 12; Kogelnik, "The History and Evolution of Radiotherapy and Radiation Oncology in Austria" (1996), 224. With the absolute support of the Social Democratic Party and the city's mayor, Karl Seitz, Tandler also established a new pavilion for the cure of tuberculosis with 300 beds and modern facilities at the municipal hospital. At the time, tuberculosis was considered the "Viennese disease" given the exceptionally high numbers of patients; see Luger, *70 Jahre Krankenhaus der Stadt Wien-Lainz* (1977), 3.

74. Urbach, "Einiges aus dem Physikalischen Laboratorium eines Krankenhauses" (1933); Meyer, "Die Vorgeschichte" (1950), 20; Bischof, *Frauen am Wiener Institute für Radiumforschung* (2000), 67. For a detailed account of Urbach's dismissal from the Radium Station see Reiter, "The Year 1938 (1995).

75. Meyer to Rutherford, January 22, 1920, AÖAW (in English).

76. Meyer, "Das erste Jahrezehnt" (1920), 11.

77. Meyer to Rutherford, February 8, 1920, M191, CUL (in English).

78. Rutherford to Meyer, February 18, 1920, AÖAW.

79. Karlik and Schmid, *Franz Serafin Exner* (1982), 144.

80. Hittner, *Geschichte des Studienfaches* (1947), 77.

81. Almanach, Akademie der Wissenschaften in Wien (1920), 180, AÖAW.

82. Meyer to Rutherford, January 22, 1920, AÖAW (in English).

83. Meyer to Rutherford, January 22, 1920, AÖAW (in English).

84. Augustus Trowbridge to Wickliffe Rose, April 2, 1925, Box 24, RAC.

85. Rutherford to Meyer, July 25, 1921, AÖAW; Rutherford to Meyer, October 17, 1921, AÖAW; Rona, *How it Came About* (1978), 21–22.

86. *Almanac of the Austrian Academy of Sciences*, (1921), 194–95, AÖAW.

87. Ruthenberg, "Friedrich Adolf Paneth (1887–1958)" (1997); Meyer, "Die Vorgeschichte" (1950), 26.

88. Bischof, *Frauen* (2000), 66.

89. Hilda Fonovits, Curriculum Vitae, AUW.

90. Fonovits, "Über die Erreichung" (1919).

91. Meyer to the Professorenkollegium der Philosophischen Fakultät der Universität, October 27, 1920, AÖAW.

92. Personalstand, Universität Wien, academic year 1920–21, AUW; Bischof, *Frauen* (2000), 67.

93. Fonovits-Smereker to Meyer, September 9, 1922, AÖAW.

94. Fonovits-Smereker to Meyer, September 9, 1922. AÖAW.

95. File Mittarbeiten/Assistenten, AÖAW.

96. Gustav Ortner, curriculum vitae, no 021, Rigorosenakt, AUW.

97. Karl Przibram, "1920 bis 1938" (1950), 27.

98. Berta Karlik, "Karl Przibram Nachruf" (1974) 380. The biographical information on Przibram comes from Karlik's Nachruf.

99. Pettersson to Meyer, November 28, 1921, AÖAW. Most of the biographical information on Pettersson comes from an interview with his daughter Agnes Rodhe on September 22, 2001, Göteborg and with Artur Svansson, an oceanographer and biographer of Otto Pettersson, on September 21, 2001, Göteborg. See also Deacon, "Hans Pettersson 1888–1966" (1966).

100. Svansson, interview by the author, September 21, 2001, Göteborg. Petterrson, "Contribution à la connaissance" (1910). See also Hans Pettersson's report to the International Education Board, April 1928, GUB.

101. Karlik, "Hans Pettersson Nachruf" (1970), 305.

102. Rodhe to Rentetzi, October 29, 2001.

103. The same problem had occupied both his advisor Angström, who had made a balance as early as 1895, and Ramsay, who measured the density of a tiny amount of radium emanation using a balance made by his students Steel and Grant. For example, Ramsay and Whytlaw-Gray determined the atomic weight of radium using a minute quantity of pure $RaBr_{2,}$ verifying earlier measurements done at the Curies' laboratory; see Pettersson, *A New Microbalance and its Use*, 1914. It is worth noticing that the quantities of radium which Ramsay had been working on also came from the Bohemian mines. In 1907, Ramsay received a loan of 300 mg $RaBr_2$ from the Austrian Academy of Sciences with the commitment to share it with Rutherford. Soon after, the arrangement brought tension between the two British laboratories. That could explain the reasons for which Pettersson later on turned to Vienna for his radioactivity research instead of Rutherford's laboratory in Cambridge.

104. Rodhe to Rentetzi, October 29, 2001. Pettersson, "Some New Instruments in Oceanographical Research" (1917); Pettersson and Angström, "Ein neues Totalimmersions-Aräometer" (1917); Pettersson, "Zur Technik der Dichtigkeitsbestimmung" (1917).

105. Rodhe, interview by the author, September 22, 2001, Göteborg.

106. Rutherford to Pettersson, June 24, 1921, GUB.

107. Pettersson to Rutherford, July 4, 1921, GUB; Rutherford to Pettersson, July 12, 1921, GUB; Pettersson to Rutherford, July 17, 1921, GUB.

108. According to Rodhe, Otto Pettersson was the one who approached the Prince of Monaco, a friend of his, and asked whether Hans could work in his laboratory. Rodhe to Rentetzi, August 11, 2003.

109. Rodhe to Rentetzi, October 29, 2001. Pettersson had already contacted Meyer in 1914 from Bornö, asking his permission to perform some of his measurements in Vienna. (Petersson to Meyer, April 26, 1914, AÖAW; Pettersson to Meyer, May 24, 1914, AÖAW).

110. Pettersson to Meyer, November 28, 1921, AÖAW.

111. Galison, *Image and Logic* (1997), 150.

112. Kautz, "Ermittlung der Halbierungszeit" (1926); Kautz studied physics and mathematics at the University of Vienna under Meyer and Gustav Jäger, and graduated in 1926 (Kautz Theodora, Rigorosentakt 9216, AUW.) She left the institute to become schoolteacher. Meyer kept in touch with her and in 1934, he informed her that the granddaughter of the president of the Academy, Eduard Suess, was her student (Meyer to Kautz, May 11, 1934, AÖAW); Bussecker, "Verflüchtigungskurven" (1928); Bussecker graduated in 1929 after studying physics and mathematics under Meyer and Gustav Jäger as well (Bussecker Erna, Rigorosenakt 10210, AUW); Weiss-Tessbach, "Mikrokalorimetrische" (1928); Schneidt, "Das Electrochemische Verhalten" (1929); Schneidt was born in Komotan, Bohemia and was a fellow student of Bussecker (Schneidt Selma, Rigorosentakt 10442, AÖAW.); Holesch, "Uber die Vardampfung" (1931). Holesch was a student of Meyer and Thirring. She graduated in 1931 (Holesch Elsa, Rigorosenakt 11257, AUW).

113. Pettersson to Meyer, December 14, 1921, AÖAW.

114. Pettersson to Meyer, June 4, 1922, AÖAW.

115. From 1910 to 1922, there had been a long debate on whether Einstein should be awarded the Nobel Prize or not. The Nobel committee, as Carl-Olov Stawström argues, was dominated by representatives of the Uppsala's strong empirical tradition; see Stawström, "Relative Acceptance" (1993), 299. Given his prestigious status within the University of Uppsala and the international community, Otto Pettersson must have known Einstein from that time.

116. Pettersson to Einstein, August 18, 1923, AEA.

117. For example, from 1923 to 1925 Pettersson received funds from the following Swedish sources: Längmanska Fonden, Lars Hiertas Minne, and the Göteborg Högskolas Oceanografiska Institution (Hans Pettersson's Finances, International Education Board, GUB).

118. Hans Pettersson's report to the International Education Board, April 1928, GUB.

119. Rayner-Canham, M. and G., *A Devotion to their Science* (1997), 18.

120. Rona, *How it Came About* (1978), 3–13.

121. The information is based on the individual Rigorosenblätter of all the women of the Institute, AUW Meyer to Dora Kautz, May 11, 1934, AÖAW; Meyer to Gertrud Wild, June 17, 1946, AÖAW.

122. Bragg to Meyer, August 29, 1930, AÖAW; Meyer to Bragg, September 1, 1930, AÖAW.

123. Rona, *How it Came About* (1978), 15.

124. Blau, curriculum vitae, GDSCA, (in English). In a self-description written later (1963) than this one (1941), Blau mentions that she received the fellowship in 1933. (Blau, curriculum vitae, Leopold Halpern Papers) From letters to Meyer it becomes obvious that the earlier date is correct. See for example, Blau to Meyer, October 4, 1932, AÖAW; Blau to Meyer, December 12, 1932, AÖAW. On Blau's time in Göttigen see Bischof, *Frauen am Wiener Institut*, (2000), 75–76.

125. Reiter, "Stefan Meyer" (2001), 119.

126. Five years younger than Meyer, Gleditsch was one of the first women who entered the field of radioactivity before the the First World War, collaborating closely with Marie Curie. Her scientific and friendly relationship with Meyer goes back to the the First World War when Gleditsch supplied radioactive materials for the research that Horovitz and Hönigschmid were performing at the institute. Besides spending time at Curie's laboratory in Paris, Gleditsch worked with Bertram Boltwood in his laboratory at Yale University for a short period of time. The exact determination of the half-life of radium brought Gleditsch to the forefront of radioactivity research and established her as a specialist in the separation of radioactive substances from minerals. In 1916, she was appointed *Dozent* at the University of Oslo. Politically aware of the difficulties women faced in their scientific careers, Gleditsch became heavily involved in the International Federation of University Women, established in 1919. See Weidler and Grzegorek, "Ellen Gleditsch" (1997); Lykknes, *Ellen Gleditsch: Professor, Radiochemist, and Mentor* (2005).

127. Gleditsch to Rona, August 19, 1934, AÖAW.

128. Gleditsch to Karlik, May 4, 1937, AÖAW.

129. In 1919, Gleditsch thanked Meyer for sending her his valuable papers and expressed hope to visit the institute in the future (Gleditsch to Meyer, April 27, 1919, AÖAW). In a letter to Gleditsch, Karlik assures her that she sent a microscope table, which should arrive soon in the mail (Karlik to Gleditsch, April 7, 1938, AÖAW).

130. Föyn, Kara-Michailova, and Rona, "Zur Frage der Künstlichen Umwandlung des Thoriums durch Neutronen" (1935); Föyn, Pettersson, and Rona, "Künstliche Umwandlung des Thoriums durch Neutronen" (1935). Föyn remained at the institute for a year and worked closely with Rona (Meyer to Gleditsch, August 18, 1934, AÖAW; Gleditsch to Meyer, August 30, 1934, AÖAW; Meyer to Rona, September 12, 1934, AÖAW; See also *Almanac of the Austrian Academy of Sciences*, (1935), 196; (1936), 213, AÖAW). Later on, he worked on radioactivity of the seawater with Pettersson, Karlik, and Rona. (Föyn, Karlik, Pettersson, and Rona, "Radioactivity of Seawater" (1939). Meyer to Gleditsch, August 18, 1934, AÖAW; Gleditsch to Meyer, August 30, 1934, AÖAW.

131. Rona to Meyer, September 9, 1935, AÖAW; Pettersson to Karlik, September 27, 1934, GUB.

132. Paneth, "Prof. Stefan Meyer" (1950), 548.

133. Lawson, "Prof. Stefan Meyer" (1950), 549.

134. Lawson, "Prof. Stefan Meyer" (1950), 549. In 1919, Lawson moved to Sheffield having already developed close relations with most of his colleagues in Vienna. As Rutherford reported to Meyer in 1920, "Lawson came to see me in Manchester and gave me detailed news of all of you which I am very glad to hear. Lawson has now got a post in Sheffield and I hope is comfortably situated. He seems a very keen and prolific worker. I appreciate very much the kind way you looked after him in difficult times." (Rutherford to Meyer, January 13, 1920, CUL). In 1918, in collaboration with Hilda Fonovits, Lawson started to work on the number of ion pairs produced by alpha particles of radium C. After he left, she carried the study over without Lawson and presented her paper to the Austrian Academy of Sciences in 1922 (Fonovits, "Die Zahl der von einem a-Teilchen" (1922), 356).

135. It was Annette Vogt who first drew our attention to a similar case of the so-called "Harnack Principle" for the employment of women in scientific institutes of the Kaiser Wilhelm Society. The principle named after the first president of the Society, Adolf von Harnack, describes the fact that each institute that belonged to the society was created in a new scientific discipline and specifically for one individual. Thus, the director of the institute was endowed with the absolute power to hire and fire his personnel. When the director was progressive, the "Harnack Principle" worked in favor of women's employment. (Vogt, "Von der Ausnahme zur Normalität?" (2003) 171.

136. Rona, *How it Came About* (1978), 25.

137. Gruber, *Red Vienna* (1991), 26.

138. Halpern, "Marietta Blau" (1997), 197.

139. Freidenreich, "Gender, Identity, and Community" (1998), 166.

140. Rossiter, *Women Scientists in America* (1984), 118. Wick was in the institute during 1929–30 and 1930–31 and returned for one year in 1936–7. During her first visit, she suffered a serious infection as did Franziska Witt and Stefan Wolf, also members of the institute; see *Almanac of the Austrian Academy of Sciences* (1930), 234. Also

in 1926–27 the institute hosted R. Hasche from New Jersey and H. Raudnitz from Prague. (*Almanac of the Austrian Academy of Sciences*, (1927), 204.

141. Farneworth to Meyer, July 8, 1925, AÖAW. Although Meyer responded positively to Farneworth's request, she probably did not come since her name appears neither in publication records nor in the almanac of the Austrian Academy of Sciences.

142. Stefan Meyer, November 28, 1931. Spare report at the file Mitarbeiten/ Assistenten, AÖAW.

143. Stefan Meyer, November 28, 1931, Mitarbeiten/Assistenten, AÖAW.

144. Pettersson to Mellbye, March 3, 1926, in Rodhe to Rentetzi, October 29, 2001.

145. November 6, 1928, Mitarbeiten/Assistenten, AÖAW.

146. For biographical information on Kara-Michailova, see Elisabeth Kara-Michailova, Rigoresenakt 5215, AUW; Tsoneva-Mathewson, Rayner-Canham, M. and G, "Elizaveta KaraMichailova" (1997); Sretenova, "Elizabeta Karamichailova" (2003); Nazarska, "Elissaveta Karamichailova" (2006).

147. November 6, 1928, Mitarbeiten/Assistenten, AÖAW.

148. Meyer to Dekan, March 21, 1933, Karlik's file, Mitarbeiten/Assistenten, AÖAW.

149. Kara-Michailova to Meyer, January 18, 1933, AÖAW.

150. *Almanac of the Austrian Academy of Sciences* (1931), 285, AÖAW.

151. January 10, 1934, Karlik's file, Mitarbeiten/Assistenten, AÖAW; December 2, 1932, Kara-Michailova's file, Mitarbeiten/ Assistenten, AÖAW.

152. *Almanac of the Austrian Academy of Sciences* (1929), 202, AÖAW.

153. Rayner-Canham M. and G., "Elizabeth Rona: The Polonium Woman," A *Devotion* (1997).

154. Rona, *How it Came About* (1978), 10. While Rona taught at the University of Budapest, she did not quit her research. In 1920, two articles came out based on her earlier work with Hevesy. Rona, "Über die Wirksamkeit" (1920); Rona, "Diffusionsgrösse" (1920).

155. Rona, *How it Came About* (1978), 10–14. See also*Almanac of the Austrian Academy of Science* (1925), 216, AÖAW.

156. Institutsverechnung 1922–32, AÖAW.

157. Maria Hornyak, Rigorosenakt, no 4925, AUW.

158. Hess and Hornyak, "Über die relative" (1920).

159. Karlik to Pettersson, July 26, 1933, GUB added emphasis.

160. Karlik to Pettersson, June 20, 1937, GUB.

161. Reiter, "The Year 1938" (1995), 198.

162. Bischof, *Frauen am Wiener Institut* (2000), 102.

163. Weiss-Tessbach, "Mikrokalorimetrische" (1928).

164. Reiter, "The Year 1938" (1995), 198. In the institute's kassa, where Meyer recorded the finances, Anni appeared to have received minor amounts of money probably for performing technical tasks (Kassa, 1933–1938, AÖAW). She remained in the institute until both she and Franz immigrated to the United States in 1939.

165. On the resolution of the controversy, see Stuewer, "The Artificial Disintegration" (1985); Hughes,*The Radioactivists*(2003); Brown, *The Neutron and the Bomb: A Biography of Sir James Chadwick* (1997).

5. FROM CAMBRIDGE TO VIENNA

The Scintillation Counter in Female Hands

Bringing money and instruments into an institute that is barely supporting its personnel might be necessary, but it is surely not sufficient for boosting it to the forefront of scientific research. The skills and the ingenuity of the experimenters are those elements that give life to the material culture of the discipline and often rework its theories. Hans Pettersson had both. He was keen in designing experiments, but he was also impulsive enough to have "big ideas"[1] and persistent enough to pursue them. In a field that only "the Devil knows what can happen anytime," Pettersson was not afraid to play along.[2]

In the early 1920s, the most challenging set of problems in radioactivity research was related to artificial disintegration. In 1919, Rutherford noticed an anomalous effect in the collision of alpha particles with nitrogen. When pure nitrogen was bombarded by fast radium C alpha particles, long-range atoms arose from the collision. Those were probably "atoms of hydrogen or atoms of mass 2. If this be the case," Rutherford argued, "we must conclude that the nitrogen atom is disintegrated under the intense forces developed in a close collision with a swift alpha particle and that the hydrogen atom which is liberated formed a constituent part of the nitrogen nucleus."[3]

In the fall of 1919, when Rutherford took over the directorship of the Cavendish Laboratory in Cambridge, he pursued his earlier studies on artificial disintegration with great zeal. As Jeff Hughes has documented, Rutherford reorganized the Cavendish Laboratory as a whole by introducing this new research program.[4] An inflow of research students, changes in the material culture, and a spatial rearrangement of the laboratory marked his arrival in Cambridge. By the end of March 1920, Rutherford had concluded that the particles from nitrogen were actually hydrogen nuclei, as he had first speculated.[5]

In the course of 1921, Rutherford tested a series of light elements for the disintegration phenomenon with the help of James Chadwick, his research student in Manchester. Chadwick had followed him to Cambridge, serving as a reliable and experienced experimenter in his team. As they both concluded, only these elements whose atomic mass was given by $4n+2$ or $4n+3$ where n was a whole number, expelled long-range disintegration particles. Seventeen other elements, including carbon and oxygen, yielded no detectable disintegration protons. In August 1921, Rutherford and Chadwick thus argued that the atomic nucleus consisted of a central core of alpha particles surrounded by protons as distant satellites.[6] The artificial disintegration experiments had a theoretical bearing that made them attractive to any ambitious researcher. Pettersson was undoubtedly one of those.

Shortly after his first visit to Vienna in 1922, Pettersson established a close collaboration with Gerhard Kirsch and worked on artificial disintegration. Their first paper came out in 1923 and a full version was published in the *Sitzungsberichte* of the Vienna Academy and in the *Philosophical Magazine*, while a short report appeared also in *Nature*.[7] The multiple publications showed the importance of their results, which differed significantly from those obtained in Cambridge. "Our results," as they both argued, "seem so far to indicate that the hydrogen nucleus is a more common constituent of the lighter atoms than one has hitherto been inclined to believe."[8] Elements such as beryllium, magnesium, and silicon were disintegratable despite the fact that Rutherford and Chadwick had stated otherwise. With that paper, Pettersson and Kirsch generated a milestone in the history of physics controversy with Rutherford's research group, one that historian Roger Stuewer has documented in detail.

Between 1923 and 1924, the debate got heated, with the two groups zealously publishing their conflicting experimental results in the most internationally prestigious journals. At stake was not only the authority of the Cavendish Laboratory in the world of radioactivity and Rutherford's theoretical satellite model. The material culture of the Cambridge group—its experimental methods, the instruments, and the politics of collaboration they embodied—were under vigorous attack. The focus was mainly on the scintillation counter.

Considering the scintillation counter as a historical document, one can see that for both groups, it reflected a number of social and epistemic meanings embodied in its construction and use.[9] With Vienna being the major link that tied together a small group of physicists and chemists, the scintillation counter enabled them to bring their laboratory to the cutting edge of radioactivity research. The scientific knowledge required for its use further sustained the work of a number of women in the institute and brought together scientists from different research teams. Demonstrating technical dexterity, female physicists manipulated the counter and played the role of instrument makers along with their male colleagues.

Still, scientific instrument makers have not been studied in their institutional context.[10] The rise of practical scientific research in the nineteenth and early twentieth centuries led to an increasing demand for specialized scientific instruments. Special contractual arrangements with local makers, the direct employment of skilled technicians on the staff of the scientific institutions, and the commercial supply from instrument-making companies satisfied those needs. In Vienna, men and women often played the role of instrument maker, saving money for the institute. Focusing on their contributions meant paying attention on how experiments were performed and how those scientists formed their identities as experimenters while actively engaged in all aspects of the experiment. In such cases, women's contributions are rarely mentioned, nonetheless.

Despite the vast literature on technology and gender, most recent studies have focused on biomedical and information technologies.[11] When the physics laboratory is addressed, the point is to demonstrate that women are often used as calculators, scanners, unskilled assistants, and cheap labor force in science.[12] The active role of female researchers at Vienna's Radium Institute, in contrast to the Cambridge group, reveals implicit assumptions about gender roles within the institute, explicit gender politics of collaboration, and most importantly institutional differences in the relation of gender and laboratory technologies.

THE EARLY DAYS OF THE SCINTILLATION COUNTER

What for Friedrich Giesel was the best way to detect the easily absorbed alpha radiation of polonium, became for William Crookes the best way to exhibit the luminosity of radium on a screen. We have already encountered Giesel when he handed the first amount of radium to Meyer back in 1899. Crookes was an exceptionally gifted chemist interested in "those areas where chemistry meets physics."[13] When he approached Giesel and asked him to recommend a medium for exhibiting radium on a screen, Giesel suggested zinc sulfide.[14] In 1903, Crookes became the first to describe the evanescent flashes appearing on the suggested screen when he brought radium close to it. Performing several experiments in order to test different factors influencing the number of scintillations, Crookes constructed a very simple apparatus. No schematic representation accompanied his publication and only a short description was given: "A blend screen was fixed near a flat glass window in a vacuum tube and a piece of radium salt was attached to an iron rocker so that the movement of an outside magnet could bring the radium close to the screen or draw it away altogether. A microscope gave a good image of the surface of the screen and in a dark room, the scintillations were well seen."[15] The first scintillation counter was already constructed and Crookes proposed to call the instrument a "spinthariscope" from the Greek word spintharis—a scintillation.[16]

What Crookes thought he was actually observing was the impact of electrons on the screen. Shortly after, Julius Elster and Hans Geitel confirmed Crookes' observations. In 1908, Erich Regener used the method to record alpha particles of polonium.[17] In his Nobel speech that same year, Ernest Rutherford compared the scintillation technique to the electrical method for counting alpha particles that he and Hans Geiger had constructed. The scintillation counter was found to be reliable. He further proved that each alpha particle produced a single visible scintillation on the screen.[18] While Rutherford was still in Manchester, his assistants Walter Makower and Geiger set up a course in training research students in the experimental techniques used in radioactivity. The scintillation method was by far the most important.[19]

In 1919, manipulating the counter for his collision experiments, Rutherford argued, "Under good conditions, counting experiments are quite reliable from day to day." Taking personal charge of the counting, he added, "Those [countings] obtained by my assistant Mr. W. Kay and myself were always in excellent accord under the most varied conditions."[20] In his Bakerian lecture in 1920, where Rutherford proposed his satellite model of the nucleus, it was apparent that he relied on the use of the scintillation method for his artificial disintegration experiments.[21] Placing great value on the technique and especially on the scintillation observers, Rutherford never forgot to acknowledge them in his papers. For William Kay, he granted an important role to the discovery of artificial disintegration by thanking him "for his invaluable assistance in counting scintillations."[22]

THE USE OF THE SCINTILLATION COUNTER IN CAMBRIDGE

In the early 1920s, the scintillation counter in its generic form was a very simple instrument. It consisted of a screen, a thin glass plate spread with an equally thin layer of zinc sulfide. When it was struck by charged particles, the screen produced light flashes. The scintillations were observed through a microscope which was specifically designed to increase the brightness of the flashes. By manipulating the microscope and its light-gathering power, the experimenter could work with weak radioactive sources and still observe a fair number of particles. A typical scintillation counting experiment could proceed as follows:

> There must be two rooms and two workers. One of the rooms must be kept in good deal darker than a photographic dark room, and in it there is one of the men who is to act as the observer . . . In this room there is a microscope and scintillation screen, and also whatever may be set up of radium appropriate to the experiment. In the neighboring room the other sits and keeps the record of the count of the scintillations. Thus it may be that what is to be counted is the total number of scintillations made on the screen in two minutes and at the end of it he will write down the number told him by the observer. When the experimental set-up is to be changed, the [observer] must first blind-fold himself, and then the light is put on in the dark room and the other man comes in and resets the instrument. He must turn out the light and go out and shut the door before the observer can uncover his eyes. Altogether it is a laborious business.[23]

Indeed, the detection of alpha particles in zinc sulfide screens by the visual observation of the individual scintillations was neither easy nor always reliable. The evidence was fragile, depending on the radioactive contamination of the counting apparatus, on the experience of the observer, and on the presence of hydrogen impurities. Besides alpha particles, the source often produced beta and gamma radiation that interfered with counting and increased the number of flashes. Additionally, weak scintillations were not always observable. The observer needed to have control not only of the counter but of his optical system as well. Rutherford, Chadwick, and Ellis affirmed this: "The superior efficiency of an experienced observer appears to be due to greater concentration, to control spontaneous movements of the eye, and to practice using external portions of the retina, thereby avoiding the insensitive fovea-centralis."[24] The issue of control stood at the center of Rutherford's research project at Cavendish in a dual sense. Control of the observers of the scintillations meant constant crosschecks of their countings and manipulation of their apparatus. To Rutherford, control had a further bearing on his entire laboratory life. Essential to his theoretical atomic model, the scintillation counter was the means by which Rutherford hoped to detect the atomic structure.

Rutherford enlisted his students "under Chadwick's careful surveillance,"[25] and he eventually assigned specific research projects to them after a short introductory course which included training in scintillation counting. The aim was to complete a Ph.D. in three years, contributing mostly to the laboratory's research and problem-solving tradition. With Rutherford's research program depending entirely on the scintillation counter, the modification of the technique became urgent.

The first step was to change the optical system by utilizing a new microscope that increased the brightness of weak scintillations. "We have found most suitable for our purpose a holoscopic objective of focal length 16mm and aperture 0.45."[26] The final magnification of the system was about 40. By choosing wide aperture, the scintillations became more brilliant and counting became easier. They also tried to obtain more powerful sources of radiation. Radium was at Rutherford's "command," a trafficking material he brought with him from Manchester.[27] Furthermore, to protect the observer from the gamma rays of the source, Rutherford and Chadwick used suitable absorbing screens and a reflecting prism. Next, they improved the scintillation screen by employing a thinner and finely powdered layer of zinc sulfide, aiming to reduce the luminosity due to gamma rays. In addition, a strong magnetic field deflected the beta rays. It was at this point that the scintillation counter developed from a clumsy technology to Rutherford's powerful vehicle for restructuring not only the material culture of his laboratory but physics in the Cavendish Laboratory as well.[28] In the hands of the Cambridge experimenters, the scintillation counter was transformed into a promising technique for maintaining the laboratory's authority in the postwar world of radioactivity.

The above changes in the material structure of the instrument simultaneously posed constraints that altered the politics of collaboration in conducting experiments. To protect against visual mistakes and exhaustion, male researchers alternated in counting scintillations for one minute each. Indeed, as Rutherford's acknowledgments show, the practice of scintillation counting in Cambridge was a male preserve and this included the task of recording data, making the necessary adjustments to the instrument, and, most importantly, observing evanescent scintillations.

Likewise, in Rutherford's group, the tasks of instrument-making and experimenting were also gendered. From the published historical studies on Rutherford's team, it appears that there were no female researchers participating in the artificial disintegration experiments. In Manchester, Rutherford had a number of female research students, but none in Cambridge.[29] In the original papers of Rutherford and Chadwick, there were no female research students mentioned as scintillation counters. However, the papers often thanked men such as Ellis, Blackett, Barton, Hirst, and Osgood. As Heilbron suggests, Rutherford's laboratory was "a masculine and, sometimes, a macho place." Rutherford himself referred to his task as a professor as "driving the boys along" although he had no objection to "a woman of great charm and ability . . . a welcome addition to any research laboratory."[30]

The Cavendish Laboratory was established in 1874 to teach the physical sciences to male students of mathematics. Despite their actively pursued petitions, women had been consistently refused degrees from the University of Cambridge both by its authorities and by its male undergraduates who feared competition for jobs and loss of freedom. The refusal of full degrees and membership was the reason for women to establish their own colleges, one in Girton outside Cambridge (1873) and one on the edge of Newnham Hall (1875). Although the two colleges were in no way considered part of the university, women had the privilege to receive some of the best academic training available through the university and take Tripos exams on an unofficial basis. In 1882 following a suggestion by Lord Rayleigh, it was decided that all classes and demonstrations should be opened to the students of the two women's colleges, including the Cavendish Laboratory. Paula Gould has concentrated on those women who entered the laboratory as research students in the late nineteenth century, arguing that they "were not isolated oddballs, figures of fun for discussion by the male workers. They were family friends, relations, educated women, keen researchers seeking experimental space."[31]

Despite having the educational advantages similar to their male colleagues, women did not gain formal university acceptance until the early 1920s. A Royal Commission was set up in 1919 to review major changes due to the war years and to recommend new policies, including women's status in Cambridge. On October 20 of that year, a grace giving women the titles of degrees with no participation in the university's government was passed. Thus, at the time Rutherford took up his professorship at Cambridge, the situation of women was very much improved. This included first a new regulation in 1923 which allowed women to have full access to the university library and borrow books and second the decision of the Royal Commission to leave it to the university itself to decide on the admission of women to full membership.

Although Rutherford favored women's full equality with men, Joseph John Thomson was in favor of maintaining a few restrictions. It is indicative that the 1921 group photographs of the Cavendish researchers show one woman out of 29 people, one out of 25 people in 1923, and two out of 39 in 1932.[32] The Cavendish Laboratory experiments, including the scintillation counter and the observation of flashes were apparently male projects. They embodied specific politics of collaboration and illustrated gender assumptions about skill in observing and manipulating the apparatus.

TECHNOLOGY TRANSFER: THE SCINTILLATION COUNTER IN VIENNA

In late 1922, when Pettersson and Kirsch performed their first artificial disintegration experiments at the Radium Institute in Vienna, they used the scintillation method exclusively. If they were to undermine Rutherford's and Chadwick's experimental results concerning the artificial disintegration of light elements, there was no alternative to the scintillation technique.[33] However, as Galison has illustrated, "Objects travel clothed in culture and human interactions. Objects are encumbered, covered with meanings, symbolisms, power, and the ability to represent but also to preserve specific elements of continuity. Yet, precisely because things come dressed with meaning, it is essential not to picture the handing down as occurring without alternation."[34] Through its transfer from Cambridge to Vienna, the scintillation counter was transformed into a different instrument, "clothed" now in the Viennese culture of radioactivity. It is those transformations I want to trace here, focusing especially on the alteration of the gender assumptions that the apparatus carried to the Radium Institute.

As a gifted instrument maker, Pettersson's first step was to improve the preparation of the radioactive source employed in the experiments. He developed a new method for the preparation of radium C that produced alpha particles of high intensity. The method involved the enclosure of dry radium emanation mixed with pure oxygen in thin-walled capillary tubes made out of quartz.[35] Later, with the help of Kirsch, Pettersson constructed a different emanation vessel "in which the substances to be examined are spread in thin layers over copper."[36] According to their results, silicon, beryllium, magnesium, and lithium yielded protons of ranges 18 cm, 12 cm, 13 cm, and 10 cm in air, respectively. Apparently contradicting Rutherford's and Chadwick's research, those elements proved to be disintegratable.

In the late autumn of 1922, two of Rutherford's students, L. Bates and J. Rogers, were assigned to study long-range alpha particles from radium C. When Pettersson and Kirsch published their first paper, Bates and Rogers already had a response in hand. On September 22, 1923, they reported in *Nature* that radium C emits not only the usual range alpha particles of 7 cm but also ones of longer ranges.[37] Thus, they argued, it could have been possible that what Pettersson and Kirsch thought they were observing were actually long-range alpha particles instead of disintegration protons.

In their reply, Pettersson and Kirsch stressed the fact that the ratio in luminosity between alpha and H-particles (protons) does not permit such a mistake as Bates and Rogers attributed to them.[38] On July 19, even before the articles were published, Rutherford sent his results to Pettersson.[39] In a friendly and grateful response on July 27, Pettersson tried to reconcile and explain the discrepancies. His line of defense was centered on the scintillation method that he and Kirsch had already modified:

> I lately had the counting box modified so as to allow the source as well as the substance of being enclosed while the pressure is varied. Preliminary experiments seem to show that with this arrangement, contamination can be avoided also at low pressures, so that the particles from carbon and other elements may be investigated down to the very shortest ranges. [40]

The modification of the counting box was not the only innovation that the Viennese experimenters added to the scintillation counter. As Pettersson said:

> Our newest microscope with the scintillation screen directly attached to the front lens by means of cedar oil (a Watson Holoscopic of n.a. 0.70 and f=12) is so superior with regard to brilliancy of the scintillations viewed through it, we feel much more confident than with the microscope previously used not only in differentiating between scintillations from alpha and from H-particles but also in counting the latter relatively near the end of their range. For this reason alone, I regard any confusion between H-particles and contamination alpha-particles as improbable.[41]

In their next publication, submitted to the *Proceedings of the Royal Society* on December 5, 1923, Bates and Rogers seemed doubtful. Without any direct reference to Pettersson and Kirsch, they suggested that the observed long-range particles were alpha particles from the contamination of the source. However, they admitted that "this evidence is far from conclusive, as these particles have never been observed alone but always when accompanied by particles of different types."[42] Faithful to the material culture of their laboratory, they used the same microscope with a holoscopic objective of numerical aperture 0.45 and 16 mm focal length that was previously used by Rutherford and Chadwick.

Specially constructed, the eyepiece consisted of two large plano-convex lenses and a smaller double convex eye lens. Such arrangements resulted in increasing substantially the field of view. Between the lines was an implicit attack to the microscope used in Vienna. By having a smaller field of view in their microscope, Pettersson and Kirsch were forced to increase the intensity of the source and thus the secondary radiation. Somewhat regretful, Bates and Rogers pointed out that "the scintillation method is the only method at present available for investigating these long-range particles."[43] They avoided any reference to the Viennese group when, in February 1924, they sent their next study to the

Proceedings of the Royal Society. More confident this time, Bates and Rogers argued that the long-range particles described in their experiment were emitted by the source itself.[44] Additionally, they had slightly altered the scintillation method by introducing polonium as the radioactive source instead of radium C. There was the significant advantage that polonium did not emit beta rays.

The task of preparing polonium was neither easy nor common knowledge among the experimenters. Chadwick, as a senior colleague and greatly interested in the project, prepared the polonium source for Bates and Rogers from a solution of radium D.[45] Eventually, the contamination of the source with radium forced the two research students to use instead the contents of old emanation tubes provided by Rutherford.

While Rutherford enlisted in his group research students who never forgot to acknowledge his support, Pettersson enrolled more experienced experimentalists.[46] Dagmar Pettersson entered the debate on April 3, 1924, presenting her first paper to a meeting in the Vienna Academy.[47] She brought to the Radium Institute technical knowledge and the chemical expertise of her recent work at the Technische Hochschule, as well as the experience she had gained working with Otto and Hans Pettersson.

DAGMAR WENDEL-PETTERSSON

Daughter of a prosperous civil engineer, Dagmar Pettersson, née Wendel, received private education and entered the University of Uppsala to study chemistry as her major and mathematics as her minor.[48] All of Dagmar's three sisters sought professional training, prompted by their father to ensure a self-supporting life. Dagmar was the eldest, born in 1888. She finished her studies in 1914 and soon after secured a position as a chemist in an agricultural laboratory in Skenja, Sweden, playing a leading role in building a new chemistry lab.[49] Shortly after she sought a new position, wishing to return to Göteborg and be closer to her parents.

In the first decade of the nineteenth century, women who sought careers in science in Sweden were very few and quite visible. Dagmar had met Hans Pettersson when both were students in Uppsala, but it was not until her return to Göteborg that they developed a relationship. Hans offered to help by asking his father to hire Dagmar as a chemist in his oceanographic station in Bornö. Otto already had a female chemist assistant and seemed to have been fairly open in accepting women in his lab.[50]

After his son's intervention, Otto employed Dagmar to measure the salinity of deep water samples as a research assistant at Börno. Two years later in 1917, Hans and Dagmar were married and when he moved to Monaco, she joined him not only as his wife but as a colleague with considerable experience in chemistry.[51] When they moved to Vienna in late 1922, she secured a position at the Technische Hochschule, Vienna's Polytechnic, working in a lab as a chemist and conducting research at the Radium Institute in support of Hans's project.

Although their 3-year-old daughter joined them in Vienna in 1924, Dagmar was able to continue her work. Her tasks were not just supplementary to her husband's research. As her diary indicates, she was involved in designing apparatus for the groups experiments and her own, counting scintillations, extracting polonium from residues from Joachimsthal radium ores, and critiquing Hans's manuscript. The fact that her husband was a scientist probably made it easier for her to combine motherhood with her scientific research.[52]

Directly addressing Bates's and Rogers's experiments, Dagmar attempted to undermine their results by altering the scintillation method. Instead of allowing alpha particles to pass through air or mica as in the Bates and Rogers experiments, Dagmar enclosed the scintillation counter in a glass tube. She constructed a device for sending alpha particles emitted from radium C to the scintillation screen through thin foils of gold or of copper which acted as primary absorbers for the ordinary a-particles.[53]

Dagmar's apparatus consisted of a glass t-tube that directed the alpha-particles emitted from the source to the scintillation screen (Z) to pass through the carrier of the radioactive preparation (T). The gold absorbing foil (F) was placed in front of the preparation (P) in order to minimize the risks of getting secondary particles other than those of the source. Additionally, mica screens of varied stopping powers were interposed between the scintillation screen and the source. The mica screens offered the advantage of placing weak radioactive sources closer to the scintillation screen. Two frames (M_1) and (M_2) with five openings each, hanging from (G_1) and (G_2) vertically in the tube, determined the different absorption levels from 0 to 16 cm equivalent of air. The pipe (H) was designed to control the pressure in the tube and a magnetic field deflected the beta rays emitted from the source.

For the preparation of radium C, Dagmar adopted Pettersson's innovative method of using thin capillary tubes filled with radium emanation and dry oxygen. The microscope used for the observation of the light flashes was, as Dagmar argued, superior to that of Rutherford's and Chadwick's. That gave her the opportunity to observe the slow, weak scintillations from H-particles. Used previously by Kirsch and Pettersson, the microscope was directly attached to the scintillation screen. It had a numerical aperture of 0.70 and the field of view was decreased from 12.5 mm^2 to 8 mm^2.[54] "It may be added," Dagmar argued, "that the use of a greatly improved microscope made it relatively easy to distinguish between scintillations from H and from alpha particles, so that they could be counted separately."[55] According to the final results, the number of alpha particles of range more than 9.2 was, to the surprise of Bates and Rogers, zero. That meant that there were no long-range alpha particles emitted from the source. Dagmar published her relevant article not only in the *Mitteilungen* of the institute but also in *Nature*.[56]

Dagmar's first ambitious articles in the world of radioactivity research were actually her last ones as well. Although she continued to work in Pettersson's group, Dagmar did not publish and, according to the recollections of her daughter, "Research work never was prominent for my mother."[57] Shortly after her publication, Rutherford and Chadwick challenged her observations, criticizing both her experimental setup and her apparatus. As they argued, it was probably the use of the absorbing foils that misled Dagmar into detecting the long-range alpha particles.[58] But her claims had clearly made an impression in Cambridge. Much later, in a letter to his father, Pettersson described Bates' interesting account of the episode:

> I was visited by my enemy, nowadays friend, former colleague of Rutherford, now with Porter in University College where I met him in May. He described very dramatically how Dagmar's letter to *Nature*—in which she opposed the results of Bates and Rogers—had hit the Cavendish laboratory like a bomb and how B[ates] had been scolded by R[utherford] in spite of his being right to certain degrees.[59]

In preparing her experiments, Dagmar collaborated closely with a young lady, giving rise to Pettersson's later claims for the "unselfish" and collegial ethos of his group.[60] "I thank Elisabeth Kara-Michailova for the zinc sulfide screen," Dagmar acknowledged, "which she produced through careful examination of different scintillation substances and methods of preparation."[61]

At the time, Kara-Michailova was 27 years old with a short but important list of publications already in her vita. At the academy meeting on May 8, 1924, Kara-Michailova presented a new technique that she and Hans Pettersson had developed for identifying the flashes between alpha and H-particles.[62] Their aim was essentially to reshape the scintillation counter by utilizing a new microscope they had constructed. The result was to deeply alter not only the counter but also the gender assumptions that accompanied its use. Where Cambridge researchers aimed for a well-defined hierarchical division of labor among the experimenters, the Viennese chose a collegial partnership. It was under these assumptions that Kara-Michailova teamed up with Pettersson in 1923.

THE KARA-MICHAILOVA/PETTERSSON COLLABORATION

By the mid 1920s, although both the Vienna and the Cambridge groups had immensely improved the scintillation counter, its main disadvantage continued to be the difficulty of distinguishing between flashes produced by different kinds of particles. As Pettersson and Kirsch suggested, the considerable differences in the relative brightness of scintillation from alpha and H-particles required more attention and exact measurements.[63] Having an expertise in measurements of luminescence, Kara-Michailova was the most appropriate collaborator for Pettersson.

In early 1924, with the help of his Swedish sponsors, Pettersson purchased a so-called *Vergleichsokular* (a comparison eyepiece) from the C. Reichert Company in Vienna.[64] The eyepiece was designed to compare the images from two microscopes in combination with two Watson holoscopic objectives. Pettersson estimated the cost of a Watson scintillation microscope, a comparison eye piece for scintillation photometry, two holoscopic objectives, and a number of scintillation screens and absorbing foils, to be 86 dollars, a considerable amount at the time.[65] In the Radium Institute, there was no previous experimental tradition with such an optical system. As a result, Pettersson purchased the whole set, altering not only the scintillation method but the material culture of the institute as well.

As this figure shows, FOF was the *Vergleichsokular*, with two identical microscopes directly built to the same eyepiece. Those were Watson holoscopic objectives with numerical aperture 0.45 each and focal length 16 mm. S was the fixture holding a reflective prism that offered the opportunity to combine the pictures from both microscopes or selectively reflect the picture from only one. The microscope on the left was set to observe the scintillations produced by H-particles on the screen (Z). The source (H) was either radium emanation or radium C and the Leyboldt magnet (M) was set to deflect the beta rays of the source. The microscope on the right was set to observe the flashes of alpha particles produced by a polonium preparation (P). The zinc sulfide screens scintillated in response to both kinds of particles, but the relative brightness of the flashes was determined by means of light-absorbing screens (F) of known absorption.

In the flow of their main text, Pettersson and Kara-Michailova did not forget to acknowledge the help of Maria Belar in measuring the absorption of the gray glasses (F). "We thank Maria Belar for the measurement of the absorption of the individual gray glasses that she conducted by the means of a Glan spectrophotometer in yellow-green light."[66] At that time, Belar was closely working with

Przibram and had just finished her thesis on the spectrophotometric method.[67] The same year, Kara-Michailova also published an article in the prestigious *Physichalische Zeitschrift* on the scintillation method and the quantitative optical differentiation between alpha and H-particles.[68] In the autumn of 1925, Kara-Michailova was forced to quit her research for a few months. Suffering from a lung infection, she moved to Merran in Tirol to recover.[69] As her publication record indicates, that interruption cost her more than a year of active work although she actually returned to the institute in the spring of 1926. Their project was by no means on secure footing and thus Pettersson was in urgent need of financial support.

FUNDING FROM THE ROCKEFELLER FOUNDATION

"I have the pleasure to state that, in my opinion, the method developed by Doctor Pettersson and his coworkers is of great scientific value and that the pursuance of this scientific work merits every assistance," Albert Einstein said of Pettersson's work in 1923.[70] The next year Pettersson tried to approach the International Education Board (IEB), securing again Einstein's recommendation.[71] The immediate response of the board was negative, but its officials were intrigued. On March 26, 1925, Augustus Trowbridge, director of the European Physical and Biological Sciences Division of the IEB, visited Vienna to meet with Meyer, Przibram, and the director of the Austrian Academy of Sciences to evaluate the situation.

Obviously, IEB could not support an institute without having a hand in it. In his report, Trowbridge first evaluated the status of the institute and its finances to Wickliffe Rose, the president of the IEB. From the institute's prewar glory, only the infrastructure and the expensive radium sources remained, but there was no soft money for research. Przibram was explicit in stating that no money was needed for assistants. Paradoxically, desperately needed were the cheapest accessories because breakable material could not be easily replaced. As Trowbridge realized, young researchers were working under "what an Occidental scientist could consider to be impossible working conditions through the lack of the most necessary materials and supplies." The main contributions came from sources that Pettersson solicited. Then, Trowbridge made his suggestion:

> In my opinion, were the board to vote yearly, for one or two years, one fifth of this sum to the Vienna Academy of Sciences for the purpose of the purchase of necessary supplies and apparatus for research, this excellent institute could continue to train

> effectively more than a dozen young research students in the physical sciences and bring them to the point where a few of the best might then be given fellowships for further study in the west of Europe or America . . . . The Radium Institute of Vienna ought to be a natural feeder from southeastern Europe to the laboratories of Madame Curie and Sir Ernest Rutherford.[72]

The International Education Board's choices in funding profoundly affected and shaped physics research in Europe. The IEB was created in 1923 as the Rockefeller Foundation's education board, with a mission both to provide fellowships for young scientists internationally and to support the infrastructure of science through capital grants for building and maintaining research institutes. The unstable political situation and the financial problems that European researchers were facing made the IEB a major actor in determining physicists' careers and the future of research programs.[73] Likewise, the future of Pettersson's research group in Vienna and his own career depended substantially on the plans and decisions of the board. While IEB was willing to help the Radium Institute, their direct interest lay in the two other important radioactivity research centers in Paris and Cambridge and they viewed Vienna as their supplier of well-trained personnel. They never envisioned Vienna as an independent center capable of being in the forefront of research.

On May 29, 1925, the board decided to offer the Radium Institute 2,000 dollars per annum for a period of not more than three years for the purchase of necessary equipment. The grant was put in the hands of the Austrian Academy of Sciences on July 1, 1925.[74] In the autumn of the same year, Pettersson applied to the board for an additional fellowship, this time for his own expenses in Vienna. As he argued, the grant "would enable me not only to pursue my own investigations but also to go on training and supervising a number of younger collaborators who were engaged in important problems related with artificial disintegration."[75]

It was at this same point that Meyer stressed in his own letter of recommendation to the International Education Board that "His [Pettersson's] impulsiveness acts most beneficially on his young collaborators here and he knows excellent how to introduce them to the methods of this difficult subject, so that, also in this respect the object of the Educational Board, that is the higher training of young workers, would be served."[76]

Pettersson was slightly over the age limit set by the board but, as Trowbridge argued, "The man is thoroughly energetic with an enormous program of work in mind, an indefatigable worker, and from this point of view in every way, compares favorably to a considerably younger man." In the usual manner of the IEB's business, Trowbridge visited Pettersson in Stockholm to interview him. Pettersson had the ambition to become a professor at the University of Stockholm, succeeding Svante August Arrhenius, the Nobel laureate in chemistry in

1903. Arhenius actually owed an old debt to Hans. Otto Pettersson was the one who first emphasized the originality of Arrhenius's dissertation in 1884 and helped him to get a docentship at Uppsala in physical chemistry, the first in Sweden in this new branch of science. Besides Arrhenius's confirmation that "He is in the first line among young men who could be thought of for a professorship in physics or physical chemistry," Pettersson's appointment would largely depend on his ability to demonstrate a similar ingenuity as his predecessor. His work in Vienna was essential in increasing the probability of getting the position.[77]

Recognizing the awkwardness of the occasion, Trowbridge approached Manne Siegbahne, professor of physics at the University of Uppsala and a second candidate for Arrhenius's position. Having this in the back of his mind, Trowbridge heard Siegbahne's critique of Pettersson with reservation: "brilliant but uncritical . . . all right as an investigator, thought not sufficiently critical, but could be no good as director." Siegbahne went on to propose that the IEB should choose a "good man" from among those who had already been connected with the Radium Institute to direct Pettersson's project. The one he specifically indicated was Georg von Hevesy who was then working with Niels Bohr in Copenhagen.[78] However, Trowbridge's final recommendation was in favor of Pettersson's fellowship. On December 2, 1925, the board decided to award 250 dollars per month for a year—much higher than the ordinary amount of 182 dollars—to Pettersson who immediately moved back to Vienna. A similar request to the IEB to fund Kara-Michailova as his assistant through a fellowship was declined.[79]

Once again in the autumn of 1926, Pettersson asked the IEB for the prolongation of his own fellowship for another year. The work that had been done in the meanwhile in Vienna and the contradictory results of the Cambridge group puzzled Trowbridge. It was obvious from the visits of other members of the IEB that there was activity and an apparent new lease on life at the institute. Pettersson's presence had definitely stimulated work there, but Rutherford's opinion was quite different. Satisfying Trowbridge's request to assess the work in Vienna, Rutherford claimed that Pettersson had certainly galvanized the institute into "at any rate, a form of life," but he and his collaborators were "experimenting in a field which is full of traps for the unwary." He continued to say that they rushed their experiments too much and ending up not contributing anything important to the field. Given their ongoing competition, Rutherford's comments were no surprise.

Marie Curie was more generous. When Trowbridge directed similar questions to her and mentioned some criticism that had arrived at this office, Curie laughed. "One is from Rutherford," she guessed, and went on to say that she had personally verified some of the results claimed in Vienna. As Curie emphasized, Rutherford's strategy was to belittle any work in his field not performed by one of his own students. She suggested the prolongation of Pettersson's fellowship, and this seemed to have made the difference in Trowbridge's final recommendation. Pettersson was eventually reappointed by the IEB for 1926, which proved to be the most critical year of his career.[80]

PUTTING THE SCINTILLATION COUNTER ASIDE

The obvious disadvantages of the scintillation counter slowly but steadily pushed both the Cambridge and the Vienna laboratories to alter their material culture. As Pettersson reported later, "The subjective character of all observations made by the scintillation method added to the strain on the eyes of the counters which it involves, has made it most desirable to develop novel methods of studying atomic fragments, less exerting and less subject to errors."[81] As a response to the Viennese threat to undermine their authority in the field, the Cambridge team employed a Wilson cloud chamber in their research. Although invented in 1895, the Wilson chamber was not incorporated into radioactivity research programs before 1923. Clinton Chaloner argues that the Wilson chamber clearly gained its impetus from the dispute over the scintillation counting procedures between the Vienna and Cambridge groups.[82]

Rutherford's response to the use of the chamber was enthusiastic and soon the Wilson cloud chamber rendered the paths of particles for the Cavendish people. Pettersson did not want to miss the opportunity of using their instrument to support his own experimental results and theoretical claims. Supported financially by the International Education Board, Pettersson ordered a Shimizu Wilson ray track apparatus on November 21, 1923. This was a reciprocating form of Wilson's instrument, which enabled more photographs per second than the initial one. Within less than a month, the Cambridge and Paul Instrument Company (later the Cambridge Instrument Company) shipped the new apparatus to the Radium Institute in Vienna.[83] A young doctoral student, Rudolf Holoubek, was assigned to study the tracks of H-particles from aluminum, carbon, and iron using the new instrument. To increase its validity, Pettersson also used the new instrument repeatedly at university lectures, showing tracks from atomic fragments to curious students.[84]

A second Cavendish method was also employed by the Vienna Institute. Georg Stetter, *Assistent* at the second Physics Institute and collaborator with the Radium Institute, constructed a mass spectrograph, adopting the principle used by Francis William Aston in England. The experiments enabled by Stetter's instrument further established that boron, carbon, aluminum, and iron were also disintegratable.[85]

A young research student, Norbert Kreidl, developed a third technique, which was an electric counter with an amplifying set of special construction to record corpuscular radiation. The innovation of the technique was that the atomic fragments were made audible based on Heinrich Greinacher's original method for producing sound in a telephone by individual particles.[86] Funded by an International Education Board grant, Kreidl spent two months in 1925 at Greinacher's laboratory in Bern for training.[87] In 1926, Stetter and Gustav Ortner took over the research, designing a method of electrical amplification of ionization currents that operated a loudspeaker. By 1928, Pettersson said the method had given excellent results with alpha particles, but the technical difficulties in making the atomic fragments audible were still unresolved.[88]

The development of a fourth technique, photographic emulsions, was assigned to Blau. The method had already been used by S. Kinoshita and M. Reinganum in the beginning of the 1910s to identify trajectories of alpha particles through emulsions. In early 1924, Blau attempted to observe recoil protons produced by alpha particles in paraffin. With weak radioactive sources, she could observe the lower energy particles, but the accuracy of the measurement was limited. The only strong source available was polonium. To prevent darkening of the plate by gamma radiation, Blau worked with polonium prepared in highly concentrated preparations by Rona. In 1925, Blau first detected the trajectories of slow protons.[89]

As Pettersson wrote to his sister on March 7, 1926, "By indescribable tenacity, she [Blau] has succeeded at an almost hopeless job I suggested to her two years ago."[90] Her first attempt was to observe recoil protons produced by alpha particles in paraffin. With weak radioactive sources, she could also observe the lower energy particles. The only strong source available was polonium which Rona knew how to prepare. As Blau describes, "To prevent darkening of the plate by gamma radiation, one worked with polonium which was prepared by Dr. E. Rona in highly concentrated preparations. After a tedious series of indefinite experiments, it finally worked in 1926, and in the following year, the method could be applied to the disintegration of various atoms with alpha particles."[91]

Besides designing new apparatus for tracing particles, the Vienna group was apparently in need of radioactive sources, preferably polonium, which was extensively used in the artificial disintegration experiments. Irene Curie was one of the few experts within the radioactivity community who could extract and prepare polonium sources. The process involved the tiresome task of the chemical separation of the element, its purification, and the final concentration on a small surface.[92] Since polonium did not emit beta particles that usually interfered in scintillation counting, its use as a radioactive source was most advantageous. Used mainly in the Wilson chamber experiments, the Viennese group was anxious to obtain the technical expertise for preparing polonium sources.[93]

In May 1926, Pettersson reported to his father. "I have now managed to get Meyer to write a letter to Curie asking to send one of our scientists, Frau Doctor Rona, chemist and specialist in polonium, to her lab for three weeks in order to learn the art from Irene Curie . . . If she is allowed to go, we have no problems next year and can make our own polonium samples."[94] Probably drawn from his own problems with the Curies, Pettersson added, "I first had an idea of going myself but desisted for the reason that a man coming to the Paris lab will be getting a much less friendly welcome than a woman."[95] His father, most likely using his authoritative way, had already tried to persuade Curie to accept his son in her lab just a few years earlier, but she refused.[96] Nevertheless, Curie accepted Rona when, not surprisingly, Pettersson succeeded in obtaining a small grant from a Swedish sponsor to fund her trip.[97]

With the credential of working with some of the most important members of the radioactivity community, Rona had entered the Radium Institute during the academic year 1924–25.[98] Adopting the methods of Pettersson's group on February 10, 1926, she presented to the Austrian Academy of Sciences her work on improved methods for measuring the absorption and range of H-rays and the use of polonium as a more suitable source than radium C. The main instrument she used was the scintillation counter. She teamed up with Blau and worked on the ionization of H-rays.[99] Four months later, after Pettersson's persistent attempts, Rona arrived at Curie's institute to get trained in preparing polonium sources.[100] On Rona's return to the Radium Institute a few weeks later, Curie was generous enough to donate to the Viennese a strong polonium source concentrated on a small silver disc of 12 mm^2. Most of the following studies on artificial disintegration performed at the institute were done using either Curie's source or Rona's preparations.[101]

THE SCINTILLATION COUNTER AS A CULTURAL HYBRID

It was in the midst of the busiest times at the Radium Institute with Pettersson's group working feverishly on new, more reliable techniques for counting particles produced by atomic disintegrations that Kara-Michailova went further into the design and construction of the scintillation counter. On May 5, 1927, she presented her next scintillation study to the Vienna Academy.[102] The focus of the study was on the brightness of scintillations produced by H-particles in relation to their velocity. As she pointed out, the most important question for the application of the scintillation method was to determine the lower limit of particles' velocity at which the scintillations were noticeable to the observer. Kara-Michailova's steps in designing her new experiment involved a noteworthy exchange of instrument parts with Stetter's mass spectrograph.

Although in search of other methods, Stetter and his colleagues were still faithful to the scintillation counter. By modifying Aston's mass spectrograph, Stetter was first to replace the photographic plates with Pettersson's and Kara-Michailova's model of the scintillation counter.[103] Purchased in 1926, Stetter's new apparatus was far too costly at close to 140 dollars. By comparison,Kara-Michailova's entire annual salary for the same year was a bit more than double that.[104] Thus, replacing the photographic plates with the old, familiar scintillation screens and a microscope was not a matter of cutting down in expenses. Rather, it was an expression of loyalty to the material culture of the institute. Most of all, it revealed the commitment of the Viennese group to an experimental tradition that trusted the coworker and his or her report instead of the static, visual representation of phenomena on photographic film.

The transfer of instrument parts, however, went both ways. "I am obliged to offer many thanks to Herr Dr. Stetter," Kara-Michailova acknowledged in her paper, "for letting me use his apparatus as well as for his help with the research."[105] Literally from her working bench to Stetter's and back again, the transformation of the instrument was indicative of a dying experimental culture. Based on the fragile eyesight of the observer, the scintillation counter was put aside as experimenters sought for more trustworthy and objective methods of research.[106] The replacement of human observers by mechanical devices indicated a shift in the material culture of the laboratory. Struggling to survive the transformation, the Viennese experimenters constructed instruments as hybrids at the intersection of the two material cultures.

By working in Pettersson's group, Kara-Michailova shared a strong network of collaborators in preparing radioactive sources such as polonium (Po), constructing scintillation screens (Z), and measuring the absorption of gray glass (G) used in determining the ratio in the brightness of the scintillations. With the comparison microscope, Kara-Michailova was able to observe the scintillations of both alpha and H-particles as she was used to in her previous experimental settings. The new feature was the magnetic field (F) which separated H-particles according to their velocity. Therefore, only rays of the same velocity fell upon the scintillation screen (Z). (K) was the radioactive source and (S) was the gap through which the rays were directed toward the magnet. This part of the device was enclosed in a vacuum.

During the preparation and the performance of her experiments, Kara-Michailova acted as the experienced and mature researcher on scintillation counters, having total control over her instruments. She employed two groups of trained observers in order to report simultaneously the scintillations produced by alpha and H-particles. The microscope in use had two Watson objectives of numerical aperture 0.70 and a single eyepiece. According to Kara-Michailova's results, the brightness of the H-particles was proportional to their residual range. The lower velocity limit for observing them was found to be less than 10^9 cm/sec.

While Kara-Michailova sought to improve the scintillation measurements, Pettersson visited Cambridge in May 1927. His long discussions with Blackett and Chadwick were centered on the scintillation method. On May 16, the discussion opened up with a reference to the properties of scintillation substances and the fatigue of the observers. As Pettersson argued, his colleagues in Vienna counted for 30 seconds or often for 20 seconds instead of one minute as the Cambridge observers did.[107] Both Pettersson and Chadwick placed an enormous importance on what essentially was Kara-Michailova's project, the possibility of distinguishing between flashes of alpha and H-particles using the scintillation counter. As it turned out, the resolution of the entire controversy was based on this distinction. After a "somewhat heated discussion" with Rutherford, Pettersson was finally satisfied that the British considered the Viennese way of working "worth looking into."[108] In the hope of arriving to a conclusive settlement, Chadwick planned a visit to Vienna in December 1927. Up to that point, Kara-Michailova, Blau, and Rona had come a long way in designing their instruments and their experiments. Their role in the wake of Pettersson's research group was indeed instrumental.

JAMES CHADWICK'S VISIT TO VIENNA

After Pettersson's visit to Cambridge in May 1927, things seemed to settle down and the relations between the two laboratories greatly improved. As director of the institute, Meyer took the initiative to thank Rutherford for the warm hospitality with which he accepted Pettersson in Cambridge. In an attempt to clear up any misunderstandings, he assured Rutherford that what was at issue was certainly not his authority in the field. "We hope that you got the impression that Dr. Pettersson and his coworkers here are doing their best to improve the knowledge founded by you on the atomic disintegration and that they are trying to further the work in the most serious attempts and not to trouble the advancements."[109] Chadwick's visit had already been arranged for December that same year.

That summer, however, and before Chadwick's arrival in Vienna, Bates unexpectedly visited the institute. The research had slowed and most of the personnel were absent, including Meyer who was spending his holidays at his summer residence in Bad Ischl, upper Austria. Pettersson had just returned from Sweden. "Yesterday, I was surprised with a visit from a gentleman with a lady," he reported later to Meyer. "That was our old enemy Dr. L. F. Bates who I had shortly met in London." As a genial host, Pettersson gave him and his wife a tour at the institute and they discussed in a "friendly tone." It was so friendly that Bates mentioned to Pettersson their Cambridge stock in capillary spent radon tubes from hospitals used in the preparation of polonium. Pettersson did not miss the opportunity. He immediately prompted Meyer to ask Rutherford for a supply of small radon tubes. In spite of their conflicts, they would have more than welcomed the radium—a valuable trafficking material—in the poorly supplied Vienna institute.[110]

Hoping to resolve the disagreements between the two laboratories, Chadwick arrived in Vienna on December 7, 1927. A scrutiny of the scintillation screens and the most basic aspects of the artificial disintegration experiments persuaded Chadwick that the problem resided in the Viennese protocol of scintillation counting.[111] As Hughes interprets Chadwick's thoughts, "There only remained the possibility that the Viennese were mistaken in believing that they could distinguish between scintillations due to alpha particles and those due to H-particles by the difference in brightness, a practice which the Viennese vehemently defended."[112] It was Kara-Michailova who "vehemently" defended her method in front of the scintillation counter and during the replication of the experiments. She was among the "young ones" who stood around "stiff legged and with bristling hair," as Chadwick vividly described it.[113]

Friday, December 9, "ended up with a fierce and very loud discussion."[114] Apparently, Chadwick interfered on Pettersson's territory and questioned his authority in his own laboratory. As he reported to Rutherford, "It is essential that I should prepare the experiment. So far, I cannot get Pettersson to agree to this."[115] Chadwick had probably suppressed the fact that when Pettersson visited Cambridge a few months earlier, he treated him as a tourist. As Hughes put it, Pettersson indeed remained a "disengaged witness" in his antagonist's territory.[116] Similarly, during the first two days of his visit, Chadwick was tenaciously kept away from the scintillation screens and the microscopes. On Monday December 12, however, Chadwick took advantage of Pettersson's absence and, treating the personnel as inexperienced students, he assumed control at the institute.[117]

> Today I arranged that the girls should count and that I should determine the order of counts. I made no change whatever in the apparatus, but I ran them up and down the scale like a cat on a piano—but no more drastically than I would in our own experiments if I suspected any bias. The result was that there was no evidence of H-particles.[118]

Chadwick's bias probably did not concern the experimental techniques, but instead the female experimenters. Here is how he described to Rutherford the testing of Schmidt's apparatus for the disintegration of carbon by polonium A-particles: "Their counters, two girls, managed to find a few [scintillations]."[119] The shift from "experienced observers," in Rutherford's language, describing scintillation counting in his lab, to "girl counters," in Chadwick's report of the work done in Vienna is more than a naïve linguistic slip in reference to the same task. [120] Instead, it indicates the gendering of skills.

The counting of scintillations, so crucial for the early experiments on artificial disintegration, was treated as a highly skilled task of experimental observations when performed by Rutherford's male team in the Cavendish Laboratory. Chadwick considered the same task, performed in the Viennese setting by the female experimenters, to be disconnected from observation and its meaning to the experimental process. As he noticed, "Not one of the men does any counting. It is all done by three women."[121] The immediate inference was that those women were mere counters. Nonetheless, as he described to Rutherford, he was the one who arranged that "the girls should count."

Apparently, the three women were Kara-Michailova, Blau, and Rona.[122] All of them were active participants in the controversy. Besides Kara-Michailova's work on the scintillation method, Rona was the polonium expert in the institute and the one who introduced Schmidt to the technique of preparing polonium sources while they worked on the penetration of polonium to metals. Further, in collaboration with Blau, Rona employed polonium to the study of ionization of H-rays. Blau had been working on the alternative method of photographic emulsions. What Chadwick saw, nevertheless, was routine counting and, as he later acknowledged, "the young women were perfectly honest."[123]

Chadwick's presupposition, though, legitimized the way historians of science Peter Achinstein and Owen Hannaway interpreted the episode in 1985. Chadwick suspected that women were informed about the experiments in advance and "they were seeing what they expected to see."[124] The minds that designed the experiments were, in Chadwick's account, separate from the eyes that recorded the results. Women's skills were limited to counting, recording, and following orders from their male colleagues.

If one asks whether Chadwick's version of the work in Vienna was entirely his interpretation, the answer is no. Pettersson seems to have played a role in this wrong account. According to Chadwick, "Pettersson says the men get too bored with routine work [scintillation counting] and finally cannot see anything, while women can go forever." According to Chadwick's later recollections, Pettersson also argued that women were more reliable than men as scintillation counters because they could not be thinking while observing. He specifically preferred women of "Slavic descent" as counters, believing that they had superior eyesight.[125] In the intense environment of the replication of the experiments, it is plausible that Chadwick and Pettersson misunderstood each other. A stiff Englishman "completely lacking a sense of humor," Chadwick had difficulties in understanding the humor of a Bohemian Swede.[126] Besides, Pettersson had used a similar argument concerning the vision of the scintillation observers for describing the work of R. L. Hasche, the young American physicist from John Hopkins University who visited Vienna in 1926. His task was to compare the different kind of microscopes used in Cambridge and Vienna for counting scintillations. "Owing to his particularly good eyes," as Pettersson stressed, "he seems to be especially qualified for that kind of work."[127]

It is also hard to reconcile Chadwick's claims either with Pettersson's anti-hierarchical style as the research leader of his group or with the respect he paid to his colleagues. Writing his report to the International Board of Education in 1928, Pettersson used language that surprises the reader even today. "Each collaborator has *his or her* particular share to take in making the practical preparations necessary for an experiment. Besides each has *his or her* particular theme for research which he pursues and where he can count on the help from one or more of his fellow workers." Appraising his experience of working in Vienna, Pettersson continues, "I have learned the very high value of teamwork for attacking the intricate problems that are found within the comparatively new field of nuclear physics."[128]

In this collegial atmosphere, Kara-Michailova had already adopted the research on scintillation counters as her main project contributing productively to Pettersson's team. As I have already argued, Kara-Michailova was actually the one who, in collaboration with Pettersson, had altered the optical system to suit the needs of the experiment. Additionally, she offered her expertise in fluorescence and lighting measurements in order to improve the method and distinguish between alpha and H-particles, an essential issue in the Cambridge-Vienna controversy. She even introduced trained observers to her experiments, proving the seniority of her position within the team.[129] As an experienced experimenter, she had absolute control of her instrument, knowing its tricks better than any of her male colleagues. Clearly, she did a great deal more than to count scintillations in front of a microscope.

Apparently, scintillation screens were easily prepared at the institute, the procedure proving as easy as spreading zinc sulfide over a glass screen and brushing it out very smoothly. Kara-Michailova was also skilled at experimenting with different scintillating substances and preparation methods. The manipulation of the numerical aperture and the field of view of the microscope were also within her capacity. Obviously, she had control over every part of the scintillation apparatus and was able to take the entire tabletop instrument apart and reconstruct it from scratch. Counting the scintillations was not a secondary, routine task for Blau, Rona, or Kara-Michailova. On the contrary, what Chadwick failed to recognize was that all these tasks demanded individual attention and defined part of the team's identity as experimenters in the institute. It appears that Chadwick was too absorbed by his own struggle supporting the Cavendish experimental results to understand the teamwork involved.

Focusing on laboratory technologies and gender in the two laboratories, it becomes apparent that there is a shift in the epistemological meaning of the concept of skill from an objectively quantifiable quality to an ideological category assigned to men and women on the grounds of gender biases. As the study of the scintillation counter reveals, the definition of a position as "skilled" depends on the gender identity of the person performing it. Therefore, skill becomes a political concept in the laboratory and plays a central role in maintaining the gender division of labor.[130]

Scientists and even historians of science are more inclined to see women as users than as designers of the technologies they encounter in their everyday laboratory practice. For example, through the eyes of the sociologist of science Jan Golinski, the women in Vienna who participated in the dispute between the two laboratories were the "support staff" that "bore the responsibility for worrying about discrepancies in observations." As he argues, "this was found to be more of an agreeable solution than blaming the physicists on either side, presumably at least in part because the technicians were not consulted and their judgment could more easily be doubted."[131] Golinski echoes Achinstein's and Hannaway's argument that, in the Radium Institute, "a division of labor had separated the eyes of the observers from the minds of the experimenters."[132] All of the women who were part of the lengthy exchange of publications in the 1920s, however, studied physics, had a background in mathematics, and had experience in experimental physics.

Kara-Michailova's effort was to secure a method that, although highly important in the 1910s and the beginning of the 1920s, was at its end as radioactivity was rapidly heading toward nuclear physics. An essential member of her research group, she had a "volcanic energy" at the peak of the controversy, as Pettersson admitted.[133] Yet passion for scientific work was not enough. With the total monthly budget of 250 dollars coming from the Rockefeller grant, Pettersson was trying to manage his entire research team.[134] It was therefore worth trying to save the cheap scintillation counter. A microscope with a couple of lenses, scintillation screens, and absorbing foils, was priced around 86 dollars, while a Wilson chamber lamp alone cost 14 dollars.[135] Low cost was not the only advantage of the scintillation counter. For Kara-Michailova and the rest of Pettersson's team, to save the method meant to retain control over the material culture of their laboratory. For instance, as a young research student, Karlik was one of those who focused on saving the instrument.

Two months before Chadwick visited the Vienna Institute in 1927, Karlik defended her thesis to Meyer and Hans Thirring.[136] Not surprisingly given the importance of the scintillation method, Karlik's topic was the dependence of scintillations on the condition of zinc sulfide and the nature of the scintillation process.[137] She described a photometric method for determining the relation between the range of alpha-particles and the brightness of the scintillations for differently prepared zinc sulfide screens. In her effort to defend the scintillation method, Karlik introduced photographic plates in her model of the counter. In order to reduce the light entering the eye through the microscope, she placed photographic plates between the objective and the eyepiece.

Especially innovative, the Viennese group was characterized by their attempts to save an old technique by taking advantage of the elements and instrument parts of new ones. Their colleagues in Cambridge were clearly shifting from the old, shaky laboratory technologies that the scintillation counter represented to the new, persuasive evidence of the photographic plates. Instead, the Viennese were hesitantly swinging between the two worlds, transferring pieces of instruments from one workbench to another and persistently defending the old physics experimental tradition to which the scintillation counter belonged. The portable, tabletop scintillation counter which almost all of the Viennese experimenters in Pettersson's group could control and construct offered them a feeling of security, not only in the fast moving world of radioactivity research, but also in the politically unstable Viennese setting.

THE POLITICS OF A "PRIVATE" RESOLUTION

When Chadwick met Meyer in the latter's office at the Radium Institute on December 14, 1927, to discuss the awkward outcome of his visit in Vienna, there was anxiety on both sites. Chadwick had demonstrated that the Viennese researchers had been performing unreliable experiments using the scintillation technique. Meyer offered, on the other hand, a public acknowledgment of Chadwick's results. In this odd situation, Chadwick counter offered a private resolution following Rutherford's earlier wish to "better discuss these divergences of view in private than in print."[138] At issue were not only the fame of the Vienna Institute and the authority of the Cambridge lab. Equally at stake was the reliability of the scintillation counter.

For at least the next two years, the scientific community remained largely unaware of the outcome of the Cambridge-Vienna controversy.[139] Although the scientific credibility of the Radium Institute and its researchers were not widely affected given the private resolution of the episode, Pettersson's team collapsed financially. The Radium Institute was already in a bad shape. Trowbridge had made it clear to Meyer even as early as June 1927 that without any additional local money, they could not renew their grant. Both Rose and Trowbridge were of the opinion that the board should "oppose emergency help being continued so that it would not become a crutch on which an institution leans."[140] Meyer, they suggested should definitely seek support from local and other sources. Taking the suggestion seriously, Meyer first addressed the Austrian Ministry of Education which promised to support the institute with 3,500 schillings and the salary for another scientific assistant, all of which amounted to almost 1,000 dollars per year. Also, a month before Chadwick came to Vienna, Meyer raised the issue of the radium that the Austrian Academy of Sciences had lent Rutherford before the war, hoping to guarantee a yearly contribution by the investment of the expected payment to another 1,000 dollars for around 15 years.[141] He wrote to Rutherford:

> Austria is so impoverished, that neither the government nor private persons or societies here are in the position to keep our institute going without financial help from abroad. For the last three years, we were assisted by the International Education Board; but as the board no longer wishes to continue the appropriations to us, our funds are threatening to run down so that we must try to get money from somewhere else if this institute is to be able to go on working. So I take the liberty of asking you if you could manage to raise the necessary funds for buying the rest of the radium that you have in Cambridge perhaps in ten or fifteen years installments? The market price of radium is at present 70 dollars for 1 mg, but I am sure that our academy would be willing to accept a lower price.[142]

In his response, Rutherford said he was willing to discuss the matter with the university authorities. He argued, though, that the amount of radium was only 250 mg and not the 304 mg that Meyer proposed.[143] Excerpts of old notes from Rutherford's "radium book" provided evidence for his claim. By offering his own data, Meyer questioned Rutherford's estimations, but since the institute was in desperate need of financial help, the former concluded, "As there are too many uncertainties and to avoid all difficulties, I immediately propose to accept your estimation of 251 mg as the basis of all further negotiations."[144]

While Chadwick was still in Vienna, Meyer had the chance to discuss the issue thoroughly with him. On December 21, 1927, immediately after Chadwick's return to Cambridge, Rutherford settled the matter. "I am very sensible of the generosity shown by the Vienna Academy of Sciences and the Austrian government in loaning me such a valuable preparation for such a long period . . . Please let me know if the price of purchase (3000 pounds) and the mode of payment by installments is satisfactory to you."[145]

Eventually, as Meyer succeeded in raising funds from local and other international sources, Trowbridge recommended a grant of 3,000 dollars for the year 1928–29 and 1,500 for the year 1929–30 which was actually the last payment from the Rockefeller Foundation.[146] "I really think," Trowbridge bragged to Rose, "we can be self-congratulatory on the outcome of this financially minor but scientifically fairly major undertaking on our part. I think this case of the Radium Institute forms a good argument for continuing and extending assistance to centers not of general major interest, but in which unique good elements may be found." What he saw as a "somewhat moribund institute" in the early 1920s, remained a peripheral center in Trowbridge's mind until the board's assistance came to an end in 1930, strongly defining the institute's scientific fate.

While the board renewed the grant to the Radium Institute for two more years, Pettersson was not as lucky. His fellowship came to an end after Chadwick's visit to Vienna. In a letter to his sister in 1928, Pettersson described his difficult life in Vienna:

> My colleagues have been touching at looking for somewhere to live for us. The Pension Atlanta would have been too costly. We are now renting two rooms in the flat of an elderly lady in Döbling, a quiet, country-like suburb of Vienna. With time, we have been able to worm ourselves into the lady's confidence, even using her kitchen for making lunches. For dinners, we have found a good, simple Gasthaus at the Silbergasse. Dinner for three amounts to less than 5 shillings (2.5 sek) so, the affluent times of Rockefeller are gone. Kara (Michailova) who helped much in getting us rooms, also has put in quite a supply of butter, marmalade, bread, eggs, and oranges, welcoming us at our arrival on Good Friday. Gerhard Kirsch's sister had put flowers in both rooms and from Blau and Rona, there were Easter eggs and a set of Dominoes for Anne.[147]

Unable to continue his research without the Rockefeller grant, Pettersson eventually returned to Sweden. According to Stuewer, "The entire investment collapsed to the ground in a few short days in December 1927. It was a severe shock."[148] The tragic picture that Stuewer draws does not do justice to the collaboration that followed Chadwick's visit to Pettersson and his colleagues in Vienna. It was not until 1936 that as Karlik wrote to him: "To think that this has been your last regular visit regarding our collaboration in atomic physics is very sad. I can hardly realize it yet. This joint work has been so much the center of my interests for the past years that it will mean quite a readjustment of my inner life to get adapted to the new circumstances." Acknowledging Pettersson's role in her career, Karlik added, "It's you who have shown me what experimental physics really is; if I look back, I must say I had only a very faint idea of it before I began working with you."[149]

Chadwick's visit, though, led Pettersson to a different kind of tragic circumstances other than the immediate collapse of his work with the Viennese. The defeat for Pettersson was personal since he lost the battle over his father's wishes. In 1928, he was refused a professorship of physics at the University of Stockholm. Instead, his father Otto arranged money for a new professorship, this time on oceanography in Göteborg.[150] Pettersson never gave up his interest in radioactivity research, but he also never acquired a position in physics. After years of attempts to avoid his father's influence, Pettersson had to rely on him once more for his academic career. It was not until 1934 that he admitted to Karlik, "I am of course aware that I am counted out before the physicists of Europe, 'suspect d'otre suspect,' a position I do not mind so much personally, but which is of course unfortunate for the work."[151]

The Cambridge group, financially and scientifically, remained on the top of the game. Early in 1928, Rutherford was aware of the importance of wave mechanics to radioactivity and succeeded in getting a new position for a teacher in theoretical physics and additional money to organize a conference on beta and gamma rays.[152] As Andrew Brown argues, "The main purpose of calling the conference was for Rutherford, Chadwick, and Ellis to make sure they had not left anything important out of their forthcoming book."[153] The conference, nevertheless, was also a forum where the Cambridge team planned to restore their authority in the field of radioactivity after the lengthy exchange of papers in the scientific press with the Viennese. Moreover, they intended to do so undisturbed by their main opponents.

Although they invited all the key figures of the radioactivity community, it was not by chance that only one participant from the Radium Institute in Vienna attended the meeting. Ewald Schmidt, even though he was not the one to work on the central theme of the conference, was the one to receive an invitation.[154] Other young researchers who had worked on radioactivity in general attended the conference, yet the young Viennese who played an essential role in the controversy such as Kara-Michailova, Karlik, and Blau, or the men of the group such as Kirsch, Holoubek, and Ortner were silently excluded. The only other person that Chadwick thought of inviting was Stetter, but, as he mentioned to Meyer, "Our funds were not sufficient."[155]

Blau especially would have been a perfect fit. She had completed a dissertation on the absorption of gamma rays and she already had a long list of publications and experience in the alternative method for detecting radiation, the photographic emulsions.[156] Thus, what the Cambridge group avoided through a public acknowledgment of the episode, they better accomplished through the politics of exclusion. It was not the credibility of the researchers that was affected but rather the possibility of their presence in the scientific community.

Suspecting this early on, Pettersson tried to prevent it in vain. As a gesture of good will, he sent Rutherford a spinthiroscope of his design for showing the hydrogen particles from paraffin. The gift was intended to temper the anxiety and tension that surrounded Chadwick's visit and ensure that Pettersson's powerful antagonists would not affect his own and his colleagues' presence in the radioactivity community. In his thankful response, Rutherford added "There are so few workers in this difficult subject that we must try and pull together and settle our differences as far as possible by private correspondence rather than by controversies in scientific journals, which in my experience do nothing but cause irritation. If you and your friends are of the same opinion, I think there should be no great difficulty in settling our differences. During my whole scientific career I have not had any serious controversy and always advise my students to be considerate where differences of opinion are involved."[157] Even though the dispute was settled in private, the group at the Radium Institute was finally dissolved. Partly because of the silent exclusions that the Cambridge team imposed on them and partly because of financial difficulties, most of the members were soon scattered to other laboratories, abandoning their teamwork on artificial disintegration.

THE LAST ATTEMPT TO SAVE THE SCINTILLATION COUNTER

In May 1928, Karlik teamed up with Kara-Michailova in a last attempt to save the scintillation counter. At a meeting in the Vienna Academy of Sciences, they presented their coauthored paper on the luminescence caused by alpha particles and its relation to their energy. For the first time, the luminescence produced by alpha particles emitted from polonium was measured by means of the photoelectric current of a rubidium cell. Their next coauthored paper appeared the same year in the prestigious *Zeitschrift für Physik*.[158] Besides discussing the experimental details of the relation between the brightness of the scintillation and the energy given up from the alpha particles of the source, they suggested a theoretical hypothesis in explaining the mechanism of the scintillation process. They were concerned with the manipulation of the instrument, preparing and gauging the scintillation screens, and experimenting with several different elements. They went one step further, suggesting that the zinc sulfide possesses distinct points that are already in an active condition before they are struck by the particles. Theory and experiment came together in this study of the scintillation counter.

At the end of 1928, Kara-Michailova was promoted to the position of *Wissenschaftliche Hilfscraft* at the institute.[159] She was obviously the mature partner in her collaboration to Karlik, although the latter's work came to the center of attention of the Cambridge group in November 1928. Julius Chariton and C. Lea, two of Rutherford's research students, raised objections to Karlik's dissertation project, published a year earlier. As they argued, there was a considerable difference between the results obtained by Karlik's and their experiments concerning the question of how the amount of light, entering the eye from an individual scintillation, affects the number of total scintillations observed. The discrepancies were probably due to "the device used for reducing the fraction of the light from a scintillation which entered the eye."[160]

Chadwick communicated the paper to the Royal Society in November 1928, but not only Karlik's results were at stake. During his visit in December 1927, Chadwick was able to show empirically that the Viennese were wrong in the number of scintillations they were claiming to count. Chariton and Lea offered a theoretical explanation of the mistake. The Viennese were actually testing the role of the numerical aperture and the magnification of the microscope designed and used by the Viennese group. It became obvious that the scintillation counter was in its last days.[161] Rona described Chadwick during his visit to Vienna in 1927 as "probably just as uncomfortable in the role of the judge as we were in that of the judged."[162] In fact it was both the scintillation technique and the skills of the women experimenters working in Pettersson's group that were under judgment.

Karlik and Kara-Michailova insisted on saving the scintillation counter and presented their last coauthored paper in July 1929 to the Vienna Academy.[163] This time, the focus was on the brightness of the scintillations produced by H-particles. Besides Karlik's two subsequent articles in 1930 related to the scintillation technique, none of the Viennese physicists followed up on scintillation counter research, and hardly any publication on it appeared in the *Mitteilungen* of the institute after 1930.[164] Eventually, both women who played a central role in improving the counter abandoned the technique as the center of their research focus. Kara-Michailova teamed up with Blau and worked on the penetrating radiation of polonium.[165] She retained the position of scientific assistant until March 1933, when she applied to the Austrian Federation of University Women for a Yarrow Scientific Research fellowship.

Stefan Meyer willingly provided her a reference letter, stating that Elisabeth had been an "independent and stimulated researcher. She has decisively cooperated especially in works on disintegration, on problems of luminescence, and also on the amount of time radium emanations remain in the human body, without her name always appearing at the publications." Soon after, Kara-Michailova was awarded the fellowship and starting in 1935, she spent four years at the famous

Girton Women's College.[166] She was employed by the Cavendish Laboratory which was then directed by Rutherford. When he passed away in 1937, he was succeeded by W. L. Bragg. In 1939, Kara-Michailova returned to Bulgaria as an associate professor in experimental atomic physics and radioactivity at the University of Sofia. As her later publication record indicates, she crossed once more from physics to medicine, carrying over her expertise in instrument making. Furthermore, she tried to transfer the collegial atmosphere of the Vienna Institute to her new department by holding weekly tea gatherings with her students and doctoral candidates.[167]

A fellowship from the International Federation of University Women allowed Karlik to spend some time away from the Radium Institute. During the academic year 1930–31, she moved to William Bragg's laboratory in London.[168] Her research interests were centered on crystallography and the use of x-rays in the study of the structure of crystals. It was her knowledge of radiophysics that Karlik brought to Bragg's laboratory, working with the crystallographers Ellie Knaggs and Helen Gilchrist.[169] The scintillation counter was in the past for both Karlik and Kara-Michailova, as it was for the research groups in Cambridge and Vienna.

WHAT IT MEANT TO BE AN EXPERIMENTER AT VIENNA'S RADIUM INSTITUTE

What it meant to be a physicist specializing in radioactivity strongly depended on the culture within which such an identity was constructed and performed. In Vienna, Pettersson initiated a new era of experimentation and transformed the meaning of "experimenter" within the institute. Introducing the research program on artificial disintegration required changes in space arrangements, the use of new experimental techniques, and the reordering of the entire laboratory. Before his arrival, physicists worked either alone or with one collaborator who was often a practicum student. In contrast, in the early 1920s, the shift to experiments on artificial disintegration investigating several elements demanded group research. It was both the competition with the Cavendish Laboratory and the nature of the experiments themselves that necessitated the changes. "I have learnt the very high value of team work," admitted Pettersson about his research in Vienna in his report to the International Education Board in 1928 and he continued:

> Our particular kind of work requires the close and continued collaboration of at least half a dozen highly specialized workers: one for preparing and calibrating the screens

> and the absorption foils used in scintillation counts; one for preparing the disintegration apparatus itself, the substances which are to be investigated in it and the gas with which it is filled; one for working the mass spectrograph and its auxiliary instruments; and one specialist on photography . . . Finally, several of the persons mentioned must act as counters in scintillation observations when the ordinary disintegration apparatus or the mass spectrographs are used.[170]

To determine the number of protons produced by the disintegration of light nuclei, the physicists counted the number of flashes that appeared on the zinc sulfide screens of their scintillation counters. The task was painstaking, and several students were usually recruited to do the counting in addition to the main experimenters themselves. For instance, in her later experiments, Kara-Michailova had employed two groups of observers who simultaneously recorded the scintillations produced by alpha and H-particles.[171] In all cases, these observers were characterized as "highly specialized."

At the same time, in order to compete with Rutherford's team, each of the Viennese researchers focused on one of the key aspects of the controversy, collaborating with other colleagues to develop and improve new experimental methods or to probe a variety of different elements. Thus, Pettersson introduced research among a cluster of scientists with varying areas of expertise, each of whom was assigned a topic of investigation that contributed to the overall effort. Although Pettersson was the leading experimenter, he encouraged teamwork and exercised minimal control over the pace and direction of research in his group. He describes its rituals in the following way: "The papers are circulated in manuscripts and read by all the coworkers and thoroughly criticized them before publication."[172] The Viennese chose peer review as a way to construe experimentation in their local context. The result was the development of close collaborations, the exchange of ideas and scientific papers among the researchers, and even the transfer of materials and of parts of instruments from one workbench to the other. For example, Karlik reported to Pettersson, "In the afternoon of that same Wednesday, Ortner quite unexpectedly presented me with 300 milliCuries of radium (something had gone wrong with their apparatus)."[173]

In contrast, to be a physicist in the British group required accepting the hierarchical structure that Rutherford's authority imposed and working on assigned projects that were designed to maintain that authority. He was formally in charge of all research students whether they were doing experimental or theoretical work and every decision went through him. Students' training was strictly organized around the accepted research methods of the laboratory and maintained the local experimental culture. As Jeff Hughes explains, "Rutherford introduced a compulsory new training regime for fledgling experimentalists based on the course established at Manchester by Walter Makower and Hans

Geiger."[174] Rutherford's reprimand of his student Bates gives us a glimpse of the highly competitive environment at the Cavendish Laboratory. As director of the laboratory, he left little doubt as to the obligations of his colleagues and students. In Mark Oliphant's words:

> With Rutherford looking over our shoulders, impatiently awaiting the outcome of the observation, the operator tended to make silly mistakes . . . . Once at the end of a particular heavy day, when the experiments had gone well, we decided to postpone development till the next morning when we were fresh and we could handle the long strip in new developer and fixer without damage. Just as we were leaving, Rutherford came in. He became extremely angry when he heard what we had decided and insisted that we develop the film at once . . . . We did our best, but the developer was almost exhausted and the fixing bath yellowed with use. In the end, he went off muttering to himself that he did not know why he was blessed with such a group of incompetent colleagues.[175]

Quite the opposite, Pettersson's manners and his shift in emphasis from the individual researcher to the research group definitely improved collegiality at the Radium Institute. In that welcoming and less competitive atmosphere, women were more readily accepted. As the cases of Dagmar Pettersson, Kara-Michailova, and Rona suggest, they exercised the same type of control over their experiments, instruments, and theories as their male colleagues. It was these different social conditions in the two laboratories that also provided different ways to make, put in use, and sustain the scintillation counter.

NOTES

1. In his interview, Arthur Svansson stressed several times Pettersson's strong personality and his impulse for "big ideas, wild ideas." As Svansson claimed, "He was not afraid to have theories that were not easy to be proved" (Svansson, interview by the author, September 21, 2001, Göteborg).

2. Pettersson to his sister E. Mellbye, March 7, 1926 (in Swedish; Agnes Rodhe Papers, translated by Rodhe).

3. Rutherford, "Collisions of alpha-particles with Light Atoms" (1919), 589.

4. Hughes, *The Radioactivists* (1993).

5. Hughes, *The Radioactivists* (1993), 20.

6. Rutherford and Chadwick, "The Artificial Disintegration of Light Elements" (1921), 61.

7. Kirsch and Pettersson, "Über die Atomzentrümmerung" (1923); Kirsch and Pettersson, "Experiments on the Artificial Disintegration" (1923), 500–12; Kirsch and Pettersson, "Long-range Particles" (1923).

8. Kirsch and Pettersson, "Long-range particles" (1923), 395.

9. For considering instruments as historical documents, see Bedini, "The Hardware of History" (2001).

10. Julian Holland, "Scientific Instrument Makers in an Institutional Context," talk given at the XXI International Scientific Instrument Symposium, National Hellenic Research Foundation 2002, Athens, Greece.

11. See for example, Berg and Mol, *Differences in Medicine* (1998); Clarke and Fujimura, *The Right Tools for the Job* (1992); Rapp, *Testing Women, Testing the Fetus* (1999). For a review of the technology and gender issue, see Faulkner "The Technology Question in Feminism" (2001); Grint and Gill (eds.), *The Gender-Technology Relation* (1995); *Technology and Culture* (1997); Wajcman, "Reflections on Gender and Technology Studies" (2000).

12. For example, as Rossiter argued in 1980, the rise of big science and the flow of significant budgets at the end of the nineteenth century enabled women to enter scientific fields as support staff or assistants at a few research centers. The most illustrative example is the entry of women in astronomy with the main task to classify photographs of stellar spectra. See Rossiter, "Women's Work in Science" (1980), and Caroline Herzenberg's and Ruth Howes' book *Their Day in the Sun: Women of the Manhattan Project* pieces together fascinating information on women and the atomic bomb. Throughout the study, it becomes apparent that those who made key decisions and shaped laboratory technologies were not the women but the men who worked at the project; see Herzenberg and Howes, *Their Day in the Sun* (1999).

13. Gay, "Invisible Resource" (1996), 314.

14. Kirby, "The Discovery of Actinium" (1971), 300.

15. Crookes, "The Emanation of Radium" (1903), 407.

16. Levy and Willis, *Radium and Other Radio-Active Elements* (1904), 50.

17. Elster and Geitel, "Über die durch radioactive Emanation" (1903); Regener, "Über die Zählung der a-Teilchen" (1908).

18. Rutherford, "The Chemical Nature of the Alpha-Particles" (1963), 143. See also Rutherford and Geiger, "A Method of Counting the Number of Alpha Particles" (1908), 1–3.

19. Makower and Geiger, *Practical Measurements in Radioactivity* (1912).

20. Rutherford, "Collision of alpha particles with Light Atoms" (1919), 551.

21. Rutherford, "Nuclear Constitution of Atoms" (1920), 14–38.

22. Stuewer, "Artificial Disintegration" (1985), 240; see also Hughes, *The Radioactivists* (1993), 62. For issues of authorship in the discovery process, see Galison, *Image and Logic* (1997), 199–200.

23. Darwin, "The Discovery of Atomic Number" (1956) quoted in Hughes, *The Radioactivists* (1993), 4.

24. Rutherford, Chadwick, and Ellis, *Radiations from Radioactive Substances* (1930), 550.

25. Hughes, *The Radioactivists* (1993), 45.

26. Rutherford and Chadwick, "The Artificial Disintegration of Light Elements" (1921), 49.

27. Rutherford, "Bakerian Lecture, Nuclear Constitution of Atom" (1920), 381.

28. For a general history of the Cavendish Laboratory and different research styles, see Goldhaber, "Reminiscence" (1993); Growther, *The Cavendish Laboratory* (1974); Larsen, *The Cavendish Laboratory* (1964); Sviedrys, "The Rise of the Physical Sciences" (1970); Falconer, "J. J. Thomson and the 'Cavendish' Physics" (1989). For physics in Cambridge from a gender perspective see Gould, "Women and the Culture of University Physics" (1997).

29. This claim is based mainly on M. and G. Rayner-Canham, *A Devotion to their Science* (1997); Hughes, *The Radioactivists* (1993).

30. Heilbron, *Ernest Rutherford and the Explosion of Atoms* (2003), 102.

31. Gould, "Women and the Culture of University Physics" (1997), 149.

32. Heilbron, *Ernest Rutherford* (2003), 102; Hunt and Barker. *Women at Cambridge: A Brief History* (1998).

33. Although important, scintillation counting was not the only method that Pettersson and Kirsch were planning to use. As they announced in their first paper, the emanation capillaries enclosing the sources used in the scintillation counter "will be used in this institute also for studying atomic disintegration by the Wilson method." Kirsch and Pettersson, "Long-range Particles" (1923), 395.

34. Galison, *Image and Logic* (1997), 435.

35. Pettersson, "Zur Herstellung von Radium C" (1923), 55–57. The decision to replace the metal tubes with pure fused silica (quartz) was probably based on Stefania Maracineanu's recommendation that glass instead of metal plates should be used in order to obtain good results for the periods of radioactive substances; see Maracineanu, "Researches on the Constant of Polonium" (1923). For more on Maracineanu and her work at the Curie's laboratory, see Popescu, M. Rayner-Canham, and G. Rayner-Canham, "Stefanie Maracineanu" (1997).

36. Kirsch and Pettersson, "Long-range Particles" (1923), 395.

37. For a long time, it was considered that alpha particles emitted by a given substance have a definite range. By studying the emitted radiation of thorium in 1906, Otto Hahn discovered that alpha rays can have different ranges for the same source. In 1919, Rutherford established the presence of particles of range 9.0 cm from radium active deposit; see Bates and Rogers, "Particles of Long Range" (1924). With their experiments, Bates and Rogers observed even longer ranges emitted from radium C such as 9.3 cm, 11.1 cm and 13.2 cm; see Bates and Rogers, "Long-Range alpha-Particles" (1923).

38. Kirsch and Pettersson, "Long-Range Particles" (1923), 687.

39. Pettersson refers to Rutherford's letter in his own on 27 July 1923, AÖAW.

40. Pettersson to Rutherford, 27 July 1923, AÖAW (in English).

41. Pettersson to Rutherford, 27 July 1923, AÖAW (in English).

42. Bates and Rogers, "Particles of Long Range" (1924), 114.

43. Bates and Rogers, "Particles of Long Range" (1924), 114.

44. Bates and Rogers, "Particles of Long Range from Polonium" (1924).

45. Bates and Rogers, "Particles of Long Range from Polonium" (1924), 360.

46. For example, in their article published in the *Proceedings* in 1924, Bates and Rogers expressed their "best thanks to Sir Ernest Rutherford who suggested this research and who gave us many helpful suggestions during its progress." See Bates and Rogers, "Particles of Long Range" (1924), 116.

47. Pettersson, Dagmar "Über die maximale Reichweite" (1924), 149–62.

48. Rodhe, interview by the author, 22 September 2001, Göteborg.

49. Rodhe, interview by the author, 22 September 2001, Göteborg.

50. Svansson, interview by the author, September 21, 2001, Göteborg. See also Svansson, *Otto Pettersson: the Oceanographer, the Chemist, the Inventor* (2006)

51. Rodhe to Rentetzi, 29 October 2001.

52. Rodhe to Rentetzi, 29 October 2001. Hans Pettersson's report on the investigation regarding artificial disintegration of elements "Atomzertrümerung" carried out during the first half of 1926 in Radium Institute and the second Physics Institute of the University of Vienna, GUB. On couples in science, see Abir-Am, Pycior and Slack, *Creative Couples in the Sciences* (1996).

53. For details on Pettersson's method of the preparation of the radium C source, see Stuewer, "Artificial Disintegration" (1985), 248–9.

54. Pettersson, Dagmar. "Über die maximale Reichweite" (1924), 153.

55. Pettersson, Dagmar "Long Range Particles" (1924), 642.

56. Pettersson, "Long-Range Particles from Radium-Active Deposits" (1924), 641–42.

57. Rodhe to Rentetzi, October 29, 2001. During her stay in Vienna and despite the fact that she had two children, Dagmar remained part of the institute. From 1922 to 1926, she appeared in the *Bericht* of the institute as a collaborator while she followed Hans on his constant trips between Vienna and Göteborg from 1922 to 1928. Dagmar and Hans hired a Swedish nanny to take care of their kids in Vienna. Sometimes, they even brought their daughter to the institute. Rodhe recalled that during her early years, she used to observe the preparation of scintillation screens at the Radium Institute when her parents were busy with their scientific research. She was even taught how to brush the mixture of alcohol and zinc sulfide from the small rectangle glasses used for scintillation counting. Rodhe, interview by the author, 22 September 2001, Göteborg.

58. Rutherford and Chadwick, "On the Origins and Nature" (1924).

59. Hans to Otto Pettersson, August 22, 1927, in Otto Pettersson–Gustaf Ekman correspondence, Regional Archives, Göteborg. I would like to thank Artur Svansson for providing and translating this letter.

60. Hans Pettersson's report to the International Education Board, April 1928, GUB. It was during those years that Dagmar built warm, friendly relations, and close collaborations with the rest of the women that eventually lasted longer than her stay in Vienna. Blau, Karlik, and Rona occasionally corresponded with Dagmar even after Pettersson's death in the 1970s. They used to recall the pleasant years they spent together in Vienna during the 1920s. As director of the institute in the 1960s and 1970s, Karlik often conveyed information about their common friends and acquaintances.

For example, Karlik reported the death of Georg Stetter's wife and Przibram's 88th birthday in Karlik to Dagmar Pettersson, December 18, 1966, GUB.

61. Pettersson, Dagmar. "Über die maximale Reichweite" (1924), 153.

62. Kara-Michailova and Pettersson, "Über die Messung" (1924).

63. Kirsch and Pettersson, "Long-Range Particles" (1923).

64. Kara-Michailova and Pettersson, "Über die Messung der relativen Helligkeit" (1924), 164.

65. Hans Pettersson's financial report to the International Education Board, 1928, GUB.

66. Kara-Michailova and Pettersson, "Über die Messung der relativen Helligkeit" (1924), 165.

67. Belar, "Spektrophotometrische Untersuchung" (1923); Przibram and Belar, "Die Verfärbung" (1923).

68. Kara-Michailova, "Quantitative optische Unterscheidung von alpha und H-Teilchen" (1924).

69. Hans Pettersson to Mellbye, March 7, 1926 (in Swedish, Agnes Rodhe Papers, translated by Rodhe).

70. Albert Einstein, November 24, 1923, Box 25/360, RAC.

71. Pettersson had even tried to acquire a grant from the General Electric Company, as he knew that they had already supported several German, British, and French scientists. Pettersson to Irving Langmuir, June 4, 1924; Langmuir to Wickliffe Rose, October 13, 1924, Box 25/360, RAC.

72. Trowbridge to Rose, April 2, 1925; Memorandum of conversation with Professor Stefan Meyer and Karl Przibram, Extract from Doctor Trowbridge's Log of visit to Vienna, Austria, March 26, 1925, Box 25/360, RAC.

73. See for example how powerful the Rockefeller Foundation was in the case of the health-related work that it funded in central Europe after the First World War. Weindling, "Public Health and Political Stability" (1993); Page, "The Rockefeller Foundation and Central Europe" (2002). For the case of physics under National Socialism in Germany, see Macrakis, "The Rockefeller Foundation" (1989); Siegmund-Schultze, *Rockefeller and the Internationalization of Mathematics* (2001).

74. W. W. Brierley to Trowbridge, July 6, 1925, Box 25/360, RAC; Hans Pettersson's report to the International Education Board, April 1928, GUB.

75. Hans Pettersson's report to the International Education Board, April 1928, GUB.

76. Meyer to Trowbridge, October 7, 1925, AÖAW (in English).

77. Trowbridge to Rose, October 14, 1925, Box 56/923, RAC.

78. Radium Institute in Vienna, G. V. Hevesy, Box 25/360, RAC; Trowbridge to Rose, October 2, 1925, Box 56/923, RAC.

79. W. W. Brierley to Otto Sylwan, December 2, 1925, Box 56/923, RAC; Pettersson to Meyer, December 4, 1925, AÖAW. Trowbridge to Rose, Memorandum of conversation, Box 25/360, RAC.

80. Trowbridge to Rose, October 5, 1925; Memorandum of conversation with Madame Curie, October 4, 1926, Box 56/923, RAC.

81. Hans Pettersson's report to the International Education Board, April 1928, GUB (in English).

82. Chaloner, "The Most Wonderful Experiment" (1997), 364.

83. The Cambridge and Paul Instrument Company to Pettersson, December 18, 1923, GUB. For more on the marketing of cloud chambers, see Chaloner, "The Most Wonderful Experiment" (1997).

84. Holoubek, "Der Nachweis" (1927); Holoubek, "Die Sichtbarmung" (1927). Hans Pettersson Report, Part II, April 1928, Box 56/923, RAC.

85. For Aston's mass spectrograph and the rest of the Cambridge techniques adopted in Vienna, see Hughes, *The Radioactivists* (1993). Also Pettersson, "On the Investigations" (1926), GUB.

86. Hans Pettersson's report to the International Education Board, April 1928, GUB; Kreidl, "Zur Verwendbarkeit des Geiger'schen" (1927). See also Greinacher, "Über die akustische Beobachtung" (1924).

87. Hans Pettersson's report to the International Education Board, April 1928, GUB.

88. Ortner and Stetter, "Die Hörbarmachung von H-Strahlen" (1927). See also Hughes, *The Radioactivists* (1993), 118. For the interference of Stetter and Ortner to Kreidl's project see Pettersson to Meyer, August 28, 1927, AÖAW. Hans Pettersson Report, Part II, April 1928, Box 56/923, RAC.

89. Blau, "Über die photographische Wirkung" (1925); Blau, "Die photographische Wirkung" (1925); Blau "Über die photographische Wirkung" (1927); Blau, "Über die photographische Wirkung" (1928); Blau, "Über photographische Intensitätsmessungen von Polonium" (1928).

90. Pettersson to Mellbye, March 7, 1926 (in Swedish, Agnes Rodhe Papers, translated by Rodhe). For a detailed account of Blau's work related to photographic emulsions, see Galison, *Image and Logic* (1997), 146–60.

91. Blau, curriculum vitae, Leopold Halpern Papers.

92. Rona, *How it Came About* (1978), 22.

93. As Pettersson reported in 1926, the use of a strong polonium source in Holoubek's experiments enabled him to take fewer photographs than Blackett did by working on the same method in Cambridge; see Pettersson, "On the Investigations" (1926), GUB.

94. Hans to Otto Pettersson, May 24, 1926 (in Swedish, Agnes Rodhe Papers, translated by Rodhe).

95. Hans to Otto Pettersson, May 24, 1926. (in Swedish, Agnes Rodhe Papers, translated by Rodhe).

96. Rodhe, interview by the author on September 22, 2001, Göteborg. Pettersson's uncomfortable relations with the Curies deteriorated when he visited their institute in Paris in 1936. Commenting on his letter that described the situation, Karlik suggested that "you must remember, too, what queer people they are. As regards Irene I don't think you should feel puzzled by anything she does. Her manners are really perfectly intolerable." Karlik to Pettersson, 9 April 1936, GUB.

97. Pettersson, "On the Investigations" (1926), GUB.

98. *Almanach der Akademie der Wissenschaften* (1925), 216, AÖAW.

99. Rona, "Absorptions und Reichweitenbestimmungen" (1926); Blau and Rona, "Ionisation durch H-Strahlen" (1926).

100. Rona to Meyer, June 3, 1926, AÖAW. The letter was sent from Paris.

101. Pettersson, "On the Investigations" (1926), GUB.

102. Kara-Michailova, "Helligkeit und Zällbarkeit der Scintillationen" (1927).

103. Stetter, "Die Massenbestimmung von H-Partikeln" (1925); Stetter, "Die Bestimmung des Quotienten Ladung/Masse" (1926); Stetter, "Massenbestimung" (1926); Stetter, "Die Massenbestimung" (1927); Stetter, "Die neueren Untersuchungen" (1927); See also Hughes, *The Radioactivists* (1993), 107.

104. Hans Pettersson's report to the International Education Board, GUB.

105. Kara-Michailova, "Helligkeit und Zählbarkeit" (1927).

106. For a historical account of the notion of objectivity, see Daston and Galison, "The Image of Objectivity" (1992).

107. "Discussion on atomic disintegration in Cambridge" May 16, 1927, GUB.

108. Pettersson to W. E. Tisdale, May 18, 1927, Box 56/923, RAC.

109. Meyer to Rutherford, May 25, 1927, AÖAW (in English).

110. Pettersson to Meyer, August 28, 1927, AÖAW; Pettersson to Meyer, August 31, 1927, AÖAW.

111. For a detailed description of Chadwick's visit, see Stuewer, "Artificial Disintegration" (1985); Hughes, *The Radioactivists* (1993); Brown, *The Neutron and the Bomb* (1997).

112. Hughes, *The Radioactivists* (1993), 112.

113. Hughes, *The Radioactivists* (1993), 134.

114. Stuewer, "Artificial Disintegration" (1985), 285.

115. Stuewer, "Artificial Disintegration" (1985), 285.

116. Hughes, *The Radioactivists* (1993), 136.

117. As Chadwick wrote to Rutherford, Pettersson's whole family came to visit him in Vienna; see Stuewer, "Artificial Disintegration" (1985), 287.

118. Stuewer, "Artificial Disintegration" (1985), 286.

119. Stuewer, "Artificial Disintegration" (1985), 286.

120. Rutherford, Chadwick, and Ellis, *Radiations from Radioactive Substances* (1930), 550.

121. Stuewer, "Artificial Disintegration," (1985), 287, excerpt from Chadwick's letter to Rutherford on 12 December 1927.

122. Pettersson to Kerr Grant, December 29, 1927, GUB.

123. Rona and Schmidt, "Untersuchungen" (1927); Blau and Rona, "Ionisation durch H-Strahlen" (1927); Blau, "Uber die photographische" (1925). Brown, *The Neutron and the Bomb* (1997), 88.

124. Achinstein and Hannaway, *Observation, Experiment, and Hypothesis* (1985); Brown, *The Neutron and the Bomb* (1997), 88.

125. Stuewer, "Artificial Disintegration" (1985), 287.

126. ona, *How it Came About* (1978), 20. This is how Rodhe characterized her father; Rodhe, interview by the author, September 22, 2001, Göteborg.

127. Pettersson, "On the Investigations" (1926), GUB.

128. Hans Pettersson's report to the International Education Board, April 1928, GUB (emphasis added).

129. Kara-Michailova, "Helligkeit und Zählbarkeit" (1927).

130. I owe my thanks to Peter Galison for his valuable suggestions on the concept of skill.

131. Golinski, *Making Natural Knowledge* (1998), 90.

132. Achinstein and Hannaway, *Observation, Experiment, and Hypothesis* (1985), x. By working with Roger Stuewer's precise historical account of the episode, the above historians and sociologist jump to a conclusion that Stuewer never argues for. He not only reports on women's scientific work but is willing to look more carefully at Chadwick's interpretation of a derogatory comment on women's work supposedly made by Pettersson. As Stuewer argues, "Pettersson valued his coworkers highly, and they him, and hence, it is not clear in what tone of voice he made these remarks." What Stuewer did nevertheless was to write the story of the dispute between the two laboratories from the point of view of the male protagonists. Rutherford, Chadwick, Pettersson, and Kirsch did have their fair share in the historical plot and, although acknowledging some of the women, Stuewer left them on the margins of his narrative. Their absence is what permits accounts such as Golinski's, Achinstein's, and Hannaway's.

133. Pettersson to Mellbye, March 7, 1926 (in Swedish, Agnes Rodhe Papers, translated by Rodhe).

134. Pettersson to Mellbye, March 7, 1926 (in Swedish, Agnes Rodhe Papers, translated by Rodhe).

135. Hans Pettersson's financial report to the International Education Board, GUB.

136. Karlik's exams for her major in physics were on October 10, 1927, and for mathematics on January 13, 1928. She finally graduated on March 8, 1928 (Rigorosenakt 9765, AUW).

137. Karlik, "Über die Abhängigkeit der Szintillationen" (1927).

138. In a letter of July 19, 1924, Rutherford suggested to Pettersson to resolve their differences in private. "Workers in this field" as he argued "are too few and too select to misunderstand one another." See Hughes, *The Radioactivists* (1993), 138–9. Stuewer, "Artificial Disintegration" (1985), 256.

139. Hughes, *The Radioactivists* (1993), 351, 206; Brown, *The Neutron and the Bomb* (1997), 88; Weiner, *Sir James Chadwick*, oral history, AIP.

140. Memorandum, conference between Wickliffe Rose and Augustus Trowbridge, June 13, 1927, Box 25/360, RAC.

141. Meyer to Trowbridge, January 20, 1928, Box 25/360, RAC.

142. Meyer to Rutherford, November 8, 1927, AÖAW.

143. Rutherford to Meyer, November 23, 1927, AÖAW.

144. Meyer to Rutherford, November 29, 1927, AÖAW (in English).

145. Rutherford to Meyer, December 21, 1927, AÖAW. Hughes argues that Meyer agreed with Rutherford's terms not only because he was financially desperate but also because he was embarrassed by the outcome of Chadwick's visit; see Hughes, *The*

Radioactivists (1993), 141. The letters exchanged before Chadwick's visit, nonetheless, show that Meyer had already agreed with Rutherford's terms. On the contrary, it was Rutherford who seems to have been embarrassed after Chadwick's visit to Vienna.

146. *Almanach der Akademie der Wissenschaften* (1930), 235, AÖAW. Trowbridge to Rose, February 7, 1928, Box 25/360, RAC.

147. Pettersson to Mellbye, April 15, 1928 (in Swedish, Agnes Rodhe papers, translated by Rodhe). As Rodhe recalled, "I can remember those dinners for two, each parent scratching morsels onto my extra plate." Rodhe to Rentetzi, October 29, 2001.

148. Stuewer, "Artificial Disintegration" (1985), 290.

149. Karlik to Pettersson, March 28, 1936, GUB (in English).

150. Rodhe, interview by the author, September 22, 2001, Göteborg; Svansson, interview by the author, September 21, 2001, Göteborg Svansson, interview by the author, September 21, 2001, Göteborg.

151. Pettersson to Karlik, September 27, 1934, GUB.

152. Hughes, *The Radioactivists* (1993), 132, 152.

153. Brown, *The Neutron and the Bomb* (1997), 96.

154. Hughes, *The Radioactivists* (1993), 152–157. Chadwick to Meyer, June 23, 1928, AÖAW.

155. Chadwick to Meyer, June 23, 1928, AÖAW.

156. Blau, "Uber die Absorption divergenter" (1918).

157. Rutherford to Pettersson, January 9, 1928, GUB.

158. Karlik and Kara-Michailova, "Über die durch alpha-Strahlen erregte Lumineszenz" (1928); Karlik and Kara-Michailova, "Zur Kenntnis der Szintillationsmethode" (1928).

159. November 6, 1928, Karlik's file, Mitarbeiter/Assistenten, AÖAW.

160. Chariton and Lea, "Some Experiments Concerning the Counting of Scintillations" (1929), 336.

161. In a letter of June 23, 1928, Chadwick asked Meyer if he would facilitate Chariton's visit to Vienna. Chariton was in his way to Russia and because of visa restrictions, he was not allowed to cross Austria without a reason. "If the authorities are assured by a resident in Vienna of the purpose of his visit, they will be satisfied," as Chadwick explained to Meyer. "His [Chariton's] work has been on the counting and scintillations and I think" as Chadwick continued "you will be interested in what he has to say" (Chadwick to Meyer, June 23, 1928, AÖAW). However, it is interesting that in Chariton's case, Chadwick does not ask for permission but somewhat haughtily assumes that his student should be welcomed in Vienna.

162. Rona, *How it Came About* (1978), 20.

163. Kara-Michailova and Karlik, "Über die relative Helligkeit" (1929).

164. Karlik, "Uber die Szintillationsfähigeit" (1930); Karlik, "Untersuchungen zur Lumineszenz" (1930).

165. Blau and Kara-Michailova, "Über die durchdringende Strahlung" (1931).

166. Meyer to the Austrian Federation of University Women, September 26, 1933, AÖAW (in English).On October 26, 1935, Kara-Michailova wrote to Meyer from Girton College, Cambridge, to thank him for the pleasant years she spent at the Institute.

She hoped for further collaboration in the future. (Kara-MIchailova to Meyer, October 26, 1935, AÖAW).

167. Kara-Michailova, "The Radioactivity of the Water Sources" (1960); Kara-Michailova and Kamburov, "Radiological and Hydrological Research" (1961), 109; Kara-Michailova, Nikolov, and Doitchinova, "The Radioactivity of Mineral Water-Springs" (1962). As Tsoneva-Mathewson, and the Rayner-Canhams argue, Kara-Michailova organized her own laboratory at the University of Sofia and many of the instruments were handmade. (Tsoneva-Mathewson; Rayner-Canham, M. and Rayner-Canham, G., "Elizaveta Kara-Michailova," (1997), 206). See also Nazarska, "Karamichailova, Elissabeta Ivanova" (2006); Stretenova, "Elizaveta Karamihailova (1897–1968)" (2003).

168. Lintner, "Berta Karlik, Nachruf" (1990), 306.

169. Karlik and Knaggs, *Tables of Cubic Crystal Structures* (1932); Karlik and Gilchrist, "Separation of Normal Longchain Hydrocarbons" (1932), 1992.

170. Pettersson, Report to the International Education Board, Part II, Box 56/923, RAC.

171. Kara-Michailova, "Helligkeit" (1929), 361.

172. Pettersson, Report to the International Education Board, April 1928, GUB.

173. Karlik to Pettersson, July 26, 1933, GUB.

174. Hughes, "Modernists with a Vengeance" (1998), 344.

175. Oliphant, "Working with Rutherford" (1984), 185–6.

6. THE AFTERMATH OF THE CAMBRIDGE-VIENNA CONTROVERSY

Radioactivity and Politics in Vienna in the 1930s

Consequences of the Cambridge-Vienna episode ranged from the entrance of other research centers into the field as the study of the atomic nucleus became a promising area of scientific investigation to the development of new experimental methods. As Jeff Hughes describes, three key groups turned to the study of atomic nucleus. Gerhard Hoffman and his student Heinz Pose studied artificial disintegration at the Physics Institute of the University of Halle using a polonium source sent by Meyer.[1] In Paris, Maurice de Broglie turned his well-equipped laboratory for x-ray research into a center for radioactivity studies and Madame Curie started to accumulate polonium for research on artificial disintegration.

The need to replace the scintillation counters with a more reliable technique also led to the extensive use of the cloud chamber in Cambridge.[2] Simultaneously, the development of electric counting methods for measuring alpha particles in Rutherford's laboratory secured quantitative investigations and prompted Stetter and Schmidt from the Vienna Institute to focus on the valve amplifier technique.[3] Essential for the work in both the Cambridge and the Vienna laboratories was the use of polonium as a strong source of alpha particles for those methods as an alternative to the scintillation technique.

Besides serving as a place for scientific production, the laboratory was definitely also a space for work where tasks were labeled as skilled and unskilled and positions were divided to those paid monthly and those supported by grant money or by research fellowships. Pettersson's departure from Vienna meant the loss of the soul of the Viennese group. That led not only to disarray in the study of atomic disintegration but also to the development of a severe financial crisis at the institute. Most deeply affected were the women of the team given the fact that they lacked stable university positions and monthly payments from the state. As soon as the flow of grant money and generous donations ended, the women financed from these sources faced professional and financial instability. It was probably this insecurity that prompted most of them to look for stipendiums and fellowships in other institutes.

As most of the women who were actively engaged in radioactivity research remained single, their personal lives were flexible and it was easy for them to travel. Undermining the stereotype of their "devotion" to science, I want to argue that these women lived through a transitional period. Meanings of sexual difference, of what a man and a woman signified, were in flux. The social democratic politics of Red Vienna that coincided with the time of Pettersson's research at the institute offered a new conceptual framework for addressing issues of gender identities. Caught in the drastic shift from traditional values to those that envisioned women and men as socially, scientifically, and politically active, the female scientists faced the dilemma of either/or concerning scientific career and personal life.

Such being the case, when Pettersson left the institute his female collaborators scattered to other European research centers, obtaining yearly fellowships and small stipendiums. The official academic politics outside of the Radium Institute kept them away from university positions and limited their career advancements. Instead of viewing this discrimination in terms of exclusion and victimization, I argue that women took their lives and careers in their hands and, forced by the circumstances, altered the boundaries of their discipline. It is at this moment that the nature of radium as a trafficking material offered them opportunities to carry their expertise in other related fields and to other cities. By doing so, they were able to acquire grant money, widen their professional network, and remain scientifically active and innovative in a world in flux.

THE WOMEN OF THE VIENNA GROUP

As the Cavendish Laboratory had no radiochemist before Chadwick left Vienna, Rona recalls that "He extended Rutherford's invitation to me to join the staff at Cavendish Laboratory." Declining the invitation, Rona chose to stay in

Vienna.[4] Given that Rona's expertise in preparing polonium sources was highly in demand by different researchers and used in a number of newly developed methods for detecting radiation, she moved from one laboratory bench to another, forming collaborations with remarkable flexibility. In 1928 with the help of Ewald Schmidt, she modified P. Bonet-Maury's method for the vaporization of polonium. In 1930, while Blau was already working on photographic emulsions, the two women combined their expertise in the study of the H-rays, using strong polonium sources and recording the tracks on photographic plates.[5] With Karlik's return from England in 1932, Rona was able to carry her knowledge of preparing polonium to another workbench. Abandoning the ordinary scintillation counter, Karlik worked on the determination of the alpha particle ranges, utilizing a photoelectric cell while she kept the fluorescent screen intact as part of the instrument. In collaboration with Rona, she applied her method to the study of the ranges of alpha particles emitted from actinium and polonium.[6]

Although Rona had previously declined Chadwick's invitation to join the British group, she did not miss the chance to visit Paris once again in 1934. That year Meyer received a telegram from the Joliot-Curies announcing their discovery of artificial radioactivity. In his usual kind manner of appreciation, Meyer invited them immediately to Vienna to present their startling work. As Rona recalls, "I had the opportunity to hear a first-hand report about this fundamental discovery which was to have such far-reaching consequences for different branches of science. The talk was given by Irene Joliot-Curie."[7] During their visit, Rona and Karlik entertained the couple in the Vienna outskirts and before the Joliot-Curies left the city, they invited her to Paris to work on artificial radioactivity.

No doubt, when Rona arrived in Paris, the atmosphere at Curie's Institut du Radium was quite different from what it had been during her first visit. "La patrone" was dying, as Frederick Holweg, a senior scientist at the institute, explained to Rona. Since December 1933, Curie had been agonizing between illness and health. In her good days, she went to the laboratory and when she was dizzy and weak, she stayed home. In mid-May 1934, she saw her lab for the last time and a few days later, she left to a sanatorium on the advice of her doctors. On July 4 of that year, Curie died from pernicious anemia at Sancellemoz.

The day the news arrived at the institute, Rona was present and ill herself.[8] As Pettersson wrote to her on July 29, 1934, "We had no idea that you were sick in France." Meyer reminded her that "The most important thing in life remains always health," and Gleditsch urged her to pay attention before it was too late. All of them must have been shocked by Marie Curie's death. Eventually, Rona recovered and returned to Vienna the following fall. She carried over to the Radium Institute experimental knowledge on artificial radioactivity, which she introduced to her colleagues in Vienna. Forming a group with Pettersson, Kara-Michailova, and Ernst Föyn, Gleditsch's assistant from Sweden, Rona studied the effects of bombarding radioactive isotopes with neutrons.[9] In 1936, she was joined by research student Elisabeth Neuninger and they investigated the artificial radioactivity of thorium.[10]

After Pettersson's departure, Blau was still working on photographic emulsions. As she described her research strategy, "The grain thickness of proton tracks was appreciably smaller than that of alpha tracks and it was evident that the photographic conditions (emulsion characteristics and development conditions) would have to be improved if high energy protons-with smaller ionization thickness-were to be observed."[11] In the following years, the method was applied to the disintegration of various atoms.

Blau also improved the quality of the processing techniques and emulsions and was able to increase the thickness of the emulsion layers. In spite of her close collaboration over the years with Rona who provided her with stronger radioactive preparations, and the investigation of the penetrating radiation of polonium in collaboration with Kara-Michailova, Blau had not yet succeeded in making fast protons visible with the photographic technique.[12] The low intensity of radiation limited the accuracy of the measurements. What proved to be decisive for Blau's career and for the success of her method was the exposure of the emulsions to cosmic radiation. In this achievement, Blau's collaborator was Hertha Wambacher.

Nine years younger than Blau, Wambacher had a similar education. She went to the same *Volksschule* (elementary school) as Blau in the first district of Vienna and entered the private Mädchen Obergymansium in 1914.[13] In contrast to Blau's Jewish family, Wambacher's was Catholic. Her father, Ferdinand Wambacher, was an industrialist and thus able to ensure her studies at the elite gymnasium and later on in the department of physics at the University of Vienna.[14] Although she enrolled in the chemistry department in the fall semester of 1922, she soon quit for health reasons. She eventually studied law and then moved into physics.[15]

According to Leopold Halpern, Wambacher did her *Praktikum* at the Radium Institute, working closely with Blau on the improvement of the emulsion technique. Since Blau did not hold any position at the University of Vienna, she was not able to officially advise a student. Blau's help, though, was indispensable to Wambacher in completing her dissertation on the impact of photographic desensitizers to the imprints of alpha, beta, and gamma rays on photographic plates.[16] In 1931, Wambacher published her work in the *Mitteilungen*, arguing mainly that the organic dye pinakryptol yellow functioned as a desensitizer on photographic emulsions while the effect was smaller with chromic acid.[17]

The first coauthored paper with Blau appeared in June 1932 and a month later, the two women were able to detect photographically protons liberated by neutrons.[18] As Galison describes, their result was "bizarre and counterintuitive."[19]

Particles liberated by neutrons did not leave an imprint unless the photographic plates where desensitized by means of pinakryptol yellow.[20] As a consequence of this first success in photographically detecting the ionization protons and explaining the effect of desensibilization, Blau was invited by the German photographic giant Agfa, "as their guest of honor" and a medal was bestowed upon her by the Photographic Association. In the fall of 1932, Blau also received a scholarship from the Association of Austrian Academic Women and spent the next six months at Robert Pohl's physics institute in Göttigen. In 1933, she accepted an invitation from Marie Curie to spend the rest of her stipendium time at her Laboratoire du radium in Paris. During her absence, Wambacher teamed with Kirsch on an investigation of neutrons from beryllium using Blau's photographic method.[21] On Blau's return in 1934, neither the institute nor Vienna was the same. The political upheavals of 1933 had deeply affected both.

CARRYING TRAFFICKING MATERIALS AND CROSSING DISCIPLINARY BOUNDARIES

Since Pettersson's first visit to Vienna, he and his family kept moving back and forth over the years from Vienna to Göteborg, resembling nomadic travelers. In the beginning, from 1922 to 1925 Pettersson spent the academic years in Sweden teaching at the Göteborg Högskola, a *Kommissionen* lectureship position in oceanography. Each spring, he went to Vienna to work with his colleagues at the Radium Institute. In 1925 the fellowship from the International Educational Board offered him the luxury to stay in Vienna most of the year, spending only the summers in Sweden. The year 1927 was decisive in his career. The end of the financial support from the International Education Board along with the repercussions of Chadwick's visit in Vienna forced Pettersson to look for an alternative solution. Svante Arrhenius died that autumn, leaving his position vacant.[22] To Pettersson, that was a chance to finally obtain a professorship, to regain his authority, and continue research on radioactivity. In the Radium Institute, he was the leader of the group working on artificial disintegration. Otherwise, he was simply a lecturer in a position made up through his father's connections.[23]

In 1928, Pettersson applied for Arrhenius's vacant position, but his application met the tenacious opposition of Manne Siegbahn from the appointment committee. Caught in the academic politics of Sweden, Pettersson lost the battle despite the supportive letters that Meyer solicited from Hevesy, Fajans, and Marie Curie.[24] To the eyes of the scientific community the controversy was still unresolved. "[T]he results of Mr. Pettersson's and his collaborators' work are still a subject of controversy,"[25] Curie wrote to Meyer and thus, she suggested, it was not fair to deny the professorship to him. She further assured Meyer that Pettersson left a great impression on her when he visited Paris. He had definitely known how to gather capable collaborators around him and inspire them with his lively interest for research. Despite Curie's affirmations, the contradiction of Pettersson's research with the work done in Cambridge and in Berlin by Bothe and Fränz was enough of an excuse for his opponents in Sweden.[26]

In the spring of 1928, Pettersson returned to Vienna once more with his family.[27] This time he had with him a few red clay samples from sea-bottom sediments which he wanted to analyze for their radium content. Rona was assigned the task. As she soon found out, "The contamination of the Radium Institute was too high to permit small amounts of radium to be determined. The needed equipment was moved to the oceanographic station in Bornö on Gullmarfjord, in south Sweden. Here, I spent many summer months, staying sometimes well into the fall."[28]

The station had been built on a property that belonged to Peterrson's father on the Bornö island right on the shoreline in 1901. "The station is located at an extraordinary favorable position-unparalleled in Europe-on a steep cliff," Otto Pettersson ensured the Swedish Hydrographic-Biological Commission. It is "at the waterfront of the wind and waves and protected by Gullmarr Fjord which forms a natural experimental basin of 3 miles length for hydrographic and biological research-instruments." A fully equipped biological and chemical laboratory, a workshop, and two big storerooms were located on the ground floor of the building where nets were usually hanging from the ceilings and oceanographic instruments were stored. In addition, two aquariums were immured in the rock. On the second floor were located a small library, a spacious dinning room, and two bedrooms.[29]

Hosting work at the boundary of oceanography and radioactivity and encompassing research from biology to hydrology and metrology, the Bornö station was developed into Pettersson's main research institute during the 1930s. Starting in 1931, Pettersson and his colleagues performed daily measurements of the temperature and salinity at different depths of the Bornö station and extracted sea samples from different depths to be analyzed for their radium content. With the intervention of Pettersson's father, the wealthy Swede Knut Mark offered funds for a new chair in oceanography at the Göteborg Högskola, which Pettersson easily obtained. "This meant," as Pettersson's daughter put it, "that my parents for the first time in their family life could afford a place of their own to live. Until then, they had 'lived in' with my maternal grandparents during winter times."[30]

To Pettersson, the new position offered a sense of professional recognition and eased his personal life. To his female collaborators in Vienna, it meant even more: a paradoxical opportunity for professional stability in temporary occupation. Carrying with them their portable apparatus, most of the women who previously worked with Pettersson in Vienna soon crossed the boundary of their discipline, analyzing sea bottom samples for their radium content in Bornö. By Pettersson's choice, laboratory technologies traveled across disciplinary lines from radioactivity to oceanography and geographically moved from Vienna to Bornö at the station which also functioned as a summer cottage of his family. The transport of materials and instruments went both ways. "Two bottles of seawater were sent at the beginning of the week to Vienna," Pettersson informed Rona.[31]

Once again, radium functioned as a trafficking material which this time linked a physics laboratory to an oceanography one. Women researched radium's properties not only in an academic site at the center of Vienna but also in a remote cottage-laboratory setting on Sweden's shoreline. During most of the winters, the station was impossible to reach from the mainland due to the bad weather. Developing a complex set of skills far from Vienna's academic environment, women's new scientific practices resembled workshop culture and old craft routines. Their research relied mainly on portable apparatus measuring the radioactivity of the samples, on handmade instruments for extracting the seawater from various depths, on boat communication with the rest of the world, and on the patronage system that Pettersson ensured.

Unable to get serious funding for their work on artificial disintegration, Rona, Blau, and Karlik visited Bornö during the summers for almost the next decade. Having fewer choices than their male colleagues, holding unstable positions in Vienna, and leading flexible personal lives as single women, the road to Bornö became a summer vacation ideally combined with serious research. With his unique talent to enlist patrons and donors, Pettersson financed his oceanographic research, secured funds for the women of the Radium Institute, and purchased the necessary apparatus for his new endeavor. Besides turning the Bornö station into a state-owned permanent center for hydrographic research, he persuaded Alice and Knut Wallenberg, Swedish industrialists, to donate money for new facilities. Adding central heating, sanitary installations, electric lights, telephones, and new equipment, Pettersson made the space of the institute available for research during the whole year for different working groups.

In 1935, Rona reported to Meyer that "This is an ideal institute for work."[32] She was probably right. Situated in close proximity to Oslo and Copenhagen, the oceanographic institute in Bornö placed women in a convenient environment for research, for scientific visits to Gleditsch's and Bohr's institutes, and for social entertainment in the Swedish Kalhuvudet, an island north of Göteborg where Pettersson's grandparents once maintained an old herring factory. Now a summer cottage, the house in Kalhuvudet regularly hosted Karlik, Rona, and occasionally Blau, all of whom became close friends with the family.[33] When, for example, Rona worked in Bornö the summer of 1935, she visited Gleditsch in Oslo and met with her old colleague Hevesy in Copenhagen who at the time was in Bohr's institute. In Kalhuvudet, she spent time with Karlik and the Petterssons.[34] The summer before, Karlik paid a visit to Copenhagen as well where she had the chance to discuss their earlier work on artificial disintegration with Bohr's team.[35] In 1935, while Blau deputized Karlik as *wissenschaftliche Hilfskraft* at the Radium Institute, the latter had the chance to spend a longer period at the oceanographic station in Sweden.[36]

For the Viennese women becoming a part of the broader radioactivists' network proved to be important during the difficult years of the political persecution by the Nazis after 1938. Yet that was not the only benefit. Working on the boundaries between oceanography and physics, the opportunity for research was also a central concern to their work. After Rona analyzed the first ocean sediments, she discovered that the radium content was high and this undermined previous results that saw a connection between the radium content and water depth.[37] To resolve the discrepancy, Rona and Pettersson embarked on a wider project, intending to perform exact measurements of the concentration of radioactive elements in seawater.

It was then that Karlik, Föyn, Rona, and Pettersson formed a group on seawater research, joined from time to time by Gleditsch.[38] They started by analyzing the radium content of seawater taken from Gullmarfjiord and the more open Swedish sea of Skagerak. During the following years, Rona and Karlik spent part of their summers in Bornö analyzing sediments. During the winters in Vienna, they kept up their collaboration with Pettersson who often sent sealed bottles of seawater for analysis to the Radium Institute.[39] Pettersson's wife Dagmar "did jobs when needed" and helped him regularly in his experiments. Their daughter recalls, "She used to do much of the routine titration on the innumerable samples of seawater coming, e.g., from the Bornö station. She liked the simple but accurate work and got paid. I remember her doing it, measuring the NaCl content by adding certain amounts of $AgNO_3$ with a drop of a color agent signaling neutrality."[40]

Concurrently, Karlik and Friedrich Hernegger, a research student at the Radium Institute, raised concerns on biological issues in relation to the uranium content of seawater. "It appears not unlikely that an accumulation of uranium in the tissue of marine organisms may occur which again may possibly account for some interesting results obtained concerning the high radioactivity of the waters in the petroleum districts in Russia."[41] Working in Przibram's research group, Hernegger had already developed an optical method for detecting and measuring small quantities of uranium based on the fluorescence phenomenon.

Owing to uranium's weak radioactivity, previously employed electrometric methods were not satisfactorily accurate. In 1933, with the support of a grant from the Rockefeller Foundation, Karlik and Hernegger acquired a glass spectrograph of high light-gathering power.[42] Expert in spectroscopic measurements due to her earlier work with Przibram and Herbert Haberlandt, Karlik joined by Hernegger, photographed the characteristic band-spectra due to uranium fluorescence and then compared those with spectra of known uranium content.[43] Supported by a stipendium from the Austrian ministry of education, Karlik performed the first experiments in Bornö in the summer of 1935 and she moved some samples to Vienna the following fall.[44]

The whole investigation became possible through the support and encouragement of both Meyer and Pettersson. As Karlik and Hernegger stressed in their article, they were hoping to enlarge their research project "through a continued collaboration between the Institut für Radiumforschung and the Oceanografiska Institutet of Göteborg Högskola."[45] The collaboration between the two institutes proved to be important for scientists in both countries. Besides the women conducting research at the Bornö station, one of Pettersson's Swedish colleagues, Börje Kullenberg, visited the Radium Institute in the end of 1938. He was the first assistant of the Svenska Hydrografisk-Biologiska Kommissionen and he appeared as the Radium Institute's collaborator in the academy's almanac in 1939. On the other hand, as their publications reveal, Rona, Karlik, and Pettersson continued their research on the intersection of radiochemistry and oceanography throughout the 1930s.[46] At the same time, they remained by and large on the margins of research on nuclear physics.

FROM RADIOACTIVITY TO NUCLEAR PHYSICS

The transformation of the experimental culture in both the Cambridge and Vienna laboratories was only the tip of an iceberg that indicated a deeper transformation in the material culture of physics and its theories during the 1930s. Granting the possibility of partial autonomy to instrumentation, experimentation, and theory, Peter Galison argues that quasi-autonomous traditions carry their own periodization.[47] In the case of radioactivity, although during the 1930s theorists, experimentalists, and instrument makers had "a life of their own," they often met in the same laboratories and collaborated, altering the boundaries and the focus of their field.

As early as 1928, the Russian theoretical physicist George Gamow, a research fellow at Niels Bohr's laboratory in Copenhagen, applied the new quantum mechanics to the nucleus. He showed that the emission of alpha particles from radioactive sources follows a tunneling process that can be explained by the wave properties of fundamental particles. This explained why the emitted particles had lower energies that classic mechanics and the work of Rutherford and Chadwick could not account for. "On the strength of this work," as Brown argues, "Gamow became a frequent and valued visitor to the Cavendish." Although primarily an experimentalist, Chadwick coauthored a paper with the theoretician Gamow in 1930 on the artificial disintegration of alpha particles. The next year, Gamow's newly published book, *Constitution of Atomic Nuclei and Radioactivity*, was devoted to the Cavendish laboratory, making tangible the intersection of theory and experiment in what was formerly the field of radioactivity.[48]

Envisioning the potential of Gamow's theoretical work in disintegration by artificially accelerated particles, John Cockcroft, who had been working at Rutherford's lab since the early 1920s, put his hands on the instruments of the field. Joined by Ernest Walton, they designed a high-voltage apparatus for producing accelerated protons which were directed at various materials. The emitted particles were recorded in a zinc sulfide screen by means of the old scintillation technique.[49] However, nuclear physics was already the focus and practice of those who worked on radioactive substances throughout the 1920s. Theory, experiment, and instrumentation fed one another while each reserved its partial autonomy and its scope of research.

In 1932, the discovery of the neutron by Chadwick based on the work done at the Cavendish certainly benefited from the radiochemical work of Irene Curie and Frederic Joliot in their institute in Paris. Their paper on the penetrating radiation from beryllium that appeared in the *Comptes rendus* gave Chadwick the answers he needed.[50] As Ernest Lawrence from the University of California at Berkley admitted, "The pioneer work of Rutherford and his school clearly indicated that the next great frontier for the experimental physicist was surely the atomic nucleus."[51] Somewhat sadly, Blau reported to Meyer from Göttingen that "For the photographic method, there is probably no time left, since all should be in the service of neutrons."[52] At the 1933 Solvay Conference in Brussels devoted to the atomic nucleus, all but the Viennese attended the meeting.[53] Shortly thereafter, in 1934, the Joliot-Curies discovered artificial radioactivity. Their Nobel Prize in Chemistry in 1935 gave them the power to control "every piece of nuclear work in France," and added to the prestige of the Institut du radium in the international scientific scene.[54]

For Enrico Fermi, the Italian physicist working in Rome, that "was a golden opportunity." His idea was to produce effects like those recorded in Paris, using neutrons instead of alpha particles. While the Italian group, formed by Emilio Segre, Edoardo Amaldi, Franco Rasetti, and Oscar D'Agostino exploited the slow neutrons, Meitner was working jointly with Hahn at the Kaiser Wilhelm

Institute for Chemistry in Berlin. As Meitner later described, "From 1934 to 1938, Hahn and I were able to resume our joint work, the impetus for which had come from Fermi's results in bombarding heavy elements with neutrons. This work finally led Otto Hahn and Fritz Strassmann to the discovery of uranium fission."[55]

Described as the "happy thirties" for physics, and as a transition from atomic to nuclear physics, the 1930s combined apparently fruitful research in physics with an increasing political horror in Europe.[56] Behind a façade of legality, Hitler seized absolute power in March 1933 and put an end to parliamentary democracy in Germany. His first and main concern was to purge Jews from public life. He immediately succeeded with the "Law for the Restoration of the Professional Civil Service." Non-Aryans were excluded from government and civil service positions including the universities. Numerous historical studies have tackled the consequences of the rise of National Socialism for German academic life, ranging from institutional and disciplinary histories to scientific biographies and comparative studies.[57] The persecution of Jews and the gender discriminatory politics of the Third Reich were, nevertheless, an acute indication of the rise of nationalist and fascist movements in Europe.[58] While still at the physics institute in Göttingen, Blau mentioned in her regular correspondence to Meyer, "On the political circumstances, one realizes here absolutely nothing, since, in principle, one should not talk about politics in the institute."[59] In Pohl's institute, scientists might have avoided facing Hitler's reality, but in Austria, the transition from Red Vienna to Austrofascismus and consequently to the Anschluss was hard to dismiss.

FROM AUSTROFASCISMUS TO THE ANSCHLUSS

Apparently,the evidence of scintillation counting and relying upon the individual observer and his or her experience was not the only uncertainty that the Viennese experimenters faced during Chadwick's visit in 1927. In the city elections on April 2 of that year, the Austrian Social Democrats experienced their greatest electoral victory of the interwar years. As Rabinbach points out, "Its [SDAP's] electoral gains were illusory as long as it controlled neither the legal structure of the state nor the instruments of power."[60] Three months later, on July 15, the police violently stamped out a massive demonstration led by socialist workers, leaving 85 dead and hundreds injured. The event marked the beginning of the SDAP's decline and the uprising of the fascists. Although the political Catholicism and fascism preceded the National Socialist ideology, as

Friedrich Stadler points out, the transition was a seamless one. In 1931, in the context of the world's economic crisis and the crash of the *Creditanstalt*, the prominent Vienna bank that financed much of Austria's industry, mobilized a wider economic and political crisis in the country.[61]

In May 1932, Austria underwent a parliamentary restructuring once again. The coalition between the Christian Socialists and the Pan-German People's party was dissolved and the new chancellor, Engelbert Dollfuss, fellow of the Christian Socialists, turned toward the Social Democrats, offering them a coalition with his party. When the Democrats responded negatively, Dollfuss made a deal with the *Heimwehr*, an antirepublican paramilitary organization supported by big business and Catholic political leaders.[62] National Socialists had already emerged as a serious political force in the city elections a month earlier and now posed a threat for the conservative Christian Socialists who held a fragile majority in the parliament.

Eventually, faced with a still strong Social Democratic party on the left and an emerging Nazi regime on the right, Dollfuss and the other Christian Socialists suspended the Austrian parliament in March 1933. Seizing the opportunity, the Nazis committed a number of serious terrorist acts, mainly in Vienna but also in the rest of the country. That gave Dollfuss the chance to ban their party. Although illegal, the Nazis continued to exist without much difficulty.[63] With Mussolini as his ally and protector, Dollfuss fought on two fronts, against the Nazis and more fiercely against the Social Democrats. All newspapers and primarily the *Arbeiter-Zeitung*, the social democratic paper, were placed under strict government censorship. Part of Dollfuss's anti-democratic politics was to drastically reduce the budget of the city of Vienna and cancel all the social reform programs that were put forward by the Social Democrats.[64] In the following months, in the context of the wider European political crisis and Hitler's rise to power in Germany, the political situation in Vienna was increasingly unstable. The SDAP was the main concern of the new fascist regime. Under Mussolini's pressure, Dollfuss proclaimed not only the end of the liberal state and the constitution of the Austrian fascist *Ständestaat*, he also tried to extinguish any opposition from the Social Democratic camp.

The obituary of Red Vienna was finally written on the streets of the city just a few days after the arrest of the mayor, Karl Seitz. For three days, from February 12 to 14, 1934, frustrated armed workers fought with government troops in an already lost fight between socialism and Dollfuss's fascist regime. Although the Social Democrats were defeated and the dissolution of all parties was a fact, the Nazis were franticly planning Dollfuss's assassination. With Hitler's approval, the Nazis attacked the chancellery in Vienna and killed Dollfuss.

In July of the same year, another Christian Socialist, Kurt von Schuschnigg, came into power. For the next two years, Austria remained independent in the midst of a serious political crisis. In 1936, as his own solution to the uprising problems, Schuschnigg signed an agreement with Hitler, giving amnesty to imprisoned Nazis and including Nazis in the government.[65] It was these acts and the undermining of the Austrian democracy that paved the way for the German invasion in 1938. The Austrian-German reconciliation promoted Hitler's plans for Austria's annexation to Germany which finally took place on March 12, 1938, when German troops marched into Vienna.[66]

DISMANTLING THE MEDIZINER-VIERTEL

In the ideological context of German nationalism and political Catholicism, Jews and Social Democrats were among the first targets of both the fascist and Nazi regimes. The antidemocratic tensions that surrounded politics were immediately reflected in Vienna's scientific community. The excellent collection of articles put together by Friedrich Stadler and Peter Weibel, *The Cultural Exodus from Austria,* draws a vivid picture of the decline of the Mediziner-Viertel.[67] Even before Hitler's arrival on the Austrian scene, the destruction of Viennese culture and science, centered at the university and its institutes, started with the Austrofascists. The Viertel, an in vivo cultural and epistemic laboratory during the 1920s, was brutally eradicated by anti-Semitic and anti-democratic attacks.

From early on, the fascists controlled teaching appointments at the University of Vienna. At the same time Nazi students distributed "black lists" of Jewish and socialist professors, demanding restrictions and dismissals. The purge of educational institutions started with the Ernst Mach Society, which was dissolved soon after the February events on the accusation that the society disseminated Social Democratic propaganda. In April 1934, Otto Neurath's Social and Economic Museum in Vienna was closed and replaced by a new institute for Austrian picture statistics headed by the *Heimwher.*

In such a political context, the murder of Moritz Schlick, a constitutive member of the Vienna Circle, and its public justification by the press in 1936 did not come as a surprise. By 1938, most of the Vienna Circle associates were forced to emigrate.[68] The threat to the Psychoanalytic Society also became apparent with fascist and Nazi propaganda where psychoanalysis was targeted as "Jewish" science. Out of the 50 official members of the society, 47 were forced to flee Austria. The fate of the psychologists and their institute was similar and even harsher. As Bernhard Handlbauer describes, the dissolution of the SDAP disrupted the work of the Viennese psychologists, most of whom were deeply involved in the educational reforms of Red Vienna.[69] Handlbauer's argument is supported by the fact that the majority of Viennese psychologists were Jewish, liberals, and among them there were many women. One of the first to be harassed by the fascist regime was the Jewish psychologist Marie Jahoda who, after her imprisonment for nine months, fled Austria in 1936.[70] Of those science institutes that hosted a high number of women during Red Vienna, the Vivarium was banned and destroyed after the Nazis seizure of power in 1938.[71]

A similar dark picture can be seen when one considers the gender and racial politics of the fascist regime in reference to women. As the statistics indicate, the number of female students at the University of Vienna dropped sharply after the civil war. In 1933–34, 1,761 Austrian women were enrolled on the philosophical faculty, 690 on the medical faculty, and 279 on the faculty of law. By 1938–39, the numbers had dropped to 768 in philosophy, 387 in medicine, and most remarkably, in the faculty of law where only 72 women were enrolled as students. (See Chart 01/6.)[72]

More specifically, during the academic year 1933–34, female students in physics accounted for the 18.1 percent of those in the faculty of philosophy. By 1938–39, the percentage had dropped to 12.5.[73] Although at the turn of the century when the University of Vienna opened up its doors to women Jewish women entered in disproportionately large numbers, by the academic year 1933–34, they represented only 2.8 percent of those enrolled. By 1938–39, there were none.[74]

The discriminatory gender politics of the Christian Socialists and their anti-Semitism was not news in 1933. During Red Vienna, any Social Democratic attempt to alter the gender politics concerning issues of abortion, birth control, or sexuality faced the tenacious and even violent resistance of the Christian Socialists.[75] At the same time, the anti-Semitism of the Christian Socialists was clearly stated even in their party's program. According to Gruber, "That the SDAP allowed such gutter politics to go unchallenged from the beginning of the republic to its end with the prominent Jews in its leadership keeping a low profile weakened the party and undercut the republic as well."[76] Overall, the ideological mechanism of the political Catholicism of the Christian Socialists and the anti-Semitism of the National Socialists absolutely destroyed the social reforms in education and, especially in the academy, those that socialists such as Otto Glöckel and Julius Tandler had put forward in Red Vienna.

THWARTING A PROMISING PROPOSAL

The failure in 1932 to establish a joint radium laboratory under the auspices of two liberal institutes in Vienna, the Vivarium and the Radium Institute, was only the most outward sign of a deeper destruction. Tandler's political power and the reform projects of the municipality were at issue much earlier than the fascist's seizure of power in 1934 and certainly long before the Nazis' arrival in Vienna. When on June 2, 1932, the directors of the Vivarium drafted a letter addressing the Austrian Academy of Sciences and presented their joint project with the Radium Institute, it was already too late for any further substantial reforms of Vienna's educational and welfare system.

The story runs as follows. In June 1932, two of the founders of the Institute for Experimental Biology, Hans Przibram and Leopold Portheim, directors of the botanical and zoological departments respectively and of the institute in general, addressed the *Kuratorium* of the Vivarium.[77] The *Kuratorium* was the scientific and administrative supervisory committee, the intermediary between the institute and the academy that among other things handled the institute's finances.[78] The directors' aim was twofold, exceeding a simple scientific request.

The first stage of the proposal included the establishment of a laboratory devoted to the study of the effects of radium on plants and animals. Of special interest was the investigation of the impact of radium on the sex hormones, a topic not sufficiently researched at the time and one that attracted great attention. Biologists interested in studying the effects of radium on organisms were also invited. The annual amount for the function of the new laboratory was estimated at 36,000 schillings. The project was specifically targeted to the physiological work on the sex hormones of Eugen Steinach who was proposed to be, according to the plan, responsible for the new laboratory.

Born in a Jewish family in Voralberg in 1861, Steinach was the son of a physician. In 1879, he moved to Switzerland at the University of Geneva to study natural sciences. In 1880, he returned to Vienna where he enrolled at the university to the medical faculty. At the beginning of his career, he worked closely with Emil Zuckerkandl, director of the first anatomical institute in Vienna.[79] As Michael Hubenstorf points out, the anatomical institute under Zuckerkandl and Tandler was populated by Jewish, liberal, socialist, and foreign students.[80] It was probably there that Steinach developed close contacts with Tandler, later Vienna's councilor for health and social welfare. In 1886, Steinach graduated with a degree in medicine from the University of Innsbruck and for the next two years served as an assistant at the Physiological Institute in Innsbruck. In 1889, Steinach became first assistant to the eminent physiologist Ewald Hering in the Physiological Institute of the German University in Prague where he served for four years. Meanwhile, he completed his habilitation in physiology and became a *Privatdozent* at the university. His rise up the academic ladder was impressive. In 1895, he was named *ausserordentlicher Professor* of physiology and two years later, *ordentlicher Professor.* Still in Prague in 1902, he created and organized an innovative laboratory, the first of its kind in German speaking countries, for "general and comparative physiology." [81]

In 1910, already 49 years old and still in Prague, Steinach published a fundamental article on the physiology of the testis and sex biology, which marked his most productive period of scientific research. The article appeared in *Zentralblatt für Physiologie* and although it emphasized the pre-existence of female and male sex, it proved that maturation and preservation of sex characteristics were definitely controlled by the internal secretion of the testis. When the testes were removed, sexual lethargy developed whereas implantation of the testes elsewhere in the body restored sexual characteristics. Actually, Steinach was the first experimenter to successfully transplant animal testis in such a way that it became vascularized and remained functional at its new site. According to Chandak Sengoopta, his experimental results challenged fundamental beliefs about masculinity and femininity and were utilized within the larger context of homosexual rights movements in early twentieth century Central Europe. "If Steinach was right, homosexuality resulted from a specific congenital anomaly. It was not a progressive disease; it could not be spread by seduction; and, far from being a crime, it was no more dangerous than psychosexual femininity."[82]

Steinach left Prague in 1912 to become director of the physiological department of the Vivarium in Vienna, where he continued his experiments on the sex glands of animals and extended them to humans with the help of the Viennese urologist Robert Lichtenstern. From the laboratory to the clinic, Steinach's aim was to control human sexuality by manipulating the sex glands. During the 1920s, a second line of his research concerned human rejuvenation. He argued that bilateral vasectomy to senile males could achieve sexual rejuvenation. Experiments first on rats and then on three human subjects seemed to prove his claims.

The "Steinach Operation" soon gained popularity in both Europe and the U.S. Rejuvenation was not only applicable to men but equally to women in whom the germinal cells of the ovary could be destroyed by low-dose radiation.[83] By 1926, when the first International Congress of Sexual Research took place in Berlin, Steinach was already a world-famous figure. Between 1921 and 1938, he was nominated for the Nobel Prize in Physiology six times. However, he never received it. Overall, his views on masculinity and femininity depended on "glandular politics" and certainly undermined traditional social certainties of his times.

Given Steinach's own research interest to irradiation experiments, the initiative to establish a radium laboratory at the Vivarium does not come as a surprise. Scientifically, the Vivarium was in a position to foster experimental biological research as, for example, the study of the effects of radium on plants and animals required. The only thing missing was the expertise in working with radioactive substances. As the directors of the Vivarium argued, a laboratory with a focus on the biological effects of radium was necessary and would be complementary to the research done in the Radium Institute given the fact that a condition Kupelweiser posed in 1910 restrained the latter from doing research on living organisms.[84] "According to the protocol, the participation in medicine for our institute," Meyer would admit in 1950, "was impossible, but at that time, the reciprocal interest was big."[85] Politically, the directors of the Vivarium could have the support of the municipality given their good connections to Tandler. As it appears in the Vivarium's annual reports published in the almanac of the Austrian Academy of Sciences, the institute received several donations from the city of Vienna over the years.[86]

What for the directors of the Vivarium was necessary, the director of the Radium Institute found to be a definite chance to enlarge the research agenda of his institute. Kinship relations between researchers and ideological proximity to the directorship and the working ethos between the two institutes promised an environment for fruitful research. The assistant of the Radium Institute, Karl Przibram, was the brother of Hans and one of those who financially supported the Vivarium when it was first established. Meyer's wife, Emilie Maas, was the niece of Leopold von Portheim.[87] Additionally, many of the researchers of the Radium Institute had already expressed their interest in working on the boundary between physics and medicine. These included Blau, Kara-Michailova, Rona, Eduard Jahoda, and Franz Urbach.[88]

At the time, Urbach was the director of the *Physikalische Laboratorium* at the municipal hospital in Lainz. Meyer himself had already published on the physical basis of radium emanation therapy in 1929. The next year, he collaborated with Erhard Suess, the son of the president of the academy, Eduard Suess, on the use of radium emanation as an indicator for diagnostics and in therapy. Meanwhile, Meyer supervised the work of Maria Renata Deinlein on the residence time of radon in the human body after drinking therapy.[89] In 1932, Meyer was invited to prepare a special issue of the *Pharmazeutische Presse* on radioactivity, in which he involved Karl Przibram, Gerhard Kirsch, and Rona.[90] Thus, the proposal for a joint research radium laboratory that would be housed in the Vivarium seemed not only reasonable but indispensable.

Besides the scientific concerns, the second stage of the proposal carried a political dimension. As the directors of the Vivarium revealed in their plan, the proposed laboratory would function as the regulator for the supplies, the dosimetry, and the handling and shipment of radium in Austrian hospitals.[91] It was one thing for the scientists to require a new research laboratory and it was quite another to envision themselves as the regulators of radium supplies in the hospitals of the entire country. To control the distribution of a valuable trafficking material for scientific laboratories, clinics, and commercial sites meant more than the movement of equipments, materials, or experts among laboratories. It meant the absolute power over laboratory and clinical work on radium in the entire country, the expansion of the budget for both institutes, and their future scientific and financial independence from state control. Under such conditions, people like Steinach could carry out their research agenda, undermining traditional social values. Magnus Hirschfeld's example leaves no doubt about the Nazis' stand toward similar lines of research. Hirschfeld was a practicing physician and founder of the Institute for Sexual Science, which was dedicated to research on human sexuality and especially homosexuality. When the Nazis took over Germany, they ransacked his institute and burnt his books and records. Hirschfeld subsequently fled to France.[92]

In Austria, the establishment of the radium laboratory at the municipal hospital in Lainz in 1930 was already a persuasive sign of the Social Democrats' political plans to alter welfare services in Vienna. Meyer's position as the consulate for the radium purchases from the municipal hospitals implied a direct connection to Tandler's political agenda. Obviously, the proposal that was put forward from the Vivarium met Tandler's ambitions as well for a centralized station for radium supplies that exceeded the borders of the city of Vienna, but given the wider political crisis of the country, the timing was not propitious.

Just five days later, a committee was invited to meet at the academy to discuss the proposed project.[93] Among them were Meyer and Schweidler from the Radium Institute, and Hans Przibram and Portheim from the Vivarium. Hans Molisch, the president of the University of Vienna and an earlier collaborator with the Radium Institute, attended the meeting along with Arnold Durig, the director of the Physiological Institute of the University of Vienna. Durig was a member of the supreme hygiene councilors and one of the first to be dismissed for political reasons after the Anschluss in May 1938.[94] Unfortunately, no further notes about the discussion appear in the protocol book of the Austrian Academy of Sciences. A letter to Molisch signed by both Prizbram and Portheim on June 6, 1932, a day earlier than the actual meeting of the committee, however, sheds some light on the issue. After informing Molisch that they had just had 400 visitors and 80 children in the aquarium, they mentioned that "Unfortunately, nothing will happen with the other project because the one who proposed it did not contact us and the available information does not shed any favorable light on it."[95]

The proposed project never took place, thwarting the possibilities of interdisciplinary research on radiobiology. Less than two months later, on July 31, 1932, Steinach retired.[96] The failure of the Vivarium and the Radium Institute to establish a joint radium laboratory could be attributed to many factors. The fact that two of the most liberal institutes in Vienna formally tried to play not only a scientific but also a political role in the country at a time of a deep antirepublican crisis is not negligible. Adding the fact that most of the researchers, especially those in the key positions of the directors, were Jewish, one could anticipate the outcome of the proposal. The unfruitful initiative was only the beginning of what was going to happen in the political life and the scientific research of the country.

FRANZ URBACH AND THE FASCIST POLITICS OF PERSECUTION

In its early days, the new fascist regime in Austria was more interested in destroying the social democratic forces of Vienna than in Jewish scientists. Still, some fit in both categories. Franz Urbach, collaborator of the Radium Institute, was among the first to be fired from his position as director of the *Physichalische Laboratorium* at the municipal hospital in Lainz. The reason was clearly political, but Urbach was also from a well-known Jewish family in Vienna.

Urbach completed his dissertation on the phenomena of luminescence after radioactive irradiation in 1926 under Przibram. He continued to work with Przibram until 1932 when he was appointed director of the *Physikalische Laboratorium* in the hospital in Lainz. According to Wolfgang Reiter, the new municipal authorities expelled Urbach from his office in 1934, accusing him of gaining the position under the political influence of his uncle Otto Urbach, an active Social Democrat.[97] Besides this odd excuse, the fascist regime knew very well that Urbach had a key position similar to the one that the directors of the Vivarium envisioned to obtain through the joint laboratory with the Radium Institute but on a smaller scale. Indeed, the *Physikalische Laboratorium* of the municipal hospital did create the space for interdisciplinary exchanges among physicists, technicians, and physicians. Combining radioactivity and medicine, the physicists crossed the boundary of their expertise and provided scientific support to doctors over issues of radium dosimetry.

Employing a director, two physicists, and a technician, the laboratory offered technical support to the physicists who worked in the relevant pavilions of the hospital. Open to medical practitioners, it further functioned as an intermediate space of collaboration between physicists and physicians for the development and improvement of methods applied in radium therapy. Besides playing the role of an information and research center, the *Physikalische Laboratorium* had another key function. It controlled the radium carriers, 400 in total, for medical use. The whole endeavor, as Urbach acknowledged, became possible through the municipality's initiative and the support of three professors at the University of Vienna: Meyer, Thirring, and Hermann Mark.[98]

For the fascists, the removal of Urbach from his key position in the hospital at Lainz was an issue of control and demonstration of power. Red Vienna and the reforms of the Social Democrats were clearly past. The municipality was in the hands of the Christian Socialists and appointments were now controlled by Schuschnigg's regime. When Urbach found himself unemployed, he turned to the Radium Institute and his close collaborators there, Meyer and Karl Przibram. In the 1935 publication of the almanac of the Austrian Academy of Sciences, Urbach appeared as the institute's collaborator, as he was before he obtained the position in Lainz.[99] The same year, working in Przibram's group, he published an article in the *Mitteilungen* on the spontaneous change of latent pictures. He remained in the institute until his immigration in 1939.[100]

HILDA FONOVITS-SMEREKER: A PUZZLING CASE

In these odd circumstances, Hilda Fonovits-Smereker, *Assistentin* of the Radium Institute from 1919 to 1922, took over Urbach's position. The politics behind this appointment are not clear. In 1932, Fonovits-Smereker was hired as an assistant director at the Radiumtechnische Versuchsanstalt, the radium station at the General Hospital of Vienna.[101] The decision was made by the federal ministry for social administration and the ministry for justice. The connection had probably been made through Meyer and the director of the *Radiumtechnische Versuchsanstalt*, Albert Fernau. That same year, Fernau was in frequent correspondence with Meyer, discussing the therapeutic value of drinking water containing radon.[102] By that point, it was mainly the personnel of the Radium Institute that prepared radium for medical purposes in the general hospital. As Fernau mentioned, Rona had recently prepared radium D for hospital use.[103]

Although no direct evidence exists, under these circumstances, Fernau probably suggested the hiring of a physicist and Meyer mentioned Fonovits-Smereker. The function of the Radiumtechnische Versuchsanstalt and thus Fonovits-Smereker's tasks were to measure and prepare radium for medical use. She worked closely with Fernau until his death in 1934.[104] Meyer informed Rona, who at the time was at Curie's Institute in Paris, "You probably have heard that poor Fernau died last August. He was not a great scientist, but he was a very nice man and had achieved a lot for the dosimetry of medical preparations and was not always acknowledged. I have not yet heard more about whether and how his institute will be further directed."[105]

At that time, Fonovits-Smereker was promoted to Fernau's position, but that was not the only change. Based on a work contract, the city of Vienna was controlled by Schuschnigg's regime. It now entrusted the Radiumtechnische Versuchsanstalt and consequently its director, Fonovits-Smereker, to perform control measurements of the radium preparations and the rest of the scientific work previously done in the Physikalische Laboratorium at the hospital in Lainz by Urbach's group. As her publication record reveals, Fonovits-Smereker published extensively in the ensuing years on medical physics and in the prestigious journal, *Strahlentherapie*.[106]

The *Oberarzt* of the radiation department, to which the Physikalische Laboratorium belonged, was Emil Maier. When the department was established under Tandler's supervision, Maier visited several sites in Europe to gain experience in radium therapy. At the end of May 1938, right after the Anschluss, Maier became a member of National Socialists (NSDAP) and on December 1 of the same year was promoted to *Primararzt*.[107] In 1941, Fonovits-Smereker and Maier were married.[108] In 1943, her habilitation was accepted at the medical faculty of the University of Vienna and she retained the position of the director until the end of her life in 1954. Her death was attributed to severe blood damage due to her work with radioactive materials.

A puzzling case, Fonovits-Smereker was the only woman who ever held the position of a second assistant at the Radium Institute in the interwar years. Following an expected pattern of women in science, she quit her career in 1922, unable to combine motherhood and scientific work. Nevertheless, she held a prominent position ten years later, this time as an assistant director at the Radiumtechnische Verschuchsanstalt in the general hospital of Vienna. In 1934, although the fascists ousted Urbach from the hospital in Lainz, she became the director of the Physikalische Laboratorium. No letters indicate any connection to the rest of the women at the Radium Institute, and no evidence seems to exist of her political positioning. Her marriage to Maier further complicates the story. However, by being able to handle a trafficking material such as radium, she was able to be evasive and to cross disciplinary boundaries to regain entry into science. Carrying her physics expertise to the general hospital and then to the hospital in Lainz, Fonovits-Smereker secured a career in the boundary zone of medical physics.

A HISTORY OF DISARRAY: THE INSTITUTE FOR RADIUM RESEARCH, 1933–38

From 1933 to 1938, the history of the Radium Institute, affected by the political upheavals, can be written as a history of disarray. Besides losing the full benefits of a position as a *wissenschaftliche Hilfskraft* in 1933, even the scientific connection to the Second Physics Institute was threatened.[109] At the end of March

1934, Gustav Jäger, Exner's successor, was forced to retire as director of the Second Physics Institute at the age of 69.[110] In a state of anxiety, Stetter reported to Pettersson that "The disaster has already come. The ministry of education has informed the faculty-in which form, I do not know-that Hofrat Jäger should retire by the end of March and that the Second Physics Institute is dissolved."[111]

On March 5, 1934, the same evening that Stetter wrote his letter, the faculty was planning to meet and discuss its strategy. The special commission hoped to gain time by keeping Jäger for one more semester and to save the institute in some form or other. Although as Stetter admitted, "they" were not optimistic about saving Jäger, "they" proposed to mobilize the American envoy, concentrating their efforts on saving the institute. "The way through the ministry of foreign affairs is the one that can be successful. Perhaps an intervention from Rockefeller, perhaps a letter from Curie or similar?"[112] Stetter never explained who "they" were and, as it becomes obvious through Karlik's correspondence with Pettersson, a number of different strategies were proposed. "I had a very long talk with Kindinger who seems to have been discussing the matter with Stetter in details . . . I hope he has some influence on the others; his ideas about the tactic seem very sound."[113]

Stetter's main concern was that the dissolution of the institute would definitely affect his research. On the basis of his work in Pettersson's group, Stetter completed his habilitation in 1927 and six years later, in October 1933, he was promoted to an *ausserordentliche Privatdozent*.[114] Since 1922, he had been working as an *ausserordentliche Assistent* at the Second Physics Institute and used its facilities for his research. With the dissolution of the institute, Stetter's professional career was at stake and his fears were well-founded. Since 1932, he had been a member of the National Socialist Teachers League and a month before Dollfuss banned the Nazi party in July 1933, Stetter joined it.[115] On top of facing the dissolution of the Second Physics Institute, Stetter was risking possible dismissal as an illegal Nazi.

Obviously, Stetter was not the only one in the institute who subscribed to the National Socialist ideology. Ortner had been a member of the National Socialist Teachers League since 1934. Kirsch, Pettersson's close collaborator during the 1920s, became a leader of a *Keimzelle* of the National Socialist Teachers League at the University of Vienna in 1933 and had been a member of the NSDAP since 1923.[116] Probably the most outspoken one of this group, Kirsch made Karlik and Pettersson nervous. Already in 1933, she expressed her aversion toward him: "I have been to the institute this morning. Kirsch has come back and now one has to face politics again. I feel so disgusted!!" A year later, it became obvious to Pettersson that Kirsch's scientific work was "probably the least dangerous occupation one could find for him." It was the time, as Karlik reported in her regular correspondence to him, that "Many things are not pleasant" in the institute, but letters did not seem to be a safe way for conveying more details.[117]

In the meantime, Karlik decided to apply for the position of *Dozent*. In the beginning of May 1936, she put together her papers and, as the procedure required, Karlik presented her file to all the members of the examination committee to ask whether they had any objection. She recounts the situation:

> Schw.[eidler] was charming and addressed me kindly in all the formalities. Th[irrin]g—to my great surprise—took the question *au serieux*. Looked up the number of *Dozenten* we had already, asked me what I meant to lecture about and who was the last Doz.[ent] appointed and when was that and so on. But finally he wound up by saying that, of course, he would not object, that he always had had an excellent impression from me and some more such compliments.[118]

At that time, Thirring was director of the Institute for Theoretical Physics and already knew Karlik and her work, since he was on the examination committee of her doctoral thesis. Schweidler had been director of the First Physics Institute since 1920. In 1936, he was also appointed director of the Second Physics Institute which had remained without a director for two years after Jäger's retirement.[119] Felix Ehrenhaft, the director of the Third Physics Institute, was next on the list. They were all housed in Boltzmangasse next door to the Radium Institute and knew Karlik from the time she started her studies on radioactivity. However, things did not go as smoothly with Ehrenhaft as with the others. "He [Ehrenhaft] was 'terribly busy' for several days (Planck was in Vienna) and could not receive me. Then at last, I was asked to come on Saturday which was a day too late for sending the application for the May meeting of the faculty. Whether he knew this or not, I don't know."[120]

Karlik was prepared for the delay and finally Ehrenhaft presented no objection. Even with their affirmations, nothing guaranteed the outcome of the final examination. She was planning to study during the summer and present her colloquium the coming October. Given that Schweidler was the dean and also positive about her application, she expected to easily satisfy the formal requirements and complete the paperwork in the next two upcoming faculty meetings before the end of the academic year. She continued to run her own experiments with uranium, part of her research project in Bornö and found herself "unexpectedly" busy with the task of performing photometric measurements for Thirring. "I don't mean to devote too much time to this job which is really not mine, but to go a little bit into the matter, I think will pay."

Accustomed to the collegial style of her earlier work at the institute, she was hoping to learn a lot in various respects, "especially things which may also be of some use for your work," Karlik wrote to Pettersson. "I am also glad the thing is of interest to Schmidt, and Blau and Wambacher will also probably profit." The arduous work and her numerous research obligations did not curb her willingness to go through with her application. As Pettersson assured her, "I have not the least doubts about your coming up out top dog in the end, but you are in a strenuous time."[121] Eventually, Karlik received the *Venia Legendi* in 1937, the formal requirement for gaining the right to teach at the university and she started teaching the physics of inert gases in 1937–38.[122]

Other than delays and an incredible amount of work, Karlik did not seem to have faced any serious objections. Blau, however, did not receive similar treatment when she asked for the permanent position of a *Dozent*. According to her brother, a professor told her that "to be a woman and a Jew was just too much."[123] Given that Karlik was able to obtain the position, Blau's case indicates that gender was not the main discriminatory factor. The fact that Blau was Jewish and probably more politically engaged than Karlik contributed to her rejection.[124] The international attention brought by her success turned out to be a disadvantage.

After her return from Paris in 1934, Blau continued her collaboration with Wambacher. The two women worked on two fronts. First, they improved the emulsion technique by thickening the photographic plates to allow a better deposit of the particle tracks. Ilford, the English photographic company, offered to produce sufficiently thick plates, but as Blau explained, "To obtain still thicker emulsion layers, new developments methods had to be worked out." Second, while still struggling to alter their apparatus to suit their experimental needs, Blau and Wambacher applied the photographic technique to neutron studies.[125] Yet their collaboration turned out to be threatening for Blau's existence in the institute and for the control over her own method. In June 1934, Wambacher had joined the National Socialist Party and around that time, she had been intimately involved with Stetter. Facilitated by the political changes in 1936, Stetter soon started to interfere in the relationship of the two women and their scientific work.

That year, following Mussolini's suggestion, Schuschnigg sought an accommodation with the Germans. An agreement signed between the two countries led to the empowerment of the Austrian Nazis. These changes were immediately reflected in the institute. As Karlik somewhat ironically reported to Pettersson, "Stetter is looking much interested in everybody's work and affairs and he is behaving like an ideal 'chef.' "[126] What he was following very closely, though, was the work done by Blau and Wambacher. The two women were supported by Victor Hess, the institute's first assistant before World War I and an expert in cosmic radiation. In 1936, they exposed their emulsions for four months on the Haferlekar, a mountain near Innsbruck.

Their research project consisted of determining the existence of heavy particles such as protons, neutrons, and alpha particles in cosmic radiation, which at the time was quite doubtful. Proton tracks, longer than anyone else had observed by that time, were apparent in a first examination of the plates. To their surprise, the two women observed in the emulsion a "contamination star" (several tracks emanating from a point) that could neither be explained by irregularities in the emulsion nor from unknown radioactive products in the handling and storage of the plates in the laboratory. "This 'star' had to originate with cosmic radiation, since we had never observed a similar phenomenon in plates, even those that had been lying in the laboratory for much longer periods of time."[127] The assumption was that the large stars originated with the disintegration of heavy particles, probably bromine or silver, and the smaller ones originated perhaps from light elements in the gelatin. Given the theoretical limitations of nuclear physics of the time, Blau and Wambacher could not determine the nature of the primary particle and the exact process of the disintegration.

These impressive results, which Galison considers the first "golden event" using emulsions, provoked the interest of the scientific community and the brutal interference of Stetter.[128] In 1937, on the basis of their discovery, the two women were awarded the Ignaz Lieben Prize of the Austrian Academy of Sciences.[129] Furthermore, through Meyer's assistance, the two women were awarded a grant from the Austrian Academy of Sciences that was to be used for balloon flights with emulsions distributed at several of the country's mountain stations.[130]

The international recognition also came with Heisenberg's immediate response. As Karlik informed Pettersson, "Heisenberg takes personally the most vivid interest in it [the new phenomenon] and is in continual correspondence with Blau and Wambacher. He has been talking about it in a conference with the Upper Ten in Bologna." While the two women were preparing a publication, Stetter approached Blau. He accused her of being unfair to Wambacher and expected her to change the order of the names on their publication. After all, Stetter argued, Wambacher was the first to look into the microscope and find the first star. Blau refused.[131]

In 1927, the skill of observing scintillations, performed by the women of the institute, had been characterized by Chadwick and historically read as routine and technical, separate from the decisive parts of the experiment. In 1937, Stetter used Wambacher's skill of observing and situating a star as the decisive part of the whole experiment performed by both women. Symptomatic of how the Viennese valued the task of observation, Stetter's argument is at the same time ironically symptomatic of his ardent anti-Semitism. During the days of Meyer's directorship, incidents like this did not occur, yet the Nazi regime imposed different practices in the institute. Although Stetter's anti-Semitism is clear, his interference implies something more. Assuming his male power, he intervened in the relation of the two women, taking control of their collaboration. Meyer, a Jew himself, was unable to play the leading role his position required.[132]

The colleagues that knew Blau, including Karlik, recognized that she was miserable after Stetter's intervention and even thinking of abandoning her research., Karlik was very close to Blau and knew that in spite of her poor health, she had been working intensely for the last months. "The enormous pleasure the work gave her actually made her feel a little stronger." As for Wambacher, "She certainly has been very diligent, too, since the summer (chiefly examining the plates in the microscope), but Etta Blau has done all the very tiresome calculating." Also, as Karlik reminded Pettersson, Blau was still the more "mature partner" between the two.

At the same time, given her affair with Setter, Wambacher was strongly attached to him. Although, according to Karlik, she recognized that his handling of the situation was not quite correct, her emotional dilemma was important. Wambacher's behavior toward Blau was extreme. Either by being rude or as enormously generous as she often was, she had turned the relationship into an uneasy one. The most suitable solution seemed to be for Blau to leave the institute for a while. Karlik turned once again to Pettersson, asking for his intervention and suggesting that he could invite Blau to his oceanographic institute, offering her a research project and a small stipendium.[133]

The solution came from another direction. Ellen Gleditsch, probably informed by Rona, took a personal interest in Blau's situation.[134] Her research assistant, Ruth Bakken, was pregnant and Gleditsch suggested that Blau replace her for three months. The solution was ideal. Away from Vienna, Blau could work with Ernst Föyn whom she already knew through her summers at Bornö. Looking forward to her visit to Oslo and under the enormous pressure from her ex-advisee and Stetter, Blau rushed to arrange research matters with Wambacher, an arrangement which was a total defeat for her. Even before Hitler's troops marched into the city, her Nazi colleague was able to take over the most interesting part of her research project.

As the agreement went, Wambacher, in collaboration with Ortner, was going to investigate the relation of the grain and density of the tracks recorded on the photographic emulsions to the energy of the particles produced by them. By measuring the grain thickness of the tracks, one could even estimate the energy of the particles that were not lying completely in the emulsion but passed through without ending it. That had the potential of identifying the particles and the total energy released in the projects, the two key points of Blau's and Wambacher's earlier work. "It is actually one of the main points started by Bl[au] in which she is particularly interested," as Karlik reported to Pettersson. "Blau kept for herself the absorption experiments. It's less promising and more tiresome and it will take months before she can examine the first plates . . . she sacrificed more than I considered right."[135]

In his usual manner of supporting his female collaborators at the Radium Institute, Pettersson proposed to present Blau's work in the Swedish press. However, it was more urgent for Blau to give a paper at Bohr's institute on her way to Oslo. Wisely enough, Pettersson foresaw that the Bohr connection could pave her way to other prominent research centers in Europe.[136] Very cautiously and discreetly, Hevesy, who at the time was in Bohr's institute, arranged the visit for March 14, 1938.[137] The connection was made by Rona, Hevesy's earlier collaborator and friend.

While at the Radium Institute, Stetter, Ortner, and Wambacher orchestrated Blau's purge and the seizure of her scientific research, Kirsch tried to present the Nazi version of the two women's collaboration to Pettersson. In January and February 1938, Kirsch was on a scientific tour in Berlin, Kiel, Oslo, Stockholm, and Göteborg, giving lectures about his work.[138] Traveling from Oslo to Stockholm, he visited Pettersson for just an evening. That was enough for Kirsch to discuss the matter and allege that Blau had exploited Wambacher in their cooperative project. Amid the threats of those who surrounded her in Vienna and those who were willing to present to the international scientific community a version of the case convenient to Wambacher, Blau was ready to leave the institute.

Karlik expected that her leave would be temporary and both Blau and Wambacher "will find some way to each other again after her return." Karlik's naiveté went so far as to suppose that "Stetter already begins to feel sorry, but his *weltanschauliche* convictions and his sympathy for and his wish to help H[erta] W[ambacher] are very strong."[139] Contrary to Karlik's predictions, Blau's temporary distance from the institute and her Nazi colleagues turned out to be a permanent struggle for existence. On March 12, Germans entered Austria in a triumphal parade. The day before Hitler gave his speech in Vienna, Blau left the city.[140]

ANSCHLUSS AND EXILE

To answer the question "What was particularly Nazi in science after 1938 in Vienna?" one should first consider what was particularly different in the Viennese scientific community before 1934. As I have argued throughout, the physicists conducting research on radioactivity were an inseparable part of the Viennese culture and the democratic politics of Red Vienna. The collegial ethos of working in the Radium Institute and the leadership style of Stefan Meyer defined the atmosphere in the institute and welcomed a number of young scientists, many women among them. Right after the Anschluss, well-preserved patterns of research and cooperation were abruptly disturbed. The so-called friendly visit[141] of the German army to Vienna was not so friendly for the Jews of the institute. Although the final decisions on the dismissed personnel were not expected earlier than April 10, the scientists at the institute were forced to swear allegiance to the Third Reich during the last week of March and at least two of them were excluded for racial reasons.[142] Meyer had already applied for a permanent retirement to the philosophical faculty on March 18 and voluntarily retreated from his academy membership in an attempt to avoid any confrontation with the Nazis and the humiliation of a dismissal from the academy.[143] Przibram's position was also threatened and, as Karlik described to Pettersson, "When I see Karl, tears come to my eyes."[144] Both Meyer and Przibram remained at the institute as "guests" until January 1939, when a hate campaign against them forced Ortner, the new director, to forbid their work at the institute.[145]

During the war, Meyer and his family retreated to his summer residence in Bad Ischl, close to Salzburg, while Przibram and his wife immigrated to Brussels.[146] Wolfgang Reiter points out that after the Anschluss, "the Radium Institute lost a quarter of its collaborators, in particular those who had shaped the profile of the institute with their scientific achievements."[147] In front of the Radium Institute, a long banner with the slogan "One Nation, One Empire, One Leader" made tangible the dramatic changes in the city and most obviously in the institute itself. Most expressive was the slogan hanging in the Physics Institute: "*Juden sind hier unerwünscht*" (Here the Jews are undesirable).[148]

While anti-Semitism in the Institute forced the Jews into exile, the promotions of the Nazi gang after the Anschluss were impressive. Besides taking over the directorship of the institute, Ortner was named *extraordinarius Professor.*[149] As Karlik explained:

> He [Ortner] is comparatively decent but perfectly happy and very pleased with what is going on. What he does not want to know, he does not know and what he does not want to think, he does not think about. So he is flourishing and made a remark a few hours ago at which I felt I should like to smack his face. George's [Stetter] beaming satisfaction with himself is sometimes almost unbearable. His psychology is as primitive as can be. To have to listen to remarks and explanations by him and his friends is the greatest strain; much worse even then to watch the distress of some friends.[150]

Ortner's friends, Stetter and Kirsch, were both promoted to the position of *ordinarius Professor* and took over the responsibilities of those who left. "Gerhard [Kirsch] is now supervising the third [Physics] Institute and is also lecturing five times a week in E.[tta Blau]'s place . . . George [Stetter] has taken over Charles's lectures."[151]

By the beginning of May, the situation deteriorated. "A number of changes have taken place here again. Mark, Thirring, Schrod., and Ludloff had to leave," Karlik wrote to Gleditsch.[152] Filling up the positions that the Jews such as Blau, Meyer, and Przibram left behind was not ambitious enough. The continuation of the research was accompanied by a plan of expansion. Supported by the German ministry for financial developments in Berlin, Stetter seemed to have played an instrumental role in establishing an institute for nuclear research as a joint program between the Second Physics Institute and the Radium Institute.[153] The Vierjahresplan-Instituts für Neutronenforschung was directed by Stetter, and Ortner was named his official substitute.

The Nazi authorities were not opposed to science, so they channeled large amounts of money to scientific research and renovations. To accommodate the changes, the Radium Institute underwent a decisive reconstruction. The whole building was cleaned and painted, the furniture was well washed, everything was put in order, and the door leading to the staircase toward the roof was bricked shut while another one was opened up. The aim was to eliminate the radioactive contamination in the institute and use the previously contaminated rooms for the sensitive Geiger-Müller counters for measurements.[154]

While the Jews of the institute were cut off from their research and forced into exile, the Nazi circle, including Max Kindinger, Josef Schintlmeister, Willibald Jentschke, Stetter, Ortner, Kirsch, and Wambacher, secured the support of the Third Reich to play a role in the development of nuclear physics.[155] In this politically polarized atmosphere, the non-Jewish anti-Nazis, such as Berta Karlik, faced a crucial dilemma. "The question is: to stay or not to stay? I have decided to stay," she admitted to Pettersson.[156]

The following table, when compared with Table 04/4 of chapter 4, illustrates the major changes in the institute's personnel as have been described above. Notice that none of the Jews, men and women alike, remained at the institute after 1938/39. (See Table 01/6.)

BERTA KARLIK

In 1938, anxious about the political circumstances, Pettersson proposed to Karlik, his closest friend and long time collaborator, a one-year fellowship in his new oceanographic institute in Göteborg. "Remember, I am reserving my first research fellowship for you and am moreover not pressing you for an answer, but in case you do not foresee any changes where you are, I shall be very happy to have you here in 1939."

The changes in Vienna were drastic. With Ortner as the new director and Przibram's purging from the position of the first assistant, Karlik took over his responsibilities and most of his research agenda. The new arrangements brought her professionally very close to Ortner, a collaboration that threatened her own research agenda and her role at the institute. "He [Ortner] is a bit of an egoist, too, in his work and there have already been a few incidents which showed me that I had to look out or he would use me as a well-qualified kuli, [which would be] most comfortable for him."[157]

The changes in the directorship marked not only the rise of anti-Semitism within the institute and the flourishing of the Nazi group but also tended to transform the role of women scientists. Out of the 17 women at the institute in 1938, only seven remained in 1939.[158] With the departure of Blau and Rona—the women most seriously engaged in the institute's research—and the decrease in the total number of women, the responsibilities of those who remained increased. Yet there was little place for research and less time for creative work. "It's not a matter of career," Karlik admitted to Pettersson. "I hope you know me well enough to realize that I don't care for that. But I want *decent* conditions to work in; not just endless drudgery work and a lot of responsibility in all sorts of silly little matters and the care of some stupid students."[159]

In May 1938, afraid that she would lose her research status at the institute, Karlik prompted Pettersson to demand back the apparatus he left in Vienna after his departure in the early 1930s. Financed by Swedish donors and the Rockefeller Foundation, the laboratory instruments used in the disintegration experiments of the 1920s officially belonged to Pettersson. During the 1930s, while he was still in close collaboration with the women of the institute, Pettersson never claimed his instruments even when he was in need to equip the station in Bornö. Wanting to ensure her access to the most important experimental apparatus for her work, Karlik asked Pettersson to leave the glass spectrograph to her responsibility. "Years ago, Gerhard [Kirsch] hinted already at taking it away from me should he once leave Vienna. Perhaps it was more to show his power over me at that time that he actually meant it—anyhow, I am not safe."[160]

While Karlik was struggling to retain her research position at the institute, she was assigned the reorganization of the library, Przibram's administrative tasks as assistant, several odd jobs in the renovation of the building, and the supervision of a number of young students besides her own. With a feeling of ambivalence and "torn to pieces," Karlik decided to stay and rejected Pettersson's offer to take up the fellowship at his Oceanographic Institute in Göteborg. "I will have to put up with many things. I want at least the possibility to do some research that interests me," but it was not only the research Karlik was interested in. As she admitted to Gleditsch, "I think perhaps some of my English friends wonder why I am not leaving Germany in protest. I have come to the conclusion that protest on the part of a German individual is quite useless at the moment and that more is done by staying and trying to improve matters from inside the country."[161]

Indeed, Karlik remained in Vienna and during the war years, she reached the peak of her career. In 1940, she officially became *wissenschaftlichen Assistent* at the Radium Institute and in 1942, she was promoted to *Diätendozentin*. In collaboration with Traude Cless-Bernert, she discovered the natural occurrence of isotopes of astatine by observation of their radioactive alpha particle decays. Bernert was supervised by Ortner and after her graduation in 1939, she remained at the institute. The two women worked extensively together until 1945. By the end of the war, after Ortner's "disappearance," Karlik became the director of the institute and retained her position until her retirement.[162] In 1956, she was promoted to the highest academic rank of an *ordentliche Professor*, the first woman in Austria to reach such a position. The Austrian Academy of Sciences elected her as a member in 1973, the second female member of the academy after Meitner.[163]

ELIZABETH RONA

Although Karlik had the choice to remain at the institute, Rona had to flee. She left for Budapest on April 7, 1938. She was so upset and disturbed by the new political status that she abandoned her research and left her latest measurements in disarray.[164] In a letter written the same day, Karlik admitted to Gleditsch, "I have just seen Elisabeth off. I am going to miss her *very* much. We got more attached to each other in those weeks then ever, but we both felt that it was time she was going; this atmosphere of departure was beginning to tear our nerves."[165] One of Rona's last papers published in the *Mitteilungen* was in collaboration with the Nazi Josef Schintlmeister, who was under Stetter's influence.[166]

It was not only the rise of the Nazis within the institute that threatened Rona, but her everyday life became troublesome as well. "She has had to provide innumerable certificates concerning taxation, etc. and everywhere she had to queue up. Every day brought new regulations that meant some more certificates. There was a very severe control of the luggage at the station, but I hope," Karlik continued, "she will get home safely."[167]

At the age of 48 and after 13 years of work at the Radium Institute, Rona was in search of a new job. Pettersson was ready to offer her a position in Bornö for three months in the autumn, replacing his assistant, Börje Kullenberg, who was going to work in the new oceanographic institute in Göteborg. Because Hungary was still independent, Rona left for Budapest where she considered working at the university.[168] Dissatisfied with the conditions there, she instead worked in industry. As she informed Meyer, "The possibility of work for the immediate future makes me worry a lot. I have found a comfortable job in the Vatur industry through the kindness of the director, Patai."[169]

She was able to retain the job until September of 1938 when the industry shifted to mere production, eliminating laboratory positions. Threatened by the political upheavals in the neighboring countries, Rona was hesitant to accept Pettersson's invitation. In the absence of any other option, the woman who was one of the most distinguished experts in polonium preparations eventually spent October through December of 1938 in Sweden, working on oceanography.[170] Her close friend Gleditsch offered her another temporary solution. She invited Rona to spend a year in Oslo, replacing a staff member in her laboratory who was on a leave of absence. "We have had much trouble in getting the permission for Dr. Rona to enter Norway. I believe, however, that by now, everything is in order," Gleditsch informed Karlik on January 17, 1939.[171]

By the end of her stipendium in Oslo, Rona returned to Budapest in 1940. Working on the boundary of physics and medicine and taking advantage of her earlier experience preparing radium for hospital use, she obtained her next one-year position at the Radium-Cancer Hospital in Budapest.[172] As she later recalled, "In 1941, I made a big decision. Hungary was threatened from two directions; on the right bank of the Danube were the Russians; on the left, the Germans. There was no future for me in Hungary."[173] After a last visit to Vienna in January 1941, Rona fled to the United States on a visitor's visa. Hunting for a job at the annual meeting of the American Physical Society, she was able to obtain her first position at Trinity College, a Catholic College for women in Washington, D.C., as a chemistry teacher.

Rona's earlier work in the intersection of radioactivity and oceanography was the vehicle for securing a joint research position at the Geophysical Laboratory at the Carnegie Institute in Washington. Rona's experience at the oceanographic laboratory in Bornö appealed to C. Piggot and W. Urry from the Geophysical Laboratory as they investigated the radioactivity of ocean sediments. A year later, she was invited to work for the Office of Scientific Research and Development (OSRD), using her expertise in preparing polonium for work related to the war effort. Obtaining security clearance, Rona revealed her method to the Canadian Radium and Uranium Company which had contracted the mass production of polonium for the OSRD.[174] Without any compensation and accustomed to the collegial ethos of researching physics during the interwar years, Rona generously offered the knowledge she obtained at the Radium Institute in

Vienna and Curie's laboratory in Paris to her colleagues in OSRD. She was fortunate to be needed for the secret work on the atomic bomb and thus, she was able to forge a new career in the United States.

MARIETTA BLAU AND HERTHA WAMBACHER

The day that the German troops marched into the city of Vienna, Blau was on her way to present her work at Bohr's Institute in Copenhagen. After a successful talk, "She was tired and rather miserable" when she finally visited the Petterssons for a few hours on her way to Oslo in mid-March.[175] In Vienna, she had left her mother and was anxious about her return but, as Pettersson assured Karlik, "Ellen Gleditsch will do her a lot of good and put her to work which is the most important point." He was planning to bring her to Sweden during the summer and carefully mobilized his connections for a more permanent solution.

Simultaneously, Karlik kept in touch with Blau but wisely screened the news she conveyed to her and hide the fact that in the meantime, her aunt had died and her mother was hospitalized with a broken leg. She once more reminded Pettersson to reclaim the instruments he had brought to the institute in the early 1920s, including the tabletop, portable objectives and microscopes that could be of use to Blau or even Rona who were searching for a research position. "There is also Etta [Blau] to think of and perhaps even Elisabeth to consider. Especially as regards Etta some help might perhaps be offered to her by the loan of instruments. Heaven knows what her fate is going to be."[176] Indeed, Blau's fate was eventful. She was 44 years old when she was forced to start a new career, first in Mexico and later on in the United States.

In a letter of April 18, 1938, Albert Einstein addressed the American Association of University Women, asking "how it may be possible to find a position for Miss Blau where she can continue her research." Easther Brunauer, associate in the International Education of University Women, responded immediately but with unfortunate news. She promised to do whatever possible but, although she was not explicit, their priority was Meitner. James Franck had already informed them that she was to lose her position at the Kaiser Wilhelm Institute. Concluding her letter, Brunauer asked Einstein to keep Meitner in mind "if you hear of any opening at a research institute where you think her line of work might be developed."[177]

Einstein's attempts to secure a position for Blau in the United States were not successful. In July, Bakken, Gleditsch's assistant, returned to the institute and Blau planned to do translations in order to survive in Sweden.[178] In the meantime, Einstein arranged a position for her at the Polytechnic School in Mexico City.[179] In November, after a delay due to formalities, Blau left Sweden first for Copenhagen to visit Bohr's institute and then to London to accept an invitation from Fritz Paneth. On her way to Mexico, the Gestapo confiscated her scientific notebooks after forcing her zeppelin down in Hamburg. As Blau later speculated, the notebooks ended up in the hands of her Nazi colleagues in Vienna.[180]

With or without Blau's scientific notebooks, Wambacher continued to use the experimental facilities of the Radium Institute while her Jewish colleague was in a desperate search for a research position and depended on Pettersson to secure some of her instruments in Vienna. The break between the two women was definite. Wambacher had been an applicant for the NSDAP party since 1934 and heavily depended on Stetter for her scientific and emotional life. As Karlik acknowledged, "H[erta] W[ambacher]'s morality inside, I believe, is in a great mess."[181] Within just two weeks of the Anschluss, Wambacher was promoted to the position of assistant at the First and Second Physics Institutes, which were now combined and directed by Stetter. The following year she received her habilitation based on her work on the "nuclear disintegration through cosmic radiation in the photographic emulsions." This enabled her to become *Dozentin* in 1940, and in the winter semester of 1941–42, she started teaching at the University of Vienna. Publications in major German journals such as the *Zeitschrift für Technische Physik* and *Physikalische Zeitschrift* accompanied her rapid promotion in the university ranks. Nevertheless, by the end of the war and although her Nazi male colleagues such as Stetter, Ortner, and Kirsch maintained the power they gained during the National Socialist period, Wambacher lost her previous advantages.[182] She died in 1950 at the age of 46.

Meanwhile, Blau was in search of a permanent position far from Vienna on another continent. In 1941, after an unfortunate research period in Mexico, she tried to enter the United States for a second time.[183] In a letter of May 21, Alvin Johnson, director of the New School of Social Research wrote to Thomas Appleget at the Rockefeller Foundation concerning her case. She had just lost her position in Mexico and was looking for employment. As Johnson concluded, "I have informed her friends that our project does not cover cases in Latin America."[184] The same day, Herbert Solow passed Blau's file on to H. Miller from the Rockefeller Foundation with the following note:

> Perhaps you will be interested to know that her friend Mrs. Szego has told me that she thinks the reason for the failure of the Polytechnic School to renew the Blau contract has to do with some not too happy political shift since the last Mexican election. Dr. Blau is a Jewish refugee from Vienna and some of her relatives were Viennese Socialists. Conceivably, she could be the victim of any of a half a dozen conflicting types of factionalism.[185]

Two days later, Solow tried to put Blau in touch with Fritz Bach, director of the General de Estadistica in Mexico, asking for advice since he did not know her personally. It was Blau's friend Szego who brought the case to Solow's attention. Within three days, Blau's request was put in the drawer. In a letter of May 26, 1941, Appleget informed Johnson that there was no possibility of assistance under their present program.[186] Bach's response to Solow on June 5 sheds light on this speedy closure of her case. Despite her contract with the Polytechnic in Mexico, Blau's payment was suspended. The official reason was the lack of money. As Bach admitted, "I believe that the reasons may be different. At the Ministry of Education, the Stalinists are still strong and she, without being a Stalinist of course, has always been in close contact with them. I do not want to take care of this matter mainly because of the kind of friends she has and besides, I do not think that I would succeed."[187]

As Blau's situation was "rather delicate," both the New School for Social Research and the Rockefeller Foundation did not take the risk of pursuing her case. Blau remained in Mexico for three more years. When she finally entered the country in May 1944, she was on leave from the Escuela Tecnica Superior until December but she never returned.[188] It was probably through the efforts of the Jewish community in Mexico that Blau was able to find her first position in industry, working for the International Rare Metals Refinery in New York.[189]

GENDER, RACE, AND SCIENCE

Ideology, as the system of ideas and representations that one holds and according to which one acts, is inscribed in the everyday practices and choices of individuals. In the relevant literature, Social Democracy has been coded as political pluralism while National Socialism has often been characterized as ideology and certainly applying a negative meaning to the term. One should not forget, however, that Social Democracy in Austria during Red Vienna carried and implemented an ideological apparatus as well, advocating and using, democratic procedures. In contrast and without doubt, totalitarian and authoritarian as they were, Nazi and fascist ideologies repressed democracy and cruelly invaded the autonomy of the individual. In the totalitarian regime that Austria was after 1933, the first system to be targeted was education.

The dismantling of the Mediziner-Viertel was only a symptom of how Christian and National Socialist ideologies tried to transform Austrian society. The purging of outspoken liberal and social-democratic faculty and staff members of the University of Vienna, the racial politics enacted especially after the Anschluss, and the use of brutal violence were part of the ideological apparatus that fascists and Nazis mobilized and used to exercise their power. That ideological apparatus was used to produce students educated in race, science, and in new population policies; to transform the university through dismissals and changes in positions; and to take total control of key positions that the Social Democrats managed to obtain on their terms. By 1938 the entire range of educational reforms and social and cultural policies of Red Vienna was destroyed on the basis of an anti-Semitic and anti-Social-Democratic propaganda and exercise of political power. The dissolution of the Vienna Circle, the Ernest Mach Society, and Neurath's Social and Economic Museum, as well as the obstruction of the research agent of the Radium Institute, the Vivarium, and a number of other "Red" institutes occurred before Hitler's arrival on the Austrian scene and was accomplished by the earlier fascist regime.

During the fascist regime, changes in the Radium Institute did not directly concern its structure. Probably, because it was an institute devoted to research and not to education, the fascists had less interest in transforming the institute's internal hierarchy and in dismissing its undesirable personnel. For strategic reasons, their interest was focused on institutions and educational establishments with direct influence on the public and the young generation of students. It is indicative that most of the institute's personnel continued research in much the same manner as before. Karlik succeeded in becoming *Dozent* and the Jewish Blau shared the Lieben Prize awarded by the Austrian Academy of Sciences with Wambacher. The purge and transformation of the university's and the academy's members had not been radical yet. The fascist regime, however, thwarted Meyer's ambitions to elevate the Radium Institute to a national regulator of radium supplies for medical use and it cut the institute off from any key role it could have had at the municipal level.

After the Anschluss, science was turned into a servant of state ideology. The fate of the Radium Institute was absolutely in the hands of those who saw in politics a chance to rise in the scientific ranks and impose their world views. Stetter, Ortner, and the rest of the institute's Nazis were able to establish their order and fulfill their ambitions. What could it mean to be a physicist in such a context? Karlik's agonizing over the question of remaining in Austria and pursuing research at the institute or leaving gives a glimpse of the dilemmas experimenters had to face.

To sharpen the question: What could it mean to be a Jewish physicist at that time and particularly, a Jewish female physicist? As Doris Bergen argues, "Any study of women as outsiders in Nazi Germany and German-occupied Europe is necessarily a discussion of race; it is not possible to separate sex from blood in Nazi ideology and practice."[190] The National Socialist ideology constructed gender as intertwined with and inseparable from race. As the Blau-Wambacher case indicates, the complexity of their story cannot be captured by reducing the historical analysis to the interplay of gender and race as two distinct factors. To reduce women to either one or to suggest that both factors added to what it meant to be a woman in Austria is to argue against the complexity of how subjectivities were and are formed. What it meant to be a woman (and a man as well) was after all the outcome of women's (and men's) location within a range of different situations such as their gender, race, nationality, religion, and ideological commitments.

NOTES

1. Hughes, *The Radioactivists* (1993), 206, 222.

2. Chaloner, "The Most Wonderful Experiment" (1997); Brown, *The Neutron and the Bomb* (1997), 88.

3. Hughes, *The Radioactivists* (1993), 157; Schmidt and Stetter, "Die Anwendung" (1929); Schmidt and Stetter, "Die Ionisastion" (1930); Schmidt and Stetter, "Untersuchungen" (1930).

4. Rona, *How it Came About* (1978), 22. To reconcile Chadwick's invitation to Rona with his account of the work in Vienna, he probably believed Rona was a good technician having a skill that he could use for the experiments in his own group. On the contrary, as Przibram was fully aware of her research, he described Rona as the chemist of his group referring to the time that she worked on a joint research project with Karlik, Haberlandt, and Przibram on the fluorescence of fluorides; see Przibram, "1920 bis 1938" (1950), 32. According to her autobiography, instead of accepting Chadwick's invitation to join the Cavendish lab, Rona visited Paris in 1928; see Rona, *How it Came About* (1978), 23.

5. Rona and Schmidt, "Eine Methode zur Herstellung" (1928). The same article appeared in the *Zeitschrift für Physik* (1928). On Bonet-Maury's method, see Bonet-Maury, "Sur la vaporisation du polonium" (1927); Bonet-Maury, "Sur la vaporisation du polonium" (1927); Blau and Rona, "Anwendung der Chamie'schen photographischen Methode" (1930).

6. Karlik, "Eine Lumineszenzmethode" (1933); Karlik and Rona, "Untersuchungen der Reichweite" (1933); Karlik and Rona, "Untersuchungen uber die Reichweite" (1934).

7. Rona, *How it Came About* (1978), 33.

8. Crossfield, "Irene Joliot-Curie" (1997), 114. Curie, *Madame Curie* (1937). Rona to Meyer, July 4, 1936, AÖAW. Already in February 1934, Pettersson mentioned Rona's blood problems in his letter, wishing quick recovery. Pettersson to Rona, February 24, 1934, GUB. The letter was sent to Vienna. Pettersson to Rona, July 29, 1934, GUB. Meyer to Rona, September 12, 1934, AÖAW. Gleditsch to Rona, August 19, 1934, AÖAW.

9. Föyn, Kara-Michailova, and Rona, "Zur Frage der Künstlichen Umwandlung" (1935), 159; Föyn, Pettersson, and Rona, "Künstliche Umwandlung" (1935), 391. Föyn came to the institute at the end of 1934 after Gleditsch's arranged his visit with Meyer. He remained there for a year and worked closely with Rona; see Meyer to Gleditsch, August 18, 1934, AÖAW; Gleditsch to Meyer, August 30, 1934, AÖAW; Meyer to Rona, September 12, 1934, AÖAW. See also Almanach der Academie der Wissenschaften, (1935), 196; (1936), 213, AÖAW.

10. Rona and Neuninger, "Beiträge zur Frage" (1936).

11. Blau, curriculum vitae, Leopold Halpern Papers. See also Blau, "Bericht über die Entdeckung" (1950).

12. Blau and Kara-Michailova, "Über eine durchdringende" (1931).

13. Hertha Wambacher, curriculum vitae, Rigorosentakt, 10860, AUW. The school attracted the girls from the Viennese upper class and bourgeoisie and, not surprisingly, included more than 35 percent Jewish pupils; see Anderson, *Utopian Feminism,* (1992), 31.

14. Bischof, *Frauen am Wiener Institut* (2000), 137.

15. Hertha Wambacher, curriculum vitae, Rigorosenakt, 10860, AUW.

16. Halpern, interview by the author, March 5, 1999, AIP. See also Blau to Pettersson, September 25, 1933, GUB.

17. Wambacher, "Untersuchungen" (1931).

18. Blau and Wambacher, "Uber das Verhalten" (1932); Blau and Wambacher, "Uber Versuche" (1932), 180. In the fall, Blau and Wambacher presented a second lengthier paper on the same topic (Blau and Wambacher, "Uber Versuche II" (1932).

19. Galison, *Image and Logic* (1997), 151.

20. A few months earlier, in February 1932, Chadwick had just discovered the neutron; see Chadwick, "Possible Existence of a Neutron" (1932), 312.

21. Blau, curriculum vitae, GDSCA (in English). Kirsch and Wambacher, "Uber die Geschwindigkeit" (1933).

22. Hughes, *The Radioactivists* (1993), 158. Meyer to Hevesy, October 30, 1928, AÖAW; Meyer to Curie, October 30, 1928, AÖAW; Meyer to Fajans, November 5, 1928, AÖAW. As Curie and Hevesy mentioned in their letters to Meyer, they were willing to support Pettersson since the controversy between his group and Rutherford was not resolved; see Curie to Meyer, November 9, 1928, AÖAW; Hevesy to Meyer, November 3, 1928, AÖAW.

23. In 1914, Gustaf Ekman offered the funds for Pettersson's position after his father's intervention. Svansson, interview by the author, September 21, 2001, Göteborg. See also Deacon, "Hans Pettersson" (1966); Pettersson, "Recent Oceanographic Research" (1933).

24. Hughes, *The Radioactivists* (1993), 158.

25. Marie Curie to Meyer, November 9, 1928, AÖAW.

26. Gütachten von Siegbahn, Meyer Nachlass, AÖAW.

27. Pettersson to Mellbye, April 15, 1928 (in Swedish, Agnes Rodhe Papers, translated by Rodhe); Almanach, Akademie der Wissenscahften in Wien, (1929), 201. AÖAW.

28. Rona, *How it Came About* (1978), 60.

29. Nornvall and Svansson, "Bornö Oceanographic Station" (1998). Pettersson, "Recent Oceanographic Research" (1933). I owe my thanks to Artur Svansson for pointing out this article to me. See also Svansson, *Otto Pettersson*, 2006.

30. Svansson, interview by the author, September 21, 2001, Göteborg. Rodhe to Rentetzi, October 29, 2001.

31. Pettersson to Rona, February 24, 1934, AÖAW.

32. Rona to Meyer, September 9, 1935, AÖAW. See also Rona, *How it Came About* (1978), 63.

33. Rodhe interview by the author, September 22, 2001, Göteborg,.

34. Rona to Meyer, September 9, 1935, AÖAW.

35. Pettersson to Karlik, September 27, 1934, GUB.

36. Meyer, June 22, 1935, Karlik's file, Mitarbeiter/Assistenten, AÖAW.

37. Rona referred to the research done by James Joly which is presented in his book *Radioactivity and Geology* (Rona, *How It Came About* (1978), 63).

38. Rona, *How it Came About* (1978), 64.

39. Hernegger and Karlik, "Uranium in Sea-Water" (1935), 5.

40. Rodhe to Rentetzi, October 29, 2001. Pettersson to Karlik, September 27, 1934, GUB; Pettersson to Karlik, December 27, 1934, GUB. Rodhe to Rentetzi, October 29, 2001.

41. Hernegger and Karlik, "Uranium in Sea-Water" (1935), 4. See also Herneggen and Karlik, "Die quantitative Bestimmung" (1935).

42. Hernegger, "Methoden für einen empfindlichen Urannachweis" (1933). Hernegger and Karlik, "Uranium in Sea-Water" (1935), 5.

43. Haberlandt, Karlik, and Przibram, "Synthese der blauen" (1933), 235; Haberlandt, Karlik, and Przibram, "Synthese der grünen" (1934), 2; Haberlandt, Karlik, and Przibram, "Zur Fluoreszenz des Flurites II" (1934); Haberlandt, Karlik, and Przibram "Zur Fluoreszenz der Flurites III" (1935; Haberlandt, Karlik and Przibram, "Zur fluoreszenz der Flurites IV" (1935); Haberlandt, Karlik, and Przibram, "Artificial Production" (1934). On the work of this group, see Przibram, "1920 bis 1938" (1950), 32–33. Karlik worked with them as experts in spectroscopic analyses.

44. Hernegger and Karlik, "Uranium in Sea-Water" (1935), 5; Meyer, June 22, 1935, Mitarbeiter/Assistenten, Karlik's file, AÖAW. Bischof, *Frauen am Wiener Institut*, (2000), 107.

45. Hernegger and Karlik, "Uranium in Sea-Water" (1935), 6.

46. *Almanach der Akademie der Wissenschaften*, (1939), 192, AÖAW.Föyn, Karlik, Pettersson, and Rona "Radioactivity in Sea-Water" (1939).

47. Galison, *Image and Logic* (1997), 799.

48. With his letter to *Nature* in November 1928, Gamow addressed the phenomenon of artificial disintegration. He explained why Pettersson's and Kirsch's work in Vienna could not be reliable and stressed his agreement with the observations of the Cavendish team. Gamow, "The Quantum Theory of Nuclear Disintegration" (1928); See also Gamow, *The Constitution of Atomic Nuclei and Radioactivity* (1931); Gamow, *My World Line* (1970); Chadwick and Gamow, "The Artificial Disintegration by a-particles" (1930), 54–55; For secondary bibliography on Gamow, see Hughes, "'Mod-

ernists with a Vengeance'" (1998); Hughes, *The Radioactivists* (1993), 208; Pais, *Inward Bound* (1986); Cockcroft, "Some Recollections" (1984); Growther, *The Cavendish Laboratory* (1974), 227–28.

49. On the history of early particle accelerators, see McMillan, "Early History of Particle Accelerators" (1979); Seidel, "The Origins of the Lawrence Berkley Laboratory" (1992). Pais, *Niels Bohr's Times,* (1991). On the work of Cockcroft and Walton, see Growther, *The Cavendish Laboratory* (1974), 227–28; Cockcroft, "Some Recollections" (1984).

50. Brown, *The Neutron and The Bomb* (1997), 103. The Joliot-Curies' paper helped Chadwick to put into context a phenomenon described by Walter Bothe and Hans Becker, who worked in Berlin. They discovered that when beryllium was bombarded by alpha particles, a low intensity radiation was emitted that could penetrate a thick surface of lead. See also Crossfield, "Irene Joliot-Curie" (1997), 112; Trenn, *Transmutation Natural and Artificial,* (1981), 75.

51. Lawrence, "The Evolution of the Cyclotron" (1965).

52. Blau to Meyer, February 18, 1933, AÖAW.

53. See the names in the photo picturing the participants of the conference in Brown, *The Neutron and The Bomb* (1997).

54. Crossfield, "Irene Joliot-Curie" (1997), 115.

55. Segre, "Nuclear Physics in Rome" (1977), 51. On Meitner see Meitner, "Looking Back" (1964), 7. On nuclear fission, see Sime, *Lise Meitner* (1996); Stuewer, "The Origin of the Liquid-Drop Model" (1994); Frisch, "Experimental Work with Nuclei" (1979).

56. Bethe, "The Happy Thirties" (1979). Goldhaber, "The Nuclear Photoelectric Effect" (1979).

57. For examples of institutional and disciplinary histories, see Beyerchen, *Scientists Under Hitler* (1977); Deichmann, *Biologists under Hitler,* (1992); Macrakis, *Surviving the Swastika,* (1993); Hoffman, "Die Physikdenkschriften" (1989). For scientific biographies, see Sime, *Lise Meitner* (1996); Cassidy, *Uncertainty: The Life and Science of Werner Heissenberg* (1992). For comparative studies, see Walker and Sachse (eds.), *Politics and Science* (2005). For a general overview of science and technology under NS, see Renneberg and Walker (eds.), *Science, Technology and National Socialism* (1994).

58. On gender and racial discriminatory politics, see for example Gellately and Stoltzfus (eds.), *Social Outsiders in Nazi Germany* (2001). On the understanding of the rise of Austro-fascism and National Socialism, see Fellner, "The Background of Austrian Fascism" (1971).

59. Blau to Meyer, February 2, 1933, AÖAW.

60. Rabinbach, *The Crisis of Austrian Socialism* (1983), 33.

61. F. Stadler, "The Emigration and Exile" (1995), 14. K. Stadler, *Austria* (1971), 123; Barker, *Austria 1918-1972* (1973), 68–70.

62. Rabinbach, *The Crisis of Austrian Socialism* (1983), 81. On the Heimwehr, see Gruber, *Red Vienna* (1991), 201.

63. Barker, *Austria 1918-1972* (1973), 74.

64. Rabinbach, *The Crisis of Austrian Socialism* (1983), 156.

65. Brook-Shepherd, *Dollfuss* (1978). Stadler K., "Austria" (1969), 109.

66. On Sunday, March 13, 1938, the *New York Times* reported that Germany had entered Austria the day before and Hitler gave a speech at Linz where he proclaimed the Anschluss; see "Hitler Enters Austria" *New York Times*, March 13, 1938, 1. On the Austrian history of this period, see Rabinbach, *The Crisis of Austrian Socialism* (1983); Stadler K., *Austria* (1971); Kindermann, *Hitler's Defeat in Austria* (1988); Brook-Schepher, *Dollfuss* (1978).

67. Stadler and Weibel, *The Cultural Exodus from Austria* (1995). See also Stadler, *Vertrieben Vernunft I* and *II* (1987) and (1988).

68. F. Stadler, "The Emigration and Exile of Austrian Intellectuals" (1995), 15; F. Stadler, "The Vienna Circle and the University of Vienna" (1995); Cartwright et all, *Otto Neurath* (1996), 83; Dahms, "The Emigration of the Vienna Circle" (1995); Feigl, "The Wiener Kreis in America" (1968).

69. Handlbauer, "The Emigration of the Viennese Individual Psychologists" (1995).

70. Mühlleitner and Reichmayr, "The Exodus of Psychoanalysis from Vienna" (1995), 99; F. Stadler, "The Emigration and Exile of Austrian Intellectuals" (1995), 20; Handlbauer, "The Emigration of the Viennese Individual Psychologists" (1995); Gardner and Stevens, *Red Vienna and the Golden Age of Psychology* (1992), 80, 160. See also Ash, "Psychology and Politics in Interwar Vienna" (1987); Ash, "Women Émigré Psychologists and Psychoanalysts in the U.S." (1995).

71. Reiter, "Zerstört und vergessen" (1999).

72. We should note, nonetheless, that the number of men enrolled in the University of Vienna also dropped drastically from 1933–34 to 1938–39. Although in 1933–34, there were 8801 men enrolled, by 1938–39 their number had dropped to 4081; see Tuma, "Die österreichischen Studentinnen" (1993), 85.

73. Tuma, "Die österreichischen Studentinnen" (1993), 88.

74. Heindl, "Die konfessionellen" (1993); Freidenreich, "Gender, Identity, and Community" (1998), 154.

75. For example, in 1925, Hugo Bettauer, a democratic novelist who championed women's rights through his sexual reform magazines, was killed. Bettauer's assassination was the result of a hate campaign of the Christian Socialists and the Chancellor Ignaz Seipel who demanded that the municipal government exercise censorship on Bettauer's writings; see Gruber, *Red Vienna* (1991), 165; F. Stadler, "The Emigration and Exile" (1995), 15.

76. Gruber, *Red Vienna* (1991),. 27.

77. The directors of the Vivarium to the Kuratorium of the Vivarium, June 2, 1932, AÖAW. Wilhelm Figdor and Eugen Steinach were the directors of the physiological and plant departments respectively but not of the whole institute.

78. *Almanach der Akademie der Wissenschaften in Wien*, Biologische Versuchsanstalt, (1914), 231, AÖAW.

79. Reiter, "Zerstört und vergessen" (1999), 605.

80. Hubenstorf, "Anatomical Science in Vienna" (2000), 1386.

81. Eugen Steinach, Rigorosentakt, AUW; Benjamin, "Eugen Steinach, 1861-1944: A Life of Research" (1945).

82. Sengoopta, "Glandular Politics" (1998).

83. Sengoopta, "'Dr Steinach coming to make old young!'"(2003).

84. The directors of the Vivarium to the Kuratorium of the Vivarium, June 2, 1932, AÖAW; Meyer, "Die Vorgeschichte" (1950), 12–13.

85. Meyer, "Die Vorgeschichte" (1950), 19.

86. In the course of Red Vienna, the Vivarium received financial support from the city; see *Almanach, Akademie der Wissenschaftern in Wien* (1922), 180; (1923), 155; (1924), 189; The mayor Karl Seitz supported the installation of general heat system in the *Vivarium*, Almanach, Akademie der Wissenschaftern in Wien (1926), 209; (1927), 205; (1928), 227.

87. Reiter, "Zerstört und vergessen" (1999), 596.

88. In 1922, Jahoda co-authored a paper with the radiologist Guido Holzknechkt on the luminosity of x-rays; see Agetter, *Guido Holzknecht* (1998), 69. At the institute, Jahoda was working at Przibram's group.

89. Meyer, "Physikalische Grundlagen" (1929), 557–80; Meyer and Suess, "Zur Verwendung" (1930); Deinlein, "Verweilzeiten" (1933); Maria Renata Deinlein, Rigorosenakt 11725, AUW. Erhard Suess was a physician and lung specialist. Meyer to Kautz, May 11, 1934, AÖAW.

90. Meyer to Kollassa, February 3, 1933, AÖAW.

91. The directors of the Vivarium to the Kuratorium of the Vivarium, June 2, 1932, AÖAW.

92. Sengoopta, "Glandular Politics" (2003).

93. Protocol book of the Austrian Academy of Sciences, 1932, AÖAW. I thank the archivist Stefan Sienell for directing my attention to the protocol book of the academy.

94. Arnold Durig, Rigorosenakt, 3, AUW.

95. Przibram and Portheim to Molish, June 6, 1932. Vivarium file, AÖAW.

96. Bundesminister to the Präsidium der Akademie, July 1, 1932; Eugen Steinach, Rigorosenakt, AUW.

97. Reiter, "The Year 1938" (1995), 198.

98. Urbach, "Einiges aus dem Physikalischen Laboratorium" (1933), 537n.

99. *Almanach der Akademie der Wissenschaften* (1935), 197.

100. Urbach, "Über eine spontane Veränderung" (1935). Reiter, "The Year 1938," (1995), 198. Urbach's wife, Anni Urbach, entered the institute in 1935; see *Almanach der Academie der Wissenscaften* (1935), 197. Daughter of the psychoanalyst Paul Federn, Freud's collaborator, Anni studied law and worked in a juvenile court. After her marriage to Urbach, she turned to physics. No publication appeared in the *Mitteilungen* under Anni Urbach's name. In the institute's kassa where Meyer recorded the finances, Anni appeared to have received minor amounts of money, probably for performing technical tasks (Kassa, 1933–38, AÖAW. She remained in the institute until their immigration in 1939.

101. Hilda Maier, Lebenslauf, undated, Personalakt, AUW.

102. As Fernau informed Meyer, Frederick Flinn, a radiologist from Columbia University in New York, visited him on August 25, 1932 (Fernau to Meyer, August 26, 1932, AÖAW). Flinn was involved in the case of Eben Byers, whose death on March 30, 1932 shocked the medical community in the United States and prompted legislation acts concerning radium products. Since 1927, Byers, an internationally known industrialist from Pittsburgh, had been drinking a radium product called Radithor as a tonic. He soon suffered from radium poisoning with symptoms similar to those of the radium girls, the famous case of the women licking their brushes while painting the dials of watches with luminous radium paint; see Clark, *Radium Girls* (1997); Rentetzi, "Women Radium Dial Painters" (2004). Flinn was involved in both cases and tried in vain various methods to remove the radium from Byers' body; see Mullner, *Deadly Glow* (1999), 114–118. Although in the radium dial painters' case, Flinn refused to recognize the radium hazards working for the corporation that hired the women, in Byer's case he issued strong warnings against the use of radioactive products for internal use. The national scandal of Byers' death prompted Flinn to travel to Vienna and also to visit Paris and Berlin hoping to discuss with the international scientific community questions concerning radon therapy (Fernau to Meyer, August 26, 1932, AÖAW).

103. Fernau to Meyer, September 19, 1932, AÖAW.

104. Hilda Fonovits-Smereker, Lebenslauf, Personalakt, AUW. A coauthored article appeared in *Strahlentherapie* in 1933 on Byers's case. Fernau and Smereker, "Über das Verbleiben" (1933).

105. Meyer to Rona, September 12, 1934, AÖAW.

106. Hilda Fonovits-Smereker, Lebenslauf, Personalakt, AUW. See for example the articles published up to 1938. Smereker and Juris, "Messung der Beta-Strahlung" (1935); Schloss and Smereker, "Zur Radiumbehandlung" (1936); Smereker, "Untersuchungen" (1937), 267; Smereker, "Dosimetrische" (1937), 676; Smereker and Juris, "Versuche über die nicht directe Ionisierung" (1938).

107. Emile Maier, Lebenslauf, Personalakt, AUW; Emile Maier, Personalblatt, AUW.

108. Hilda Fonovits-Maier, Heiratsurkunde, Personalakt, AUW.

109. Meyer to Dekan, March 21, 1933, Karlik's file. Mitarbeiten/Assistenten, AÖAW; January 10, 1934. Berta Karlik's file, Mitarbeiten/Assistenten, AÖAW.

110. Hittner, *Gescichte des studienfaches* (1949), 210. The reasons for his retirement are not stated.

111. Stetter to Pettersson, March 5, 1934, GUB. Karlik conveyed the same information to Pettersson four days later without giving any specific reasons for the decision of the ministry (Karlik to Pettersson, March 9, 1934, GUB).

112. Stetter to Pettersson, March 5, 1934, GUB.

113. Karlik to Pettersson, March 9, 1934, GUB.

114. Kommissionsbericht n. 59, Georg Stetter, Personalakte, AUW; The Dean to the ministry of education, December 20, 1933, n. 65, Georg Stetter, Personalakte, AUW.

115. Galison, *Image and Logic* (1997), 153.

116. Galison, *Image and Logic* (1997), 153.

117. Karlik to Pettersson, September 13, 1933, GUB, (in English); Pettersson to Karlik, September 9, 1934, GUB, (in English); Karlik to Pettersson, April 15, 1934, GUB, (in English).

118. Karlik to Pettersson, May 12, 1936, GUB (in English).

119. Berta Karlik, Rigorosenakt 9765, AUW. Hittner, *Geschichte des studienfaches* (1949), 79.

120. Karlik to Pettersson, May 12, 1936, GUB, (in English).

121. Karlik to Pettersson, May 12, 1936, GUB, (in English); Pettersson to Karlik, August 3, 1936, GUB.

122. Bischof, *Frauen am Winer Institut* (2000), 108; Lintner, "Berta Karlik Nachruf" (1990), 304; Vorlesungen an der Universität zu Wien, 1937/18, AÖAW.

123. Halpern, "Marietta Blau" (1993), 57; Galison, *Image and Logic* (1997), 149.

124. For some hints on Blau's political engagement, see Miller to Solow, May 21, 1941, GDSCA; Solow to Bach, June 5, 1941, GDSCA; Interview of Arnold Perlmutter to the author, 1999, deposited in the AIP.

125. Blau, curriculum vitae, Leopold Halpern Papers For a detailed description of their experimental work, see Galison, *Image and Logic* (1997), 152.

126. Karlik to Pettersson, February 11, 1936. GUB. (in English).

127. Marietta Blau, curriculum vitae, Leopold Halpern Papers.

128. Galison, *Image and Logic* (1997), 154.

129. In her self-description written around 1963, Blau mentions that she received the Haitinger Prize of the academy in 1936. However, according to the records of the academy, Blau received the Lieben Prize in 1937; see *Almanach der Academie der Wisssenschaften* (1939), 136; Bischof, *Frauen am Wiener Institut* (2000), 79; Pettersson to Karlik, 7 June 1937, GUB. From 1865 to 1937, Blau and Wambacher were the only women besides Lise Meitner (1925) who received the Lieben Prize. The only women from 1905 to 1938 who received the Haitinger Prize of the academy were Rona and Karlik, both in 1933; see *Almanach der Akademie der Wissenschaften* (1939), 141.

130. Marietta Blau, curriculum vitae, Leopold Halpern Papers. Given the political events of the 1938, the two women never actually used the grant.

131. Karlik to Hans Pettersson, December 30, 1937, GUB, (in English). I was not able to locate Heisenberg's letters to Blau and Wambacher.

132. On Meyer's personal difficulties under the Nazis, see Reiter, "Stefan Meyer" (2001); Sime, *Lise Meitner* (1996), 287–88.

133. Karlik to Pettersson, December 30, 1937, GUB. The attempt to find a solution to Blau's problem was Karlik's initiative and nobody at the institute knew the content of her correspondence to Pettersson, not even Blau.

134. Karlik to Pettersson, February 3, 1938, GUB.

135. Karlik to Pettersson, February 3, 1938, GUB.

136. Pettersson to Karlik, February 17, 1938, GUB.

137. Blau to Bohr, March 5, 1938, NBA. Blau confirms to Bohr the information she got from Rona on the arranged colloquium in Copenhagen.

138. January 20, 1938, Gerhard Kirsch, Rigorosenakt, 92, AUW.

139. Pettersson to Karlik, February 20, 1938, GUB; Karlik to Pettersson, February 24, 1938, GUB (in English).

140. "Hitler Enters Austria in Triumphal Parade," *New York Times*, March 13, 1938; Blau to Paneth, March 21, 1938. AGMPG.

141. Karlik to Pettersson, March 14, 1938, GUB.

142. Karlik to Pettersson, March 30, 1938, GUB.

143. Reiter, "Stefan Meyer" (2001), 122; Sime, *Lise Meitner* (1996), 287–88.

144. Karlik to Pettersson, March 14, 1938, GUB.

145. Karlik, "1938 bis 1950" (1950), 35.

146. Karlik, "1938 bis 1950" (1950), 35.

147. Reiter, "The Year 1938" (1995), 195.

148. Karlik to Pettersson, April 9, 1938, GUB; Karlik to Pettersson, March 19, 1938, GUB.

149. Galison, *Image and Logic* (1997), 159.

150. Karlik to Pettersson, July 2, 1938, GUB (in English).

151. Karlik to Pettersson, May 1, 1938, GUB (in English).

152. Karlik to Gleditsch, May 2, 1938, AÖAW.

153. Karlik, "1938 bis 1950" (1950), 36. As Galison points out, Stetter was a member of the commission that met in May 1938 to consider the restructuring of physics in Vienna; see Galison, *Image and Logic* (1997), 158n.

154. Karlik to Pettersson, July 20, 1938, GUB; Karlik to Gleditsch, October 16, 1938, AÖAW.

155. As Karlik reported to Pettersson, Max Kindinger was involved in the NSDAP with great enthusiasm after the Anschluss; see Karlik to Pettersson, March 19, 1938, GUB. For Jentschke and Schintlmeister, see Bischoff, *Frauen am Wiener Institut*, (2000), 140.

156. Karlik to Pettersson, July 20, 1938, GUB (in English).

157. Pettersson to Karlik, March 30, 1938, GUB (in English); Karlik to Pettersson, July 20, 1938, GUB, (in English).

158. *Alamanch der Akademie der Wissenschaften*, (1938), 193–4; (1939), 192–93.

159. Karlik to Pettersson, July 20, 1938, GUB, (in English).

160. Karlik to Pettersson, May 1, 1938, GUB, (in English).

161. Karlik to Pettersson, May 11, 1938, GUB; Karlik to Pettersson, July 20, 1938, GUB; Karlik to Gleditsch, September 11, 1938, AÖAW, (in English).

162. Berta Karlik, Curriculum vitae, 1949, Rigorosenakt, 104, AUW; Karlik and Bernert, "Über eine vermitete" (1942), 685; Karlik and Bernert, "Zur Frage" (1942); Karlik and Bernert, "Über die Entemanierung" (1942), 267; Karlik and Bernert, "Über eine dem Element" (1943); Karlik and Bernert, "Das Element 85" (1943), 51; Karlik and Bernert, "Eine neue natürliche alpha Strahlung" (1943), 298; Karlik and Bernert, "Ein weiterer dualer Zerfall" (1943), 492; Karlik and Bernert, "Das Element 85" (1944), 44; Karlik and Bernert, "Über swei neue Alpha-Strahlungen" (1944), 2–3; Karlik and Bernert, "Entsehung" (1945), 34–35; Karlik, "1938 bis 1950" (1950), 37–38.

163. On Karlik's scientific career and life, see Bischoff, *Frauen am Wiener Institut* (2000), 101–119; Lintner, "Berta Karlik Nachruf" (1990). Vogt, "Women Members of the Academies of Science" (2000).

164. Karlik to Pettersson, April 9, 1938, GUB; Karlik to Pettersson, May 11, 1938, GUB, (in English).

165. Karlik to Gleditsch, April 7, 1938, AÖAW (in English).

166. Rona and Schintlmeister "Untersuchung der Alphastralung" (1938), 49–62. Schintlmeister worked closely with Stetter starting in 1934 (Karlik to Pettersson, March 9, 1934, GUB). During the National Socialist period, Schintlmeister became Stetter's assistant and was promoted to *Dozent* for experimental physics in 1939 while he was a member of the NSDAP (Bischof, *Frauen am Wiener Institut* (2000), 140).

167. Karlik to Gleditsch, April 7, 1938, AÖAW.

168. Pettersson to Karlik, March 30, 1938, GUB (in English); Karlik to Pettersson, July 27, 1938, GUB.

169. Rona to Meyer, July 7, 1938, AÖAW. It is not clear from Rona's letter what kind of industry Vatur was.

170. Rona to Meyer, October 3, 1938, AÖAW.

171. Gleditsch to Karlik, January 17, 1939, AÖAW (in English); Rona, *How it Came About* (1978), 42–43.

172. Rayner-Canham M. and G., "Elizabeth Rona" (1997), 214.

173. Rona, *How it Came About* (1978), 53.

174. Rona, *How it Came About* (1978), 56–7.

175. Pettersson to Karlik, March 30, 1938, GUB; Blau to Bohr, March 5, 1938, NBA.

176. Pettersson to Karlik, March 30, 1938, GUB, (in English); Karlik to Pettersson, May 11, 1938, GUB, (in English). Karlik sent a microscope table to Gleditsch in Oslo, on March 25, after Blau's departure from Vienna (Karlik to Gleditsch, April 7, 1938, AÖAW). She also planned to send a counter for Blau and Föyn to repair and work with it (Karlik to Gleditsch, May 2, 1938, AÖAW). Karlik to Pettersson, May 1, 1938, GUB, (in English).

177. Einstein to McHall, April 18, 1938, AAUW; Brunauer to Einstein, April 22, 1938, AAUW.

178. Karlik to Pettersson, 11 May 1938, GUB, (in English); Karlik to Pettersson, July 17, 1938, GUB, (in English).

179. Galison, *Image and Logic* (1997), 155.

180. Blau, curriculum vitae, GDSCA.According to Halpern, the Gestapo confiscated Blau's scientific notebooks in 1938 as she was leaving Germany from Hamburg. The fate of those notebooks is not clear. However, Stetter's and Wambacher's later publications indicate a relation of Blau's missing scientific notes. As Galison suggests "although we may never be able to confirm this, we can know something of Wambacher's attitudes in the years of Nazi rule." (Galison, *Image and Logic* (1997), 157). Karlik's letters to Pettersson suggest that Wambacher and her Nazi colleagues seized Blau's research much earlier than the Gestapo's confiscation.

181. Galison, *Image and Logic* (1997), 157; Karlik to Pettersson, December 30, 1937, GUB, (in English). Karlik to Pettersson, February 24, 1938, GUB, (in English).

182. Wambacher, "Kernzertrümmerung" (1940); Wambacher, "Mehrfachzertrümmerung" (1938); Wambacher, "Mehrfachzertrümmerung" (1938); Wambacher,

"Wirkung" (1939); Stetter and Wambacher, "Neuere Ergebnisse" (1939); Wambacher, "Höhenstrahlung und Atomkernbau" (1940). Bischof, *Frauen am Wiener Institut* (2000), 139.

183. For Blau's work in Mexico, see Galison, *Image and Logic* (1997), 155–6.

184. Johnson to Appleget, May 21, 1941, GDSCA.

185. Solow to Miller, May 21, 1941, GDSCA.

186. Solow to Bach, May 23, 1941, GDSCA; Appleget to Johnson, May 26, 1941, RAC.

187. Bach to Solow, June 5, 1941, GDSCA.

188. Blau to Venegas, April 14, 1944, E.S.I.M.E. Blau asked the director of E.S.I.M.E. to grant her a leave of absence from May 1 to May 31, December 1944.

189. The *Comite Central Israerita de Mexico* contacted the World Jewish Congress in New York as well as the Canadian Government Trade Commissioner asking for a possible position on behalf of Blau (Glikowski to Tartakower, no date, CDICAM; Lisker to the Canadian Government Trade Commissioner, February 10, 1942, CDICAM). They also financially supported Blau when her mother was sick and in need of constant attention during the difficult years of 1942–43. (Blau to Comite Central Israelita de Mexico, November 3, 1942, CDICAM; Glikowski to Blau, May 17, 1943, CDICAM).

190. Bergen, "Sex, Blood, and Vulnerability" (2001), 273.

7. MARIETTA BLAU ON THE MARGINS OF NUCLEAR AND PARTICLE PHYSICS

With the end of the Second World War, the first attempt to bring the scintillation counter back into nuclear research coincided with the passage from what Galison has described as the image to the logic tradition in physics. He argues, "What transformed the scintillator's flash and Cerenkov's glow into basic building blocks of the logic tradition was the electronic revolution begun during the war. When attached to the new high-gain photomultiplier tubes and strung into the array of amplifiers, pulse-height analyzers, and scalers that emerged from the Rad Lab and Los Alamos, then and only then did the scintillator and Cerenkov radiation became part of the material culture of postwar physics."[1] Interestingly enough, one of the first to suggest the use of a photomultiplier in combination with the scintillation counter was Blau. Nurtured in the material culture of the Radium Institute before the war, she sought possibilities for professional existence in saving the scintillation counter. It was through this instrument that Blau mingled the competing prewar and postwar cultures in physics research.

FROM EXPERIENCED EXPERIMENTER TO INDUSTRIAL DESIGNER

Although Blau found herself "at the periphery of the American research establishment," as Arnold Perlmutter has argued, it still led her "to an explosion of creative activity."[2] Her first position in the United States was at the physics department of the International Rare Metals Refinery, Inc, a corporation located in Mount Kisco, New York. The company processed pitchblende ore for the African Metals Corp to extract uranium and for the Ministry of Economic Development to recover radium and uranium. Blau teamed up with B. Dreyfus in combining the use of a photomultiplier tube to a scintillation screen for the measurement of alpha ray sources.[3] As the references to Kara-Michailova's and Karlik's work show, the driving force in designing the device was Blau. Drawing on her work and that of her colleagues at the Radium Institute in Vienna from more than a decade ago, Blau relied on her past to secure her present.

In 1933, after abandoning the ordinary scintillation counter, Karlik worked on the determination of alpha particle ranges, utilizing a photoelectric cell while she kept the fluorescent screen as the intact part of the instrument.[4] Yet, as Blau explained, Karlik's method was seldom used because of the limited range of measurements of the ordinary photocells and the lack of adequate and constant alpha-standards. "These two inconveniences," Blau continued, "have been remedied recently, thanks to the appearance of the multiplier photo-tube, and of good standards."[5]

The continuity of technology traced from Dagmar Pettersson's version of the scintillation counter in 1923 to Karlik's 1933 counter equipped with a photocell and Blau's 1945 detector endowed with photomultiplier is striking. Karlik replaced the microscope used in the early type of the scintillation counter attached directly to the scintillation screen with a photocell that Blau eventually turned into a photomultiplier.[6]

By 1933, the shift from the microscope to the photocell was an attempt to save the scintillation counter by excluding the observer. Instead of the fragile and unreliable human optical system, Karlik introduced a sensitive electric device for the detection of the scintillations. The method was barely noticed, however, not only because of its limitations but mostly due to the fact that the scintillation counter had already failed in the experimental tradition of the institute. The end of the Vienna and Cambridge controversy also brought about the end of the counter. Political upheavals soon followed in the country and greatly affected life at the Radium Institute, particularly by determining specific research directions that did not include the scintillation counter.

After the end of the war, the second shift, this time from the ordinary photocells to photomultipliers in Blau's experimental practice, was not just a simple replacement between two pieces in an instrument. The transformation was a more profound, conceptual one for both the experimenter and her instrument. From a research-oriented position in the Radium Institute in Vienna, Blau's occupation shifted to industrial physics in the postwar United States. Working in Meyer's institute, Blau was challenged by the scientific community to legitimize her theories and instruments. In her new position, she was struggling for her mere existence. Her new concerns were the "wide range of applications" and the possibilities the photomultiplier scintillation counter offered for "quick industrial measurements."[7] It is relevant that around the time Blau was hired at the International Rare Metals Refinery, Inc, the company had shifted to primarily producing radium for commercial and medical use.

Putting together the photomultiplier with a fluorescent screen and using strong polonium sources, Blau and Dreyfus had in fact described the first electrically modified scintillation counter. In the scientific literature, there was no other previous reference to such a device for the detection of radioactive emissions.[8] During the war, Samuel Curran and W. Baker had assembled a detector of alpha particles using a photomultiplier, but their report was classified as part of the Manhattan Project and in collaboration with the Radiation Laboratory in California. Although issued on November 17, 1944, it was not published until February 1948.[9] Certainly unaware of the previous detector, Blau and Dreyfus were the first to describe the method in the open literature. Used primarily as a detector of alpha particles, however, the multiplier phototube was not limited to alpha or beta measurements, as the authors argued. "In a later article, we will describe its application to the measurements of strong neutrino sources."[10]

Instead of exploring the capacities of her new device, Blau was forced to shift her research in another direction. Working for competitive industrial corporations in the 1940s was not quite the same as conducting research in the welcoming atmosphere of the Radium Institute in the early 1920s. More precisely, the corporations that Blau worked for from 1944 to 1948 were deeply involved in the manufacture of nuclear weapons, the commerce of uranium and radium, and the industrial uses of radium.[11]

Under the pressure of producing industrial devices, Blau never published her promised work on the measurements of neutrino sources. Her next article, written in 1946 and co-authored by I. Feuer, was on the production of radioactive light sources.[12] In the beginning of 1948, Blau moved to the Gibbs Manufacturing and Research Corporation and with R. Carlin, she published on the industrial applications of radioactivity. A number of radioactive devices serving as resistors, electrostatic voltmeters, leveling systems, and micrometers took up

Blau's creative time. It is not by chance that they were advertised as "representative examples of the forerunners of a wide range of industrial applications."[13]

In her effort to find an appropriate research position, Blau moved again within the next few months, this time to the Canadian Radium and Uranium Corporation (CRUC) as head of the research laboratory.[14] Her research experience at Holzknecht's Radiological Institute in Vienna in the early 1920s was now put to use. Carrying over her knowledge of medical physics to the Radium Corporation, Blau designed a photomultiplier scintillation counter for medical use. In a paper published in 1948 and co-authored with J. Smith, she argued that "with the increased availability of radioactive isotopes for medical, biological and industrial research, the problem of suitable instrumentation and consequently units of measurement has presented itself. There is the need for a practical and rugged instrument for routine measurements covering a wide unit in which to express beta radiation."[15] Designed for "persons not very familiar with radioactive measurements," Blau's scintillation counter was a convenient and practical instrument for wide use in hospitals and medical laboratories. Despite the fact that she was the first to design and suggest medical applications of the photomultiplier scintillation counter, Blau remained a peripheral figure, trapped in the competitive world of industrial physics.

Just one year earlier, the efforts to design scintillation counters by replacing the human agent with a reliable and efficient photomultiplier were already at their peak. J. Coltman and Fitz-Hugh Marschall from the Westinghouse Research Laboratories described a photomultiplier scintillation counter for detecting and measuring alpha, beta and gamma rays, and high-energy electrons and neutrons. Kuan Han Sun from the same lab proceeded Blau in extending the detector to neutron measurement.[16]

Shortly after, Hartmunt Kallman, from the Kaiser Wilhelm Institute for Physical Research in Berlin, and his student Immanuel Broser greatly advanced the technique by using naphthalene as a fluorescent screen.[17] Kallmann's expertise on the photomultiplier scintillation counters served as a passport to the United States. In 1948, he moved to the U.S. Army Signal Corps Laboratory in Belmar, New Jersey, as a research fellow. In 1949, he was appointed director of the Radiation and Solid State Laboratory at New York University's Physics Department.[18] The zenith of the photomultiplier era came with the work of Kallmann's student Robert Hofstadter, who left Princeton University to work with Kallmann in New York. As Perlmutter points out, "The development of scintillation counters by Robert Hofstadter was a critical component of his experiments on the scattering of (then) high-energy electrons (600 ev) from heavy protons and heavy nuclei during the 1950s. He received a Nobel Prize in Physics for his work in 1961."[19]

By the end of the 1940s, Blau had already lost her chance to play a central role in the uses and applications of the scintillation counter. In 1947, Blau was still not satisfied with her work at the Canadian Radium and Uranium Corporation and made several attempts to find a position at a university or a research institution. In November 1947, both the Agricultural and Mechanical College of Texas and North American Aviation Inc expressed their interest to hire her. Boris Pregel, director of the CRUC, promised her "Although you seem to doubt it, we will be doing scientific work. You know my interest in scientific work and I will not give up easily."[20]

Even as Pregel was trying to persuade her to stay at his corporation, Blau received a two-year contract from Columbia University to conduct research on improvements in the use of photographic emulsions for the study of high-speed particles. The research was initiated by the newly established Atomic Energy Committee and was carried out at both Columbia and the Brookhaven National Laboratory. Blau's work was supposed to deal directly with the usefulness of the cyclotron which was then being constructed at Brookhaven, but since the accelerator needed about a year to be ready, Blau was assigned to Columbia laboratories.[21] Her contract actually began on January 1, 1948.

But as Blau remained faithful to the experimental tradition of the 1920s and nostalgic for her work at Meyer's Institute, she was unable to continue her research in the new settings. From the cozy Radium Institute to the impersonal thirteenth floor of the Pupin Physics Laboratories of Columbia, Blau had gone a long way. When she finally moved to Brookhaven two years later, she had only restricted access to the high-energy physics facilities and found herself removed from any cooperating group. Once again, Einstein took the initiative to help her. In a letter of January 5, 1954, he reminded Samuel Goudsmit, director of the lab that "It is well known that Marietta Blau has shown really original achievements. However, it would be very unfortunate if such a personality would be condemned to inactivity due to the shortage of scientific tools . . . Of course, Marietta Blau does not know about this letter."[22] Goudsmit's response left no doubt that the days of independent work were already gone:

> The difficulties encountered by Dr. Marietta Blau can easily be formulated but are hard to solve. Physics has changed so drastically from the days of simple experimentation that group work has become an unfortunate necessity. Miss Blau's temperament is not adapted to the type of regimentation which occurs nowadays when only intense cooperations make it possible to obtain meager results from a tremendously expensive piece of apparatus.[23]

It was at that point that Blau was marginalized in the world of science. In contrast, male scientists such as Kallmann, Broser, and Hofstadter acquired privileged positions in prestigious research universities and centers instead of industrial laboratories. For Kallmann and his research students, the shift from small to big science and the growth of large-scale research came smoothly, giving them a chance to adjust to the new status of physics research. With the war over, the scintillation counter became a powerful instrument in a number of different disciplines. The new technology was essential for high-energy physics research, weapons control and guidance systems, civilian mass communication, and medicine.[24] In 1949, portable scintillation counters were developed for fieldwork in geology and for the detection of uranium and radium ores.[25] As Hans-Jörg Rheinberger has shown, the production of the liquid scintillation counter in 1953 "opened new epistemic dimensions for radioactive experimentation in biology and medicine."[26] In the 1970s, the instrument was finally transformed into a generic technology in molecular biology and medical laboratories. Interestingly, a report issued in 2003 by the Human Resources Development group in Canada concerning medical radiation technologists showed that 77 percent of the personnel who operate scintillation counters and other kinds of radiation detection equipment were women.[27]

THE NOBEL PRIZE AND THE CULTURE OF POSTWAR PHYSICS

On January 8, 1950, while Blau was struggling to survive in the world of high-energy physics, Erwin Schrödinger addressed the Nobel Committee of Physics. "I herewith propose that this year's Nobel Prize for physics be given to Marietta Blau and Hertha Wambacher for experimental work conducted and published in common." Stressing the importance of their method, namely the photographic emulsions technique, he continues, "The method has become of increasing importance in the last years both for studying those nuclear explosions and for revealing details of the nature of ultimate particles and their decay into one another. It has become one of the most important tools for promoting our knowledge of the constitution of matter." As Schrödinger argued, Blau and Wambacher were certainly worthy candidates for the prize as they had priority for using the method and of discovering the first "stars," results of the disintegration process of nuclear particles originated in cosmic radiation.[28]

That year, the committee had to evaluate 28 nominees, among them important figures such as Hans Bethe (Nobel Laureate in 1967), Arnold Sommerfeld who had been nominated 81 times between 1901 and 1950 (although he never won), Pyotr Kapitza (Nobel Laureate in 1978), and Cecil Powell. In general,

Nobel prizes in the physical sciences are awarded in a three-stage process which involves the committees for physics and chemistry, the relevant section of the academy, and a final meeting of the whole academy. Each nominee is examined in a special report prepared by a committee member.

Three of the five members of that year's committee belonged to the Siegbahn school of X-ray spectroscopy. The chairman was Manne Siegbahn. He was a Nobel Laureate in physics in 1924 for his research in X-ray spectroscopy and Pettersson's earlier critic. Siegbahn had dominated Swedish physics with his authoritarian style since the 1920s, influencing university appointments and research orientations. Svante Arrhenius referred to him as the "small pope" of Swedish science.[29] Siegbahn's former student and chair of experimental physics at Stockholm University, Erik Hulthén, and his successor at Uppsala University, Axel Edvin Lindh, were also members of the committee. The fourth experimentalist was Gustaf Ising, the prominent Swedish physicist who proposed the first accelerator that used time-dependent fields in 1924. Thanks to Rolf Wideroe, Irsing's concept was tested and demonstrated through the design of the first RF linear accelerator (LINAC) in 1927. The fifth member, Ivar Waller, was the only theoretician in the committee and a professor of theoretical physics at Uppsala University. The remaining member was Carl Olof Gudmund Borelius, a professor of physics at the Royal Swedish School of Technology. He was called in as an additional member to participate in discussions and advise on decisions that year. Borelius had been a member of the Royal Swedish Academy of Sciences since 1942.[30]

The evaluation of Blau and Wambacher was conducted by Lindh as the most appropriate to provide information on their experimental work and guide the committee members to their final recommendation. His thorough and largely neutral account of their writings indicates that the two women were indeed the first to use the photographic method to demonstrate cosmic radiation and the particle effects of the disintegration. Lindh, however, diminished their achievement by stressing two things: first, the improvements of the photographic plates that the Ilford laboratory had accomplished, and second, the fact that disintegration stars had been previously observed by other physicists such as Carl Anderson and his student Seth Neddermeyer by means of the cloud chamber. Lindh therefore argued that the work of the two Viennese women did not add up to a Nobel Prize.[31]

Blau had previously acknowledged the role of the photographic giant Ilford in thickening emulsion layers on the photographic plates to make the entire particle tracks visible. Despite Ilford's help, she insisted that still thicker emulsion layers had to be obtained as the thinness of the plates continued to put geometric constraints on the capture of the particles' paths. In 1932, Blau and Wambacher discovered that pinacryptol yellow sensitizes photographic emulsions to energetic protons. It was not until 1935 that the Ilford laboratories developed independently photographic emulsions sensitive to energetic protons without preliminary immersion in pinacryptol yellow.[32] Although important, Ilford's contribution was not enough by itself to ensure Blau's and Wambacher's experimental success.

Concerning Lindh's second argument, in the summer of 1936, Anderson and Neddermeyer had mounted their cloud chamber on an old flatbed truck and drove it to the top of the Pike's Peak. During their month and a half of experimenting, they took about 10,000 photographs, but, as Maurice Shapiro explained in a later interview to Laurie Brown, in the course of a whole summer of expansions of the cloud chamber, Anderson's and Neddermeyer's results were rather limited. Their photographic plates revealed clusters of tracks but would usually not see the interaction. Whereas the chamber discovered particles and processes only during the brief moments of exposure, the photographic method had the advantage of registering continuously. It was around the same time that Shapiro was advised to look into Blau's and Wambacher's work.

> I became aware that this technique [photographic emulsion] had been used with some success in collecting quite impressive accounts of data statistically on the "stars" and that this photographic emulsion technique had the marvelous attribute of being a completely visual technique in that the stars would have the vertex and would be generated somewhere in the middle of this emulsion. Then when the particle emerged, they were all visible.[33]

In his 1941 scrupulous review of the photographic emulsion method, Shapiro made it clear that Blau and Wambacher were the first to observe the simultaneous ejection of several particles from a nucleus on plates exposed to cosmic radiation. Two years later, commenting on the history of the method, Cecil Powell also acknowledged that "notably" Blau and Wambacher have employed the method for experiments in nuclear physics.[34] So why did Lindh present this two-pronged argument against an award for Blau and Wambacher?

It is all too easy to argue that Lindh and that year's Nobel committee discriminated against the two female scientists, but one thing is clear. From 1901 to 1950, the Nobel population—those people who were nominated for the prizes in physics and chemistry—was predominantly male with only eight female candidates, of whom just three became prize winners. Lindh's report does not indicate a direct gender bias and Siegbahn's previous opposition to Meitner's Nobel Prize award can only be a fragile indication of systematic gender discrimination.[35] But the subtle discrimination against female scientists that occurred after the end of the Second World War and the changing culture of high-energy physics are powerful parts of Blau's and Wambacher's history.

In a splendid royal ceremony on December 10, 1950, the Nobel Prize was finally awarded to Cecil Powell, the physicist who turned Blau's technique into an "extremely effective aid," in Lindh's words, for the study of mesons. Throughout its report, the Nobel committee seemed intent all along on awarding the prize to someone who was doing follow-up work to the previous year's prize winner, Hideki Yukawa, on the existence of mesons. Powell had been proposed by the largest number of nominators, among them Enrico Fermi. As he had experimentally demonstrated Yukawa's theory, he appeared to be the most obvious candidate. As the committee bluntly put it, "Powell partly developed the photographic method, deployed earlier, into an outstanding, useful tool for the study of nuclear particles and processes and partly by using this method, has made foundational discoveries concerning mesons and their properties. Powell has thereby laid the grounds for a new era in research on mesons."[36]

Located at his laboratory at the University of Bristol, Powell utilized Blau's work in Vienna to undertake experiments both in cosmic radiation and, by using the Bristol high-tension generator, to produce a powerful beam of needed particles. Starting in 1938, he expanded his method by dividing the emulsion into sectors and assigning each sector to a different group of scanners, all of whom were female observers. In a wonderfully written account, Galison has demonstrated how Powell turned an old technique of an unsystematic cosmic ray research into a tool for the detailed analysis of accelerator-produced as well as cosmic ray nuclei.[37] By 1947, Powell and his collaborators had discovered the existence of primary and secondary mesons, determined their masses, and explored their properties.

Through the work at Bristol, the reliability of the photographic method increased as the emulsions and the optical equipment for analyzing particle traces were improved. One of the nominators captured the significance of Powell's work: "His special claim to consideration is, in my view, the fact that discoveries of fundamental importance can still be made with the simplest apparatus—in this case, special nuclear emulsions developed under his general direction and microscopes."[38]

Although the method was still based on the same simple principle that Blau and Wambacher had put it to use earlier in radioactivity research, Powell had deeply transformed the culture that surrounded its use. By the time he was nominated for the Nobel Prize, Powell was directing a large group of female scanners and experimenters, he had strong university affiliations and powerful industrial connections, and he managed significant material sources. As Galison put it, Powell's lab was "a cottage industry" with its links to chemists, theorists, and scanners.

In contrast, Blau, the unsuccessful Nobel candidate, tried to continue her emulsion work during the 1940s. However, she never obtained a stable professional affiliation to a university and moved constantly from one industrial site to another, working primarily with the commercial uses of radium. Even when working in the field of photographic emulsions at the Brookhaven Laboratory, Blau did most of the scanning herself despite the common practice to have the trained scanners, usually women, perform the actual scanning of bubble chamber images.[39] Throughout the 1940s and 1950s, she remained largely peripheral, working for industrial firms and having only restricted access to high-energy physics machinery.

A study of the Nobel nominations in physics from 1901 to 1950 reveals an interesting trend that sheds light on Blau's case. The vast majority of the nominees—67 percent—were based in university teaching departments and laboratories. Only 10 percent worked in institutes of technology, and hardly anyone was an industrial physicist.[40] Thus, Blau's possibilities of obtaining an award were closely interrelated to a number of more nebulous considerations such as the committee's stance toward university rather than industrial physics research or a cutting edge, specific discovery rather than a life's work. These considerations were closely interrelated to women's limited possibilities for research and experimentation during the 1940s and 1950s.

THE END: BLAU'S RETURN TO VIENNA

In 1955, Blau moved once again from the Brookhaven Laboratory this time to the University of Miami. As Harry Robertson, the head of the physics department explained to Blau, they were in need of teachers for graduate and undergraduate courses who can also carry on research programs. They did not possess, however, any facilities for high-energy nuclear physics. Ingenious as usual and always having to put up with practical constraints in her experimental work, Blau proposed research on "elementary particles and hyperfragments with nuclear emulsions, exposed to the machines in Brookhaven and Berkeley. The work could be done in collaboration with students and the exposed emulsions could be used as tools in teaching nuclear physics." This plan would only require, as she put it, "several good microscopes and one scattering microscope."

A year later, in collaboration with younger colleagues at the department, Blau applied for a research grant to the National Science Foundation, a project that was actually funded by the Air Force Office of Scientific Research. Setting up from scratch the missing facilities for high-energy physics research at her department, Blau was assured of a three-year grant for the study of photographic emulsions exposed to the Bevatron at the University of California and the Cosmotron at Brookhaven. She was able to remain active in the world of emulsions and the Air Force satisfied its intense interest in what hazards lie in the space their jets and missiles were soon going to penetrate. [41]

But times had changed not only concerning possibilities for experimental work but also concerning the politics of science. On May 2, 1957, Morton Miller, the dean of the University of Miami, addressed the U.S. Department of State at Blau's request with an unusual question. Blau had been recently nominated for the Leibnity Medallion by the Academy of Sciences in Berlin, honoring her achievements in physics. Once again, Schrödinger had placed the nomination,[42] but the Second World War had not only displaced most of the prominent European scientists such as Blau, it had also created a deep gulf of animosity between East and West.

The Academy of Sciences, as Miller argued, resided in East Germany in the Soviet sector of Berlin. Schrödinger was also associated with the academy and obviously these facts could pose serious issues in accepting an award from a communist scientific institution. Working with a research grant from the U.S. Air Force, Blau also did not want to risk her position. The immediate answer from the Department of State was definitely expected:

> This government does not, as you know, recognize the regime which controls the Soviet zone of Germany. By the same token, the Department of State deprecates any action which tends to clothe this regime or any of its agencies with legality or respectability. Acceptance of a medal from the East German Academy of Sciences would, it seems to us, constitute such an action. In addition, it is quite likely that Dr. Blau's name would be misused by the East Germans in a way which might be embarrassing to her. There is no genuine scholarship in the Soviet-controlled part of Germany, and it therefore seems beyond doubt that the offer is being made to her solely for propaganda purposes.[43]

In its boldness, the paragraph reflects the ways postwar science was intertwined with politics. Blau, caught in the shift from small to big science and stood at the nodal point between prewar internationalism and postwar nationalism concerning scientific culture. The era in which the Curies had refused to patent their discovery of radium and polonium were gone. When Blau declined the honor, she was already 63 years old. Because of her prolonged exposure to radioactivity, she started to develop cataracts, requiring an operation. At the end of her life, facing financial and health problems, Vienna seemed the most suitable destination. In March 1960, she requested a release from her university contract. Having no retirement rights or health insurance, she left Miami at the end of May 1960.[44]

In Vienna, a number of old colleagues tried to gather funds for her, and Schrödinger put her up for the Schrödinger Prize which she received in 1962. Poor, disconnected from any major scientific network, and bitter with several members of the Radium Institute for accepting the Nazi Stetter as one of the heads of the Physics Institute after the end of the war, Blau distanced herself from serious research and from old friends such as Karlik. She died on January 27, 1970, at Lainz hospital, lonely and unknown to the international physics community.

A CONSIDERABLE SHIFT IN TRAFFICKING MATERIALS

In writing this account of the men and women who worked at Vienna Radium Institute, I have been trying to emphasize the history of radium. This book has focused on a major trafficking material and the gendered experimental practices that it generated in the early twentieth century. Among the many sites in which radium appeared—scientific, medical, technological, and industrial—I have focused on Vienna's physics and chemistry laboratories. I have paid attention to the city's medical institutions and to those other sites where scientists—mainly women—moved, carrying radioactive materials. I have been concerned with the controversies that radium spawned in physics and chemistry over its properties and with the conflicts that occurred when it was connected to the atomic structure. Briefly, I have argued that by being a trafficking material, radium brought together men and women in scientific laboratories; enforced novel laboratory designs; constituted new scientific networks among chemists, physicists and physicians; created opportunities for professional careers and unprecedented collaborations; dictated the design of new techniques and instruments; imposed new experimental practices; and changed laboratory culture. It prevailed in medicine and forced physicians to become familiar with radium technologies, share their power with physicists within their hospitals, and collaborate with them in order to produce new medical instruments and standardized radium doses for cancer treatments.

In its early days, radium built a whole new domain of knowledge at the intersection of physics and chemistry and at crossroads of the laboratory, the lecture hall, the workshop, and industrial and commercial sites. This scientific object became a valuable commodity with a life of its own, inundating the shelves of beauty shops and apothecaries. Radium mattered in the public's eyes both through commercial advertisements and physics laws. On the one hand, the industry created a powerful market for the new material with the surprising properties and on the other hand, scientists built careers and networks working with radium as a therapeutic agent and a means for probing nature.

Radium's rapid expansion in the scientific world relied on a complex system of chemistry and physics technologies, on standardization techniques, and on the institutionalization of the new field of radioactivity. As we saw in the case of Vienna, designing a new institute to host radium research meant designing according to the new experimental culture of radioactivity, taking into account the interdisciplinary nature of the field and the gendered experiences of those working in the institute's laboratories. The architecture of the Radium Institute materialized the identities that the physicists working on radioactivity wished to acquire. Built as a separate building from the physics and chemistry institutes, the Radium Institute legitimized research on radioelements as a scientific specialty, which was still dependent on physics and chemistry.

The new building provided and stabilized professional identities for those physicists whose new disciplinary identities were at risk of blending with well-established ones. At the same time, the internal spatial arrangements of the institute reflected the ways through which the architecture of the laboratory acknowledged that physicists were gendered. The separate sanitary installations for men and women indicated the existence of both in the institute. Moreover, by having a work bench for themselves, women gained a sense of belonging. They were no longer transients in the field. Here, the city played an additional role.

As a context for intellectual work, Vienna, and especially the Mediziner-Viertel, was linked to physicists' self-images and their interdisciplinary practices. The urban reconstruction of the city becomes sociologically interesting for it provides the space to study the social and political negotiations that shaped the concrete realization of scientific institutions. Implementation of the decision to move from the ramshackle Physics Institute in Türkenstrasse to the new natural-science quarter across from the Josephinum took physicists more than three decades. Moreover, it involved endless negotiations on the nature of their discipline, the exact location of their institutes in relation to the exchange of radioactive materials, their professional practices, and the identities they wanted to portray through the architecture of their buildings.

The location of the institute at the center of the cultural and epistemic life of Vienna worked especially in women's favor. In one sense, the few women in the early-twentieth-century science became visible in the relatively small scientific community of Vienna. Several times a day, women crossed Währingerstrasse in order to attend classes and participate in laboratory courses. The face-to-face interactions in the Mediziner-Viertel affected their social lives and contributed to their visibility in the community. For example, it was not hard for Meitner to recall that Horovitz had been a chemistry student when Hönigschmid asked her to suggest a possible collaborator to him.

Although few in absolute numbers, the women who entered the Radium Institute during its first decade were surprisingly numerous considering that the University of Vienna had started admitting them less than two decades before. In another sense, the crossing of Währingerstrasse facilitated the crossing of disciplinary boundaries as well. From physics to medicine and chemistry to physics, women transferred their knowledge among disciplines, shifting and shaping the boundaries of radioactivity. More importantly, they literally transferred radium preparations, needles, and tubes from the physics laboratory to the clinic.

In the interwar period, radium inaugurated significant changes in laboratory organization. By welcoming different researchers with diverse research projects, the laboratory space had been reassigned and the material basis for experimental work had changed dramatically. During the 1920s, scintillation counters, Shimizu-Wilson ray track apparatus, photographic emulsions, and powerful microscopes become part of the material culture of the Radium Institute. After Pettersson's arrival in Vienna, polonium sources, spent radon needles from the neighboring hospitals, and photographic plates lay around the workbenches and occupied the interest of the experimenters.

The patterns of experimentation were transformed as well. Although several researchers worked closely together before the 1920s, those groups were limited to two people. For instance, Hönigschmid teamed up with Horovitz, Ludwig Flamm worked with Mache, and Hess with Lawson. With Pettersson's arrival, the groups grew appreciably and complicated the patterns of collaboration. "Our particular kind of work," Pettersson noted in 1928, "requires the close and continued collaboration of at least a dozen highly specialized people."[45] At the same time, Przibram gathered around him a number of students and occasionally attracted experienced researchers from Pettersson's team. Karlik and Kara-Michailova, for example, often shifted to Przibram's research projects and co-authored papers with him.

Experimental scientists such as the Viennese "radioactivists," continued to deal with material units and constantly needing radium in their daily bench work. Even in the midst of the Cambridge-Vienna controversy, Pettersson thought of asking Rutherford for the capillary-spent radon tubes from hospitals used in the preparation of radium and polonium that were useless to the affluent British researchers but essential for the work in Vienna. In addition, Przibram made it clear to the Rockefeller Foundation that funding for an assistant was less important than money "for new apparatus, for replacement of breakage, for liquid air (or for a liquid air machine which could be a source of income to them through their ability to sell it to physicists and chemists, all of whom need air and now have to purchase it from a concern on the outskirts of Vienna)."[46] Thus, materials are not fixed entities, but they mean what they mean as far as they can be placed in a local experimental context and used in specific experimental practices.

For the period preceding the outbreak of the Second World War, radium was also connected to several attempts to exercise political power. The plan of Meyer and the directors of the Vivarium to create a centralized station for controlling radium supplies for the whole country is such an example. In the midst of the 1930s political unrest, controlling radium supplies meant securing research autonomy. In a similar vein, radium transformed medical practices and the organization of work in the clinic. The application of radium therapy replaced that of radiotherapy or rontgentherapy (the use of x-rays in medical treatments). This transformation was particularly evident in Tandler's attempts to bring in radium therapy to municipal hospitals. His venture introduced reorganization of medical work while creating vocational opportunities especially for the female scientists who, unable to foster academic careers, looked for other prospects.

Finally, in configuring and reconfiguring trafficking materials, scientists were forced to shape new experimental practices, which were often heavily gendered. Work on radium became connected with the gendering of experimental skills as in the case of the Viennese women who designed and handled the scintillation counter. For them, radium also worked as a vehicle for crossing disciplinary boundaries and moving from physics to oceanography. Trading their expertise to prepare radium sources and analyze radium samples became another way for women to finance scientific research and professionally survive when grant money ended and political unrest arose.

Nonetheless, since the mid-1930s, changes in nuclear physics had repercussions both on materials and instruments and on experimenters and their practices. The development of new technologies such as the particle accelerators and experiments for the production of artificial radioelements or radioisotopes with these new machines forced radium into disuse. Likewise, its decline pushed many of those who knew how to prepare it and handle it into despair, especially the women. Immediately after the Second World War, women such as Blau who had been nurtured in a small-scale laboratory found themselves far from cutting-edge research, unable to adapt to highly hierarchical research groups and having limited access to experimental machines. As their expertise in experimenting with radium lost significance, they lost their ability to survive in the field as well. The shift from radium to radioisotopes marked an important shift not only in the kind of trafficking materials but also in the type of experimental practices each of them embodied.

Radioisotopes are the product of irradiation in the cores of nuclear reactors. If it is suitably adapted, any reactor could irradiate isotopes. Although production of radioisotopes had started in the prewar period, it did not become widespread until after the Second World War, dictating a new experimental tradition. Making use of large instruments such as the cyclotron, physicists, chemists, and biologists started to manufacture material substances, generating new markets and practices. In contrast to radium, radioisotopes were relatively cheap and easy to make. They were used in radiodiagnostics, new cancer treatments, in thyroid conditions, in nuclear medicine as tracer material, and in biology for investigating organic metabolisms and visualizing elusive molecular processes.

The Atomic Energy Commission, which in 1948 offered Blau a promising contract for working at Columbia's laboratories and later at the Brookhaven cyclotron, was the agent which carried out the most extensive production and distribution of isotopes to "friendly nations" for research and medical purposes in the late 1940s. After the end of the Manhattan Project, the AEC also decided that it could use the nuclear piles that had produced plutonium for the bomb as a source of radioisotopes for biomedical research and therapeutic purposes in the local U.S. market. The reactors at Oak Ridge where Rona found employment after she emigrated from Europe could produce, for example, 200 millicuries of carbon-14 in a few weeks for about $10,000, an amazing income for the laboratory.[47]

The big machines brought big money and opened opportunities for prestigious research positions. The way the Los Alamos Laboratory, for example, advertised research positions in the mid 1950s leaves no doubt that women were not welcomed as experimenters in this new "big" science: "Excellent opportunities now exist at Los Alamos for qualified *men* wishing to further their scientific careers . . . The future resides in *men* with imagination" (emphasis mine).

At the same time, readers were exposed to subtle images about women's role as the privileged wives who could "live and raise a family" in "an ideal small city" where the laboratory was located.[48] In a best-case scenario, women's role in the large laboratories shifted from active experimenters to scanners, calculators, and assistants.[49] Coupled with the fact that in the mid-1950s, the proportion of all science and engineering degrees awarded to women remained under 10 percent, the images above—no matter which way we view them—remained unappealing to female scientists.[50]

Throughout this book, I have tried to convey a sense of how trafficking materials and gendered experimental practices are tied together in a specific local context, how they can eventually shape an entire laboratory culture, and how the handling of materials constructs gender differences in scientific practice. My concern has been with a history of traces of materials and the study of the people who actually produced, traded, manipulated, consumed, and used them. In the hands of early-twentieth-century physicists, radium evolved into a powerful agent. In the period following the Second World War, as radium lost its power and radioactivity research came to an end, physicists, and especially women physicists, eventually experienced the same fate.

NOTES

1. Galison, *Image and Logic* (1997), 454.

2. Perlmutter, "Marietta Blau's Work After World War II" (unpublished manuscript), 2.

3. Blau and Dreyfus, "The Multiplier Photo-Tube" (1945).

4. Karlik, "Eine Lumineszenzmethode" (1933). Karlik presented the modified scintillation counter on February 23, 1933 at a Vienna Academy meeting. During the same meeting, she also presented her work with Rona on the use of the instrument for the study of ranges of alpha-particles emitted from actinium and its products; see Karlik and Rona, "Untersuchungen der Reichweite" (1933).

5. Blau and Dreyfus, "The Multiplier Photo-Tube" (1945), 246.

6. Karlik's device was similar to the one that Adolf Krebs developed in 1941. Krebs was a staff member of the Kaiser-Wilhelm Institute for Biophysics in Frankfurt since 1937. In 1947, he became director of the division of Radiobiology of the U.S. Army Medical Research Laboratory at Fort Knox. Krebs, "Ein Demonstrationsversuch" (1941). See also Rheinberger, "Putting Isotopes to Work" (1999), 7; Krebs, "Early History of the Scintillation Counter" (1955).

7. Blau and Dreyfus, "The Multiplier Photo-Tube" (1945).

8. Perlmutter, "Marietta Blau's Work," (unpublished manuscript), 2.

9. Curran and Baker, "Photoelectric Alpha-particle Detector" (1948), 116; Curran and Baker, A photoelectric Alpha-Particle Detector, U.S. Atomic Energy Commission Rpt. MDDC 1296, November 17, 1944, declassified September 23, 1947. See also Galison, *Image and Logic* (1997), 455; Perlmutter, "Marietta Blau's Work" (unpublished manuscript), 2; Rheinberger, "Putting Isotopes to Work" (1999), 7.

10. Blau and Dreyfus, "The Multiplier Photo-Tube" (1945), 248.

11. The Energy Employees Illness Compensation Act of 2000 ("Act" Public Law 106–398) signed by president Bill Clinton in December 7, 2000, established a program to provide compensation to individuals who developed illnesses as a result of their employment in nuclear weapons production. The two corporations that Blau worked for between 1944 and 1948, the International Rare Metals Refinery Inc (listed as Atomic Weapon Employer) and the Canadian Radium and Uranium Corporation, both in Mount Kisco, an hour outside New York, are included in the long list of covered facilities (Department of Energy 6450–010p). The Canadian Radium and Uranium Corp was founded in 1943 and collected radium from airplane industries and watch dials; see Hughes, B. "U.S. Begins Compensating Workers Exposed to Toxic

Substances," *The Journal News*, August 20, 2001. Boris Pregel and his brother Alexander, Russian bourgeois who lived in Paris, came to New York in the early 1940s and established the company as one of the main uranium providers. Alexander was the administrative vice-president and both brothers looked after refugee scientists after World War II. Elisabeth Rona worked for them as well (all the information related to the Canadian Radium and Uranium comes from my personal communication to Vilma Hunt, retired professor of Environmental Health at Harvard School of Public Health. Her information is based on extensive interviews with the Pregels).

12. Blau and Feuer, "Radioactive Light Sources" (1946). Blau and Feuer constructed a device for using the fluorescent effect of radioactive radiation and especially that of the highly ionizing alpha radiation as a light source. Blau's contributions were based on Karlik's 1933 paper and the use of photocell, this time for transforming the alpha particles into light. The application of the radioactive light sources was enormous. For example, they were used for the standardization of the color of luminous compounds for television purposes. They were also utilized as a source of light for instruments previously painted with luminous compounds dangerously mixed with radioactive material. See, for example, the case of radium dial painters and the use of radium paint in dials of watches and instruments. Rentetzi, "Women Radium Dial Painters" (2004).

13. Blau and Carlin, "Industrial Applications of Radioactivity" (1948), 82.

14. News of the University of Miami, 1 April 1960, AUM.

15. Blau and Smith, "Beta-ray Measurements and Units" (1948), 67.

16. Marshall and Coltman, "The Photo-Multiplier Radiation Detector" (1947), 528.

17. Broser and Kallmann, "Über die Anregung" (1947); Broser and Kallamn, "Über den Elementarprozess" (1947). The sensitivity of naphthalene, the first organic and large volume scintillator, made Kallman's counter more efficient than the previous ones.

18. Rheinberger, "Putting Isotopes to Work" (1999), 8; Perlmutter, "Marietta Blau's Work" (unpublished manuscript).

19. Perlmutter, "Marietta Blau's Work" (unpublished manuscript).

20. J. C. Potter to Blau, August 25, 1947, CLP; H.P. Yockey to Blau, September 4, 1947, CLP; Boris Pregel to Blau, November 3, 1947, CLP.

21. George Pegram to Blau, December 23, 1947, CLP.

22. Einstein to Goudsmit, January 5, 1954, AEA.

23. Goudsmit to Einstein, February 11, 1954, AEA.

24. Rheinberger, "Putting Isotopes to Work" (1999), 7; Galison, *Image and Logic* (1997), 454–63; Mayneord and Belcher, "Scintillation Counting and its Medical Applications" (1950), 259.

25. Pringle, "The Scintillation Counter" (1950).

26. Rheinberger, "Putting Isotopes to Work" (1999), 2.

27. Medical Radiation Technologists, Web site http://jobfutures.ca/noc/print/3215.html.

28. Schrödinger to the Nobel Committee for Physics, January 8, 1950, NARSAS.

29. Sime, Lise Meitner (1996), 290. Crawford, Sime, and Walker, "A Nobel Tale of Postwar Injustice" (1997).

30. I would like to thank the archivist of the NARSAS, Maria Asp Romefors, for her help in identifying the members of the Nobel Prize committee.

31. Report by Lindh on Blau and Wambacher, Uppsala, July 1, 1950, NARSAS. I would like to sincerely thank Anders Stephanson for providing a brief translation of both Lindh's and the Nobel committee's reports from Swedish.

32. Blau, curriculum vitae, Leopold Halpern Papers. Shapiro, "Tracks of Nuclear Particles" (1941).

33. Maurice Shapiro interview to Laurie Brown, May 30, 1978, AIP.

34. Shapiro, "Tracks of Nuclear Particles in Photographic Emulsions" (1941). Powell, "Applications of the Photographic Method" (1943)

35. Crawford, Elisabeth. "Nobel Population 1901–1950: Anatomy of a Scientific Elite" *Physics World*, 2001 http://physicsweb.org/articles/world/14/11/7. Crawford, Sime and Walker, "A Nobel Tale" (1997).

36. Committee report, Nobel Committee for Physics, September 25, 1950. NARSAS.

37. Galison, *Image and Logic* (1997), 161–186.

38. Presentation speech by Axel Lindh, member of the Nobel Committee for Physics http://nobelprize.org/nobel_prizes/physics/laureates/1950/press.html.

39. Luke Yuan to Harry Robertson, November 11, 1955, AUM.

40. Crawford, "Nobel Population" (2001).

41. Robertson to Blau, October 5, 1955, AUM; Blau to Robertson, November 1, 1955, AUM. Although her salary at the Brookhaven was $7,200, she finally accepted the teaching position at the University of Miami for only $5,500 and became an associate professor under the pressure of applying for a research grant to the National Science Foundation in 1956. See memoradum, Charles Doren Tharp, June 6, 1956. AUM; see Collier, Bert. "Woman Scientist Probes the Atom" Miami Herald, October 5, 1958, AUM.

42. Hans Thirring had also nominated Blau for a Nobel Prize in 1955, but again she was not successful. Thirring to the Nobel Prize Committee, January 29, 1955, CLP.

43. E. Ramsauer, Department of State to Morton Miller, May 7, 1957, AUM; Morton Miller to the U.S. Department of State, May 2, 1957, AUM.

44. Blau to Charles Tharp, March 31, 1960, AUM; Charles Tharp to Blau, April 4, 1960. AUM.

45. Hans Pettersson's report to the International Education Board, April 1928, AUG.

46. Augustus Trowbridge, memoradum of conversation with Stefan Meyer and Karl Przibram, March 26, 1925, Box 25/360, RAC.

47. Krige, "Atoms for Peace, Scientific Internationalism and Scientific Intelligence" (2006); Creager, "Tracing the Politics of Changing Postwar Research Practices" (2002).

48. Advertisements that appeared in the *Journal of American Chemical Society* (JACS) (either in the front or the last page of each issue, not numbered) in 1955, 77(21); 1955, 77(23); 1956, 78(3); 1955, 77(21).

49. See Galison, *Image and Logic* (1997), 199–200.

50. LaFollette, "Eyes on the Stars" (1988).

BIBLIOGRAPHY

Achinstein, Peter and Hannaway, Owen, eds. *Observation, Experiment, and Hypothesis in Modern Physical Science.* Cambridge, MA: MIT Press, 1985.

Albrecht, Eleonore. "Über die Verzweigungsverhältnisses bei RaC, AcC, ThC und die Zerfallskonstantender C"-Produkte." *Sitzungsberichte Akademie der Wissenschaften in Wien, Mathematisch-naturwissenschaftliche Klasse,* Abteillung IIa, 128 (1919): 925–944.

Alcoff, Linda and Elizabeth, Potter. *Feminist Epistemologies.* New York: Routledge, 1993.

Alth, Gerhart. *50 Jahre Strahlentherapie Lainz.* Wien, 1981.

Anderson, Hariett. "Feminism as a Vocation: Motives for Joining the Austrian Women's Movement." In *Vienna 1900: From Altenberg to Wittgenstein* edited by Eduard Timms and Ritchie Robertson. 73–86. Austrian Studies 1, Edinburg University Press, 1992.

Anderson, Harriet. *Utopian Feminism: Women's Movement in the fin-de-siecle Vienna.* New Haven: Yale University Press, 1992.

Andraschko, Elisabeth. "Elise Richter-einer Skizze ihres Lebens." In *Durch Erkenntnis zu Freiheit und Glück: Frauen an der Universität Wien* edited by Waltraud Heindl and Marina Tichy, 221–231. Schriftenreihe des Universitätsarchivs, Band 5, Wien: Universität Wien, 1990.

Angetter, Daniela. *Guido Holzknecht: Leben und Werk des Pioniers der österreichischen Röntgenologie.* Wien: Werner Eichbauer Verlag, 1998.

Antony, Louise and Witt, Charlotter. eds. *A Mind of One's Own: Feminist Essays on Reason and Objectivity.* Boulder: Westview Press, 1993.

Appadurai, Arjun. "Introduction: Commodities and the Politics of Value." In *The Social Life of Things: Commodities in Cultural Perspective* edited by Arjun Appadurai. 3–63.Cambridge: Cambridge University Press, 1986.

Arias, Ingrid. "Die ersten Ärztinnen in Wien" in *Töchter des Hippokrates: 100 Jahre akademische Ärztinnen in Österreich* edited by Birgit Bolognese-Leuchtenmüller, and Sonia Horn. 55–78. Wien: Pressestelle und Verl. Der Österr. Ärztkammer, 2000.

Ash, Mitchell. "Psychology and Politics in Interwar Vienna: The Vienna Psychological Institute, 1922–1942." In *Psychology in Twentieth-Century Thought and Society* edited by Mitchell Ash and William Woodward. 143–164. Cambridge University Press, 1987.

Ash, Mitchell. "Women Émigré Psychologists and Psychoanalysts in the United States." In *Between Sorrow and Strength: Women Refugees of the Nazi Period* edited by Sibylle Quack. 239–264. Cambridge: Cambridge University Press, 1995.

Auer, Barbara. *Das Physikalische Institut der Universität Heidelberg an der Albert-Überle-Strasse* 7. MA Thesis, University of Heidelberg, 1983.

Auer, Barbara. *Das Physikalische Institut der Universität Heidelberg.* Heidelberg, 1984.

Badash, Lawrence. "The Suicidal Success of Radiochemistry." *The British Journal for the History of Science,* 12 no. 42 (1979): 245–256.

Badash, Lawrence. *Radioactivity in America: Growth and Decay of a Science.* Baltimore: The Johns Hopkins University Press, 1979.

Badash, Lawrence. *Rutherford and Boltwood: Letters on Radioactivity,* New Haven, 1969.

Baker, Thorne. "Industrial Uses of Radium: Possibilities in Agriculture, Medicine, and Other Directions" *Engineering Magazine Co,* 49 (1915): 597.

Bandhauer-Schöffman, Irene. "Zum Engagement der österreichischen Frauenvereine für das Frauenstudium." In *Durch Erkenntnis zu Freiheit und Glück: Frauen an der Universität Wien* edited by Waltraud Heindl and Marina Tichy. 49–78. Schriftenreihe des Universitätsarchivs, Band 5, Wien: Universität Wien, 1990.

Barea, Ilsa. *Vienna.* New York: Alfred Knopf, 1966.

Barker, Elisabeth. *Austria 1918–1972.* London: MacMillan, 1973.

Barthes, Roland. "Myth Today." In *Mythologies.* 109–159. London: Paladin Grafton Books, 1973.

Bates, L. and Rogers, J. "Long-Range alpha-Particles from Radium Active Deposits." *Nature,* 112 (1923): 435–436.

Bates, L. and Rogers, J. "Particles of Long Range from Polonium." *Proceedings of the Royal Society,* A, 105 (1924): 360–369.

Bates, L. and Rogers, J. "Particles of Long-Range Emitted by the Active Deposits of Radium, Thorium and Actinium." *Proceedings of the Royal Society,* A, 105 (1923): 97–116.

Becker, Barbara. "Dispelling the Myth of the Able Assistant: Margaret and William Huggins at Work in the Tulse Observatory." In *Creative Couples in the Sciences* edited by Helena Pycior, Nancy Slack and Pnina Abir-Am. 98–111, New Brunswick: Rutgers University Press, 1996.

Bedini, Silvio. "The Hardware of History." *Professional Surveyor.* (December 2001): 10–14.

Belar, Maria. "Spektrophotometrische Untersuchung der Verfärbungserscheinungen durch Becquerelstrahlen." *Sitzungsberichte Akademie der Wissenschaften in Wien, Mathematisch-naturwissenschaftliche Klasse*, Abteillung IIa, 132 (1923): 45–54.

Beller, Steven. *Rethinking Vienna* 1900. New York: Berghahn Books, 2001.

Benetka, Gerhard. "The Vienna Institute of Psychology: Obituary for a Once Important Research Institution." In *The Cultural Exodus from Austria* edited by Friedrich Stadler and Peter Weibel. 127–131. New York: Springer-Verlag, 1995.

Benjamin, Harry. "Eugen Steinach, 1861–1944: A Life of Research." *The Scientific Monthly*, 61 no. 6 (1945): 427–442.

Bequerel Henri. "Sur les Radiations Invisible Emises par les Corps Phosphorescence." *Comptes Rendus des Seanses de l'Academie des Sciences, Paris*, 122 (1896): 501–3.

Bequerel, Henri and Curie, Pierre. "Action physiologiques des rayons du radium." *Comptes Rendus des Seanses de l'Academie des Sciences, Paris*, 132 (1901): 1289–91.

Berg, Marc and Mol, Annemarie. *Differences in Medicine: Unraveling Practices, Techniques, and Bodies*. Durham: Duke University Press, 1998.

Bergen, Doris. "Sex, Blood, and Vulnerability: Women Outsiders in German-Occupied Europe." In *Social Outsiders in Nazi Germany* edited by Robert Gellately and Nathan Stoltzfus. 273–293. Princeton: Princeton University Press, 2001.

Berth, Marianne. "Die Stellung der Frau im Recht." In *Frauenbewegung, Frauenbildung und Frauenarbeit in Österreich* edited by Martha Stephanie Braun et al. 95–103. Wien, 1930.

Berven, Elis et al. *The First Fifty Years: Radiumhemmet 1910–1937 and King Gustaf v Jubilee Clinic 1938–1960*. Stockholm: Acta Radiologica Supplementum 250, 1965.

Bethe, Hans. "The Happy Thirties," In *Nuclear Physics in Retrospect: Proceedings on a Symposium on the 1930s* edited by Roger Stuewer. 11–31. Minneapolis: University of Minnesota Press, 1979.

Beyerchen, Alan. *Scientists under Hitler: Politics and the Physics Community in the Third Reich*. New Haven: Yale University Press, 1977.

Beyler, Richard. "'Imagine a Cube Filled with Biological Material': Reconceptualizing the Organic in German Biophysics, 1918–1945." In *Fundamental Changes in Cellular Biology in the 20th Century* edited by Charles Galperin, Scott Gilbert, and Brigitte Hoppe. Proceedings of the XXth International Congress of History of Science. 38–46. Belgium: Brepolis, 1997.

Bischof, Brigitte. "Olga Steindler (1879–1933)" *Koryphäe* 29 (2001): 4–5.

Bischof, Brigitte. "The 'Marie Curie Syndrome,' the Role of Mentors and Romanticism or Why Were There so Many Women In Radiaoctivity Research in Vienna?" In *Women Scholars and Institutions Proceedings of the International Conference: Women Scholars and Institutions, Commission Women in Science of the IUHPS/DHS* edited by S. Strbanova, I. Stamhuis, and K. Mojsejova. 639–658. Prague, 2004.

Bischof, Brigitte. *Frauen am Wiener Institut für Radiumforschung*. MA Thesis, University of Vienna, 2000.

Bischof, Brigitte. *Physikerinnen: 100 Jahre Frauenstudium an den Physikalischen Instituten der Universität Wien*. Broschüre zur Ausstellung von Brigitte Bischof. University of Vienna, 1998.

Blau, Marietta and Carlin, J. R. "Industrial Applications of Radioactivity." *Electronics,* 21 no. 1–4 (1948): 78–82.

Blau, Marietta and Dreyfus, B. "The Multiplier Phott-Tube in Radioactive Measurements." *The Review of Scientific Instruments,* 16, no. 9 (1945): 245–248.

Blau, Marietta and Feuer, I. "Radioactive Light Sources." *Journal of the Optical Society of America,* 36 no. 10 (1946): 576–580.

Blau, Marietta and Kara-Michailova, Elisabeth. "Über die durchdringende Strahlung des Poloniums." *Sitzungsberichte Akademie der Wissenschaften in Wien, Mathema tisch-naturwissenschaftliche Klasse,* Abteillung IIa, 140 (1931): 615–622.

Blau, Marietta and Rona, Elisabeth. "Anwendungen der Chamié'schen photographischen Methode zur Prüfung des chemischen Verhaltens von Polonium." *Sitzungsberichte Akademie der Wissenschaften in Wien, Mathematisch-naturwissenschaftli che Klasse,* Abteillung IIa, 139 (1930): 275–279.

Blau, Marietta and Rona, Elisabeth. "Ionisation durch alpha-Strahlen." *Sitzungsberichte Akademie der Wissenschaften in Wien, Mathematisch-naturwissenschaftliche Klasse,* Abteillung IIa, 135 (1926): 573–585.

Blau, Marietta and Smith, J. E. "Beta-ray Measurements and Units." *Nucleonics,* 2 no. 6 (1948): 67–74.

Blau, Marietta and Wambacher, Hertha. "Über Versuche durch Neutronen ausgelöste Protonen photographisch nachzuweisen." *Sitzungsberichte Akademie der Wissenschaften in Wien, Mathematisch-naturwissenschaftliche Klasse,* Abteillung IIa, 141 (1932): 180.

Blau, Marietta and Wambacher, Hertha. "Über Versuche durch Neutronen ausgelöste Protonen photographisch nachzuweisen II" *Sitzungsberichte Akademie der Wissenschaften in Wien, Mathematisch-naturwissenschaftliche Klasse,* Abteillung IIa, 141 (1932): 617–20.

Blau, Marietta and Wambacher, Hertha."Über das Verhalten konloser Emulsionen gegenüber Alpha Partikeln." *Sitzungsberichte Akademie der Wissenschaften in Wien, Mathematisch-naturwissenschaftliche Klasse,* Abteillung IIa, 141 (1932): 467–474.

Blau, Marietta. "Bericht über die Entdeckung der durch kosmische Strahlung erzeugten Sterne." *Sitzungsberichte Akademie der Wissenschaften in Wien, Mathem atisch-naturwissenschaftliche Klasse,* Abteillung IIa, 159 (1950): 53–57.

Blau, Marietta. "Die photographische Wirkung von H-Strahlen aus Paraffin und Aluminium." *Zeitschrift für Physik,* 34 (1925): 285–295.

Blau, Marietta. "Über die Absorption divergenter γ-Strahlung." *Sitzungsberichte der Kaiserl. Akademie der Wissenschaften in Wien, Mathematisch-naturwissenschaftlic he Klasse,* Abteillung IIa, 127 (1918): 1253–1279.

Blau, Marietta. "Über die photographische Wirkung naturliche H-Strahlen." *Sitzungsberichte Akademie der Wissenschaften in Wien, Mathematisch-naturwissensc haftliche Klasse,* Abteillung IIa, 136 (1927): 469–480.

Blau, Marietta. "Über die photographische Wirkung naturliche H-Strahlen." *Sitzungsberichte Akademie der Wissenschaften in Wien, Mathematisch-naturwissensc haftliche Klasse,* Abteillung IIa, 134 (1925): 427–436.

Blau, Marietta. "Über die photographische Wirkung von H-Strahlen aus Paraffin und Atomfragmenten." *Zeitschrift für Physik*, 48 (1928): 751–764.

Blau, Marietta. "Über photographische Intensitätsmessungen von Poloniumpräparaten." *Sitzungsberichte Akademie der Wissenschaften in Wien, Mathematisch-n aturwissenschaftliche Klasse*, Abteillung IIa, 137 (1928): 259–268.

Blau, Marietta. "Über photographische Untersuchungen mit radioactiven Strahlungen." In *Zehn Jahre Forschung auf dem Physikalisch-Medizinischen Grenzgebiet*, edited by F. Dessauer. 390–98. Leipzig: Georg Thieme, 1931.

Blaufox, Donald. "Bequerel and the Discovery of Radioactivity: Early Concepts." *Seminars in Nuclear Medicine*, 26 no. 3 (1996): 145–154.

Bonet-Maury, P. "Sur la vaporisation du polonium." *Comptes Rendus hebdomadaires des seances de l' Academie des Paris*, 184 (1927): 1376–1378.

Bonet-Maury, P. "Sur la vaporisation du polonium." *Comptes Rendus hebdomadaires des seances de l' Academie des Paris*, 185 (1927): 204–6.

Borngässer, Barbara. "Architecture from the Turn of the Nineteenth Century to the Present Day." In *Vienna: Art and Architecture* edited by Rolf Toman. 272–357. Cologne: Könneman, 1999.

Boudia, Soraya. "The Curie Laboratory: Radioactivity and Metrology." *History and Technology*, 13 (1997): 249–265.

Boudia, Soraya. *Marie Curie et son laboratoire: Sciences et industrie de la radioactivitè en France*. Paris, 2001.

Bowker, Geoffrey and Star, Susan Leigh. *Sorting Things Out: Classification and its Consequences*. Cambridge, Mass.: MIT Press, 1999.

Brinda-Konopik, Nelly. "Robert Wilhelm Bunsen und die Chemische Schule an der Universität Wien (Alma Mater Rudolphina)." In *Bunsentagung 1992 Wien*. 2–13. Wien: Fredinand Berger, 1992.

Brook-Schepherd, Gordon. *Dollfuss*. Westport, Conn.: Greenwood Press, 1978.

Broser, von Immanuel and Kallmann, Hartmut. "Über den Elementarprozess der Lichtanregung in Leuchtstoffen durch alpha-Teilchen schnelle Elektronen und gamma-Quanten II." *Zeitschrift für Naturforschung*, 2a (1947): 642–650.

Brown, Andrew. *The Neutron and the Bomb: A Biography of Sir James Chadwick*. Oxford: Oxford University Press, 1997.

Burcham, W. E. "Rutherford at Manchester, 1907–19." *Contemporary Physics*, 5 (1964): 304–8.

Burian, Wilhelm. "History of the Vienna Psychoanalytical Society." *Newsletter of International Psychoanalysis*, 8 no. 2 (1999).

Bussecker, Erna. "Verflüchtigungskurven von RaB und RaC, die auf Gold bei einfachem und zweifachem Rücksto;sz niedergeschlagen sind." *Sitzungsberichte Akademie der Wissenschaften in Wien, Mathematisch-naturwissenschaftliche Klasse*, Abteillung IIa, 137 (1928): 117–126.

Cahan, David. "'The Geopolitics and Architectural Design of a Meteorological Laboratory: The Physikalisch-Technische Reichsanstalt in Imperial Germany." In *The Development of the Laboratory: Essays on the Place of Experiment in Industrial Civilization* edited by Frank James. 137–154. Basingstoke, Hampshire: Macmillan Press, 1989.

Cahan, Davis. *An Institute for an Empire: The Physikalisch-Technische Reichsanstalt 1871–1918*. Cambridge: Cambridge University Press, 1989.

Cartwright, Nancy; Cat, Jordi; Fleck, Lola and Uebel, Thomas. *Otto Neurath: Philosophy Between Science and Politics*. Cambridge: Cambridge University Press, 1996.

Cassidy, David. *Uncertainty: The Life and Science of Werner Heissenberg*. New York: W. H. Freeman and Company, 1992.

Cercignani, Carlo. *Ludwig Boltzmann: The Man Who Trusted Atoms*. Oxford: Oxford University Press, 1998.

Cernuschi, Claude. *Re/Casting Kokoschka: Ethics and Aesthetics, Epistemology and Politics in Fin-de-siecle Vienna*. London: Associated University Presses, 2002.

Chadwick, James and Gamow, George. "Artificial Disintegration by alpha-particles." *Nature* 126 (1930): 54–55.

Chadwick, James. "Possible Existence of a Neutron." *Nature*, 129 (1932): 312.

Chaloner, Clinton. "The Most Wonderful Experiment in the World: A History of the Cloud Chamber." *British Journal for the History of Science*, 30 (1997): 357–74.

Chariton, J. and Lea, C. "Some Experiments Concerning the Counting of Scintillations Produced by Alpha Particles. Part III. Practical Applications." *Proceedings of the Royal Society*, 122 (1929): 335–352.

Chase, C. "American Literature on Radium and Radium Therapy prior to 1906." *American Journal of Radiology*, 2 (1921): 29–34.

Clark, Claudia. *Radium Girls, Women and Industrial Health Reform, 1910–1935*. Chapel Hill, N.C.: University of North Carolina Press, 1997.

Clarke, Adele and Fujimura, Joan. *The Right Tools for the Job: at Work in Twentieth-Century Life Sciences*. Princeton, N.J.: Princeton University Press, 1992.

Cockcroft, J. "Some Recollections of Low Energy Nuclear Physics." In *Cambridge Physics in the Thirties* edited by J. Hendry. 74–80. Bristol: Adam Hilger, 1984.

Coen, Rachel. *A Scientific Dynasty: Probability, Liberalism, and the Exner Family in Imperial Austria*. Ph.D. Thesis. Harvard University, 2004.

Collins, James. "The Design Process for the Human Workplace." In *The Architecture of Science* edited by Peter Galison and Emily Thomson. 399–412. Cambridge Mass.: MIT Press, 1999.

Cooper, Annabel and Law, Robin and Malthus, Jane. "Rooms of their Own: Public Toilets and Gendered Citizens in a New Zealand City, 1860–1940." *Gender, Place and Culture*, 7, no. 4 (2000): 417–433.

Crane, Verner. "The Club of Honest Whigs: Friends of Science and Liberty." *William and Mary Quarterly*, 23 no. 2 (1966): 210–233.

Crawford, Elizabeth; Sime, Ruth and Walker, Mark. "A Nobel Tale of Postwar Injustice." *Physics Today*, (September 1997): 26–32.

Creager, Angela N.H. "Tracing the Politics of Changing Postwar Research Practices: the Export of 'American' Radioisotopes to European Biologists." *Studies in the History and Philosophy of Biological and Biomedical Sciences*, 33 (2002): 367–88.

Crookes, William. "The Emanations of Radium." *Proceedings of the Royal Society*, 71 (1903): 405–408.

Crossfield, Tina. "Irene Joliot-Curie: Following Her Mother's Footsteps." In *A Devotion to their Science: Pioneer Women of Radioactivity edited by* Rayner-Canham, Marelene and Rayner-Canham, Geoffrey. 97–123. Philadelphia: Chemical Heritage Foundation and Montreal: McGill-Queen's University Press, 1997.

Crowther, J. G. *The Cavendish Laboratory 1874–1974*. London: MacMIllan, 1974.

Curie, Eve. *Madame Curie*. New York: Doubleday and Company Inc, 1937.

Curie, Marie. "Radium and Radioactivity." *Century Magazine*. 1904: 461–466.

Curie, Pierre and Curie, Marie. "Sur une substance nouvelle radioactive contenue dans le pecheblende." *Comptes Rendus des Seanses de l' Academie des Sciences, Paris*, 127 (1898): 175–178.

Curie, Pierre; Curie, Marie, and Bèmont, Gustav. "Sur une nouvelle substance fortement radio-active contenue dans le pechblende." *Comptes Rendus des Seanses de l' Academie des Sciences, Paris*, 127 (1898): 1215–7.

Curran, Samuel and Baker, W. R. "Photoelectric Alpha-Particle Detector." *The Review of Scientific Instruments*, 19, no. 2 (1948): 116.

Dahms, Hans-Joachim. "The Emigration of the Vienna Circle." In *The Cultural Exodus from Austria* edited by Friedrich Stadler and Peter Weibel. 57–79. Wien, New York: Springer Verlag, 1995.

Darwin, C. G. "The Discovery of Atomic Number." *New Zealand Science Review*, 14, (1956): 102–108.

Daston, Loraine and Galison, Peter. "The Image of Objectivity." *Representations*, 40 (1992): 81–128.

Daston, Lorraine. ed. *Biographies of Scientific Objects*. Chicago: University of Chicago Press, 2000.

Daston, Lorraine, ed. *Things That Talk: Object Lessons from Art and Science*. New York: Zone Books, 2004.

Davis, J. L. "The Research School of Marie Curie in the Paris Faculty, 1907–14." *Annals of Science*, 52 (1995): 321–355.

de Grazia, Victoria and Furlough, Ellen. *The Sex of Things: Gender and Consumption in Historical Perspective*. Berkeley: University of California Press, 1996.

Deacon, G. E. "Hans Pettersson 1888–1966." *Biographical Memoirs of Fellows of the Royal Society*, 12 (1966): 405–421.

Deichmann, Ute. *Biologists under Hitler*. Cambridge, Mass.: Harvard University Press, 1992.

Deinlein, Maria Renata. "Verweilzeiten der Radiumemanation (Radon) im menschlichen Körper bei Trinkkuren." *Sitzungsberichte Akademie der Wissenschaften in Wien, Mathematisch-naturwissenschaftliche Klasse*, Abteillung IIa, (1933): 127–134.

Dierig, Sven; Lachmund, Jens; Mendelsohn, Andrew. *Science and the City*, Osiris 18, Chicago: Chicago University Press, 2003.

Diethe, Carola. *Aspects of Distorted Sexual Attitudes in German Expressionist Drama with particular Reference to Wedekind, Kokoschka and Kaiser*. New York: Peter Lang, 1988.

DiSantis, David and DiSantis, Denise. "Radiologic History Exhibit." *Radiographics*, 11 no. 6 (1991): 1121–1138.

Dutreix, Jean and Dutreix, Andree. "Henri Bequerel (1852–1908)." *Medical Physics*, 22 no. 11 (1995): 1869–1875.

Dutreix, Jean and Tubiana, Maurice and Pierquin, Bernard. "The Hazy Dawn of Brachytherapy." *Radiotherapy and Oncology*, 49 (1998): 223–232.

Elster, Julius and Geitel, Hans. "Über die durch radioactive Emanation erregte scintillierende Phosphoreszenz der Sidot-Blende." *Physikalische Zeitschrift*, 4 (1903): 439–440.

Esden-Tempska, Carla. "Civic Education in Authoritarian Austria," *History of Education Quarterly*, 2 (1990): 187–211.

Exner, Alfred and Holzknecht, Guido. "Die Pathologie der Radium Dermatitis," *Sitzungsberichte der Kaiserl. Akademie der Wissenschaften in Wien, Mathematisch-na turwissenschaftliche Klasse*, Abteillung IIa, Band cxii, (1903): 155–162.

Exner, Alfred. "Über die Bahandlung von Ösophagus-Carzinomen mit Radiumstrahlen." *Wiener Klinische Wochenschrift*, 4 (1904): 96.

Exner, Franz."Le nouvel institut pour le radium a Vienne." *Le Radium*, 6 (1910): 244.

Exner, Sigmund. "Einige Beobachtungen über die durch Radiumstrahlen in den thierischen Geweben erzengte Phosphorescenz." *Centralblatt für Physiology*, xvii, (1903): 177–179.

Falconer, Isobel. "J. J. Thomson and 'Cavendish Physics'." In *The Development of the Laboratory: Essays on the Place of Experiment in Industrial Civilization* edited by Frank James. 104–117. Basingstoke, Hampshire: Macmillan Press, 1989.

Faulkner, Wendy. "The Technology Question in Feminism: A View from Feminist Technology Studies." *Women's Studies International Forum*, 24 (2001): 79–95.

Feather, Norman. *Lord Rutherford*. London: Priory Press, 1940.

Feigl, Herbert. "The Wiener Kreis in America." In *Inquiries and Provocations: Selected Writings* 1929–1974 edited by Herbert Feigl and Robert Cohen. 57–69. Dordrecht, Holland: Reidel, 1980.

Fellner, Fritz. "The Background of Austrian Fascism." In *Native Fascism in the Successor States, 1918–1945* edited by Peter Sugar. California: Clio Press, 1971.

Fellner, Günter. "The Emigration of Austrian Historians." In *The Cultural Exodus from Austria* edited by Friedrich Stadler and Peter Weibel. 174–186. Wien, New York: Springer Verlag, 1995.

Fernau, Albert and Fonovits-Smereker, Hilda. "Über das Verbleiben radioaktiver Substany im Organismus bei Radium-Emanationstrinkkuren" *Strahlentherapie* 46 (1933).

Fischer, Isidor. *Geschichte der Geselschaft der Ärtze in Wien 1837–1937*. Wien: Verlag von Julius Springer, 1938.

Flevaris, Spiros. *Universität Wien*. MA Thesis, Technical University of Graz, 2001.

Fonovits, Hilda. "Über die Erreichung des Sättigungsstromes für a-Strahlen im Plattenkondesator." *Sitzungsberichte Akademie der Wissenschaften in Wien, Mathemat isch-naturwissenschaftliche Klasse*, Abteillung IIa, 128 (1919): 761–793.

Fonovits-Smereker, Hilda. "Die Zahl der von einem a-Teilchen von RaC erzeugten Ionenpaare." *Sitzungsberichte Akademie der Wissenschaften in Wien, Mathematisc h-naturwissenschaftliche Klasse*, Abteillung IIa, 131 (1922): 355–363.

Forgan, Sophie and Gooday, Graeme. "Constructing South Kensington: The Buildings of Politics of T. H. Hixley's Working Environments." *British Journal for the History of Science*, 29 no. 103 (1996): 435–468.

Forgan, Sophie. "Bricks and Bones: Architecture and Science in Victorian Britain." In *The Architecture of Science* edited by Peter Galison and Emily Thomson. 181–208. Cambridge, Mass.: MIT Press, 1999.

Forgan, Sophie. "Context, Image and Function: A Preliminary Enquiry into the Architecture of Scientific Societies." *British Journal for the History of Science* 19 (1986): 83–113.

Forgan, Sophie. "Eine angemessen Häuslichkeit? Frauen und die Architecture der Wissenschaft im 19. Jahrhundert." In *Zwischen Vorderbühne und Hinterbühne Beiträge yum Wandel der Geschlechterbeyiehungen in der Wissenschaft vom 17. Jahrhundert bis zur Gegenwart* edited by W. Theresa, Bielefeld. 137–58. 2003.

Forgan, Sophie. "The Architecture of Display: Museums, Universities and Objects in Nineteenth-Century Britain." *History of Science*, 32 (1994): 139–162.

Forgan, Sophie. "The Architecture of Science and the Idea of a University." *Studies in History and Philosophy of Science*, 20 no. 40 (1989): 405–434.

Fox, B.W. "The History of Radium in Medicine in Manchester," *Clinical Oncology*, 10 (1998): 115–124.

Föyn, Ernst; Kara-Michailova, Elisabeth and Rona, Elisabeth. "Zur Frage der Künstlichen Umwandlung des Thoriums durch Neutronen." *Sitzungsberichte Akademie der Wissenschaften in Wien, Mathematisch-naturwissenschaftliche Klasse*, Abteillung IIa, 72 (1935): 159.

Föyn, Ernst; Karlik, Berta; Pettersson, Hans, and Rona, Elisabeth. "Radioactivity of Seawater." *Göteborgs Kungl. Vetensk. och Vitterhets-Samhälles Handlingar.* 6 (1939): 1–44.

Föyn, Ernst; Pettersson, Hans and Rona, Elisabeth. "Künstliche Umwandlung des Thoriums durch Neutronen." *Naturwissenschaft*, 23 (1935): 391.

Freidenreich, Harriet Pass. "Gender, Identity, and Community: Jewish University Women in Germany and Austria." In *In Search of Jewish Community: Jewish Identities in Germany and Austria, 1918–1933* edited by Michael Brenner and Derek Penslar. 154–175. Bloomington: Indiana University Press, 1998.

Friedmann, Friderike. "Experimentelle Bestimmung der Schwankungen in der Reichweite der einzelnen a-Teilchen." *Sitzungsberichte der Kaiserl. Akademie der Wissenschaften in Wien, Mathematisch-naturwissenschaftliche Klasse*, Abteillung IIa, 122 (1913): 1269–1289.

Frisch, Otto. "Experimental Work with Nuclei: Hamburg, London, Copenhagen." In *Nuclear Physics in Retrospect: Proceedings on a Symposium on the 1930s* edited by Roger Stuewer. 65–79. Minneapolis: University of Minnesota Press, 1979.

Fürth, Ernestine von. "Geschichte der Frauenstimmrechtsbewegung." In *Frauenbewegung, Frauenbildung und Frauenarbeit in Österreich* edited by Martha Stephanie Braun et al. 65–83. Wien, 1930.

Galison, Peter and Jones, Caroline. "Trajectories of Production: Laboratories, Factories, Studios." In *Laboratorium* edited by Ulrich Obrist and Barbara Vanderlinden. 205–210. Dumont: Antwerpen Open, Roomade, 2001.

Galison, Peter and Thomson, Emily. *The Architecture of Science.* Cambridge, Mass.: MIT Press, 1999.

Galison, Peter. "Constructing Modernism: The Cultural Location of *Aufbau.*" In *Origins of Logical Empiricism Minnesota Studies in the Philosophy of Science, Vol. XVI* edited by Ronald Giere and Alan Richardson. 17–44. Minneapolis: University of Minnesota Press, 1996.

Galison, Peter. "Marietta Blau: Between Nazis and Nuclei." *Physics Today,* 50 no. 11 (1997): 42–48.

Galison, Peter. *Image and Logic.* Chicago: The University of Chicago Press, 1997.

Gamow, George. "The Quantum Theory of Nuclear Disintegration." *Nature,* 112 (1928): 805–806.

Gamow, George. *My World Line. An Informal Autobiography.* New York: Viking Press, 1970.

Gamow, George. *The Constitution of Atomic Nuclei and Radioactivity.* Oxford: Clarendon Press, 1931.

Gamper, Martina. "'. . . so kann ich nicht umhin zu wendern, dass nicht mehr Ärztinnen da sind . . .' Die Stellung weibliche Ärzte im 'Roten Wien'. " In *Töchter des Hippokrates: 100 Jahre akademische Ärztinnen in Österreich* edited by Birgit Bolognese-Leuchtenmüller and Sonia Horn. 79–96. Wien: Pressestelle und Verl. Der Österr. Ärztkammer, 2000.

Gardner, Sheldon and Stevens, Gwendolyn. *Red Vienna and the Golden Age of Psychology, 1918–1938.* New York: Praeger, 1992.

Gay, Hannah. "Invisible Resource: William Crookes and his Circle of Support, 1871–81." *British Journal for the History of Science,* 29 (1996): 311–336.

Gay, Peter. "Sigmund Freud: A Brief Life in Sigmund Freud." In *Five Lectures on Psych-Analysis,* Sigmund Freud. xi–xxix, New York: Norton, 1989.

Gay, Peter. *Freud: A Life for Our Time.* New York: Norton, 1988.

Geehr, Richard. *Karl Lueger: Mayor of fin-de-siècle Vienna.* Detroit: Wayne State University Press, 1990.

Gehmacher, Johanna. "Men, Women, and the Community Borders: German-Nationalist and National Socialist Discourses on Gender, 'Race,' and National Identity in Austria, 1918–1938." In *Nation, Empire, Colony: Historicizing Gender and Race* edited by Ruth Roach Pierson and Nupur Chaudhuri. 205–219. Bloomington: Indiana University Press, 1998.

Gellately, Robert and Stoltzfus, Nathan. *Social Outsiders in Nazi Germany.* Princeton and Oxford: Princeton University Press. 2001.

Gerward, Leif and Rassat, Andre. "La decouverte des rayons gamma." *Comptes Rendus des Seanses de l' Academie des Sciences, Paris,* 1 no. 7 (2000): 965–973.

Gieryn, Thomas. "Two Faces on Science: Building Identities for Molecular Biology and Biotechnology." In *The Architecture of Science* edited by Peter Galison and Emily Thomson. 423–455. Cambridge, Mass.: MIT Press, 1999.

Gieryn, Thomas. *Cultural Boundaries of Science: Credibility on the Line*. Chicago: University of Chicago Press, 1999.

Goldhaber, Maurice. "Reminiscence from the Cavendish Laboratory in the 1930s." *Annual Review of Nuclear and Particle Science*, 43 (1993): 1–25.

Goldhaber, Maurice. "The Nuclear Photoelectric Effect and Remarks on Higher Multiple Transitions: A Personal History." In *Nuclear Physics in Retrospect: Proceedings on a Symposium on the 1930s* edited by Roger Stuewer. 83–110. Minneapolis: University of Minnesota Press, 1979.

Golinksi, Ian. "Conversations on Chemistry: Talk About Phlogiston in the Coffee House Society, 1780–1787." In *Discussing Chemistry and Steam: The Minutes of a Coffee House Philosophical Society 1780–1787* edited by Trevor Levere and Gerard Turner. 191–205. Oxford: Oxford University Press, 2002.

Golinski, Jan. *Making Natural Knowledge: Constructivism and the History of Science*. Cambridge: Cambridge University Press, 1998.

Golitschek, Fritz and Elbwart, Edler. *Der Neubau des Physikalischen Institutes der Wiener k.k. Universität*. Wien, 1915.

Gordin, Michael. *A Well-Ordered Thing: Dmitrii Mendeleev at the Shadow of the Periodic Table*. New York: Basic Books, 2004.

Gould, Paula. "Women and the Culture of University Physics in Late Nineteenth Century Physics." *British Journal for the History of Science*, 30 (1997):127–49.

Greinacher, H. "Über die akustische Beobachtung und galvanometrische Registierung von Elementarstrahlen und Einzelionen." *Zeitschrift für Physik*, 23 (1924): 361–378.

Grint, Keith and Gill, Rosalind. eds. *The Gender-Technology Relation: Contemporary Theory and Research*. London: Taylor and Francis, 1995.

Growther, James. *The Cavendish Laboratory 1874–1974*. New York, 1974.

Gruber, Helmut. *Red Vienna: Experiment in Working-Class Culture 1919–1934*. New York: Oxford University Press, 1991.

Haberlandt, Herbert; Karlik, Berta and Przibram, Karl. "Artificial Production of the Blue Fluorescence of Fluorite." *Nature*, 133 (1934): 99–100.

Haberlandt, Herbert; Karlik, Berta and Przibram, Karl. "Synthese der blauen Fluoritfluoreszenz." *Sitzungsberichte Akademie der Wissenschaften in Wien, Mathematisc h-naturwissenschaftliche Klasse*, Abteillung IIa, 1933: 301–2.

Haberlandt, Herbert; Karlik, Berta and Przibram, Karl. "Synthese der grünen Tieftemperaturfluoreszenz des Flurites." *Sitzungsberichte Akademie der Wissenschaften in Wien, Mathematisch-naturwissenschaftliche Klasse*, Abteillung IIa. 71 (1934): 1–2.

Haberlandt, Herbert; Karlik, Berta and Przibram, Karl. "Zur Fluoreszenz des Flurites II." *Sitzungsberichte Akademie der Wissenschaften in Wien, Mathematisch-naturw issenschaftliche Klasse*, Abteillung IIa, 143 (1934): 151–161.

Haberlandt, Herbert; Karlik, Berta and Przibram, Karl. "Zur fluoreszenz der Flurites IV; über einen Urannachweis in Fluorit und über die Tieftemperaturfluoreszenz."

Sitzungsberichte Akademie der Wissenschaften in Wien, Mathematisch-naturwissenschaftliche Klasse, Abteillung IIa, 144 (1935): 135–140.

Haberlandt, Herbert; Karlik, Berta and Przibram, Karl. "Zur Fluoreszenz der Flurites III; das Linienfluoreszenzspektrum." *Sitzungsberichte Akademie der Wissenschaften in Wien, Mathematisch-naturwissenschaftliche Klasse,* Abteillung IIa, 144 (1935): 77–85.

Hacohen, Malachi Haim. *Karl Popper-The Formative Years, 1902–1945*. Cambridge: Cambridge University Press, 2000.

Hahn, Otto. "Friedrich Giesel." *Physikalische Zeitschrift,* 12 (1928): 353–357.

Haitinger, Ludwig and Ulrich, Karl. "Bericht über die Verarbeitung von Uranpecherzückständen." *Sitzungsberichte der Kaiserl. Akademie der Wissenschaften in Wien, Mathematisch-naturwissenschaftliche Klasse,* Abteillung IIa,, 117 (1908): 619–630.

Halpern, Leopold. "Marietta Blau (1894–1970)." In *Women in Chemistry and Physics: A Bibliographical Sourcebook* edited by Louise Grinstein, Rose Rose and Miriam Rafailovich. 57–64. Westport: Greenwood Press, 1993.

Handlbauer, Bernhard. "The Emigration of the Viennese Individual Psychologists." In *The Cultural Exodus from Austria* edited by Friedrich Stadler and Peter Weibel. 122–126. Wien, New York: Springer Verlag, 1995.

Hannaway, Owen. "Laboratory design and the Aim of Science, Andreas Libavius versus Tycho Brahe." *Isis,* 77 (1986): 585–610.

Harvie, David. "The Radium Century." *Endeavour,* 23 no. 3 (1999): 100–105

Hayter, Charles. *An Element of Hope: Radium and the Response to Cancer in Canada, 1900–1940*. Montreal: McGill-Queen's University Press, 2005.

Heering, Kurt-Jürgen. *Das Wiener Kaffeehaus*. Frankfurt am Main: Insel Verlag, 1993.

Heidegger, Martin. *Poetry, Language, Thought*. New York: Harper Colophon Books, 1971.

Heilborn, J. L. *Ernest Rutherford and the explosion of Atom*. Oxford: Oxford University Press, 2003.

Heilmann, Hans-Peter. "Radiation Oncology: Historical Development in Germany." *International Journal of Radiation Oncology, Biology, Physics,* 35 (1996): 207–217.

Heimann, Bertha. "Über die Lebensdauer des Thoriums." *Sitzungsberichte der Kaiserl. Akademie der Wissenschaften in Wien, Mathematisch-naturwissenschaftliche Klasse,* Abteillung IIa, Band cxxii, (1914): 1369–1372.

Heindl, Waltraud. "Die konfessionellen Verhältnisse. Jüdische und katholische Studentinnen." In *"Durch Erkenntnis zu Freiheit und Glück" Frauen an der Universität Wien (ab 1897)* edited by Waltraud Heindl and Marina Tichy. 139–49. Wien: Universität Wien, 1990.

Heindl, Waltraud. "Zur Entwicklung des Frauen-studiums in Österreich." In *"Durch Erkenntnis zu Freiheit und Glück" Frauen an der Universität Wien (ab 1897)* edited by Waltraud Heindl, and Marina Tichy. 17–26. Schriftenreihe des Universitätsarchivs, Band 5, Wien: Universität Wien, 1990.

Heise, Ulla. *Kaffe und Kaffeehaus: Die Geschichte des Kaffees*. Insel Verlag, 2002.

Heise, Ulla. *Kaffee und Kaffee Haus: Eine Kulturgeschichte*. Zurich: Olms Presse, 1987.

Hernegger, Friedrich and Karlik, Berta. "Die quantitative Bestimmung sehr kleiner Uranmengen und der Urangehalt des Meerwassers." *Sitzungsberichte Akademie der Wissenschaften in Wien, Mathematisch-naturwissenschaftliche Klasse,* Abteillung IIa, 144 (1935): 217–25.

Hernegger, Friedrich and Karlik, Berta. "Uranium in Seawater." *Göteborgs Kungl. Vetensk. och Vitterhets-Samhälles Handlingar,* 4 no. 12 (1935): 1–15.

Hernegger, Friedrich. "Methoden für einen empfindlichen Urannachweis in Quellwässern und Quellsedimenten." *Sitzungsberichte Akademie der Wissenschaften in Wien, Mathematisch-naturwissenschaftliche Klasse,* Abteillung IIa. 70 (1933): 15–18.

Herrberg, H. and Wagner, H. *Wiener Melange: Frauen zwischen Salon und Kaffeehaus.* Vienna: Edition Ebersbach, 2002.

Herzfeld, Karl. "Über die Schwankungen der Reichweite bei den einzelnen a-Teilchen." *Physikalische Zeitschrift,* 13 (1912): 547–550.

Hess, Victor and Hornyak, Maria. "Über die relative Ionisation von alpha-Strahlen in verschiedenen Gasen." *Sitzungsberichte Akademie der Wissenschaften in Wien, Mathematisch-naturwissenschaftliche Klasse,* Abteillung IIa, 129 (1920): 661–672.

Hess, Victor. "Persönliche Erinnerungen aus dem ersten Jahrzehnt des Instituts für Radiumforschung." In *Festrift des Institutes für Radiumforschuung Anlässlich seines 40jährigen Bestandes (1910–1950).* 43–45. Wien, 1950.

Hess, Victor. "Über den Ionenwind." *Sitzungsberichte Akademie der Wissenschaften in Wien, Mathematisch-naturwissenschaftliche Klasse,* Abteillung IIa, 128 (1919): 1029–1079.

Hessenbruch, Arne. "Rutherford's 1901 Experiment on Radiation Energy and His Creation of a Stable Detector." *Archive for History of Exact Sciences,* 54 (2000): 403–420.

Hevesy, Georg von and Rona, Elizabeth. "Die Lösungsgeschwindigkeit molekularer Schichten." *Zeitschrift für Physikalische Chemie,* 89 (1915): 294–305.

Himstedt, Franz. "Radioactivity" in *Annual report of the board of regents of the Smithsonian Institution showing the operations expenditures and conditions of the institution,* 1906: 117–130.

Hittner, Lotte. *Geschichte des Studienfaches Physik an der Wiener Universität in der Letzten Hundert Jahren.* Ph.D. Thesis: University of Vienna, 1949.

Hliniak, Andrzej; Reinfuss, Marian and Skolyszewski, Jan. "An Outline of the History of Radiation Oncology in Poland." *International Journal of Radiation Oncology, Biology, Physics,* 35 (1996): 799–802.

Hofer, Veronica. "The Beginnings of Biological System Theory in the Context of Scientific Modernism in Vienna: The Lines of Discourse and Scientific Communities Forming the Intellectual Background for Ludwig von Bertalanffy" (unpublished manuscript)

Hoffman, Dietrich. "Die Physikdenkschriften von 1934/36 und zur Situation der Physik im fascistischen Deutschland." *Wissenschaft und Staat,* 1989: 185–211.

Hoffmann, Christoph. "The Design of Disturbance; Physics Institutes and Physics Research in Germany, 1870–1910." *Perspectives on Science,* 9 no. 2 (2002): 173–195.

Holmes, Frederick and Levere, Trevor. *Instrument and Experimentation in the History of Chemistry*. Cambridge, MA: MIT Press, 2000.

Holoubek, Roudolf. "Die Sichtbarmung von Atomtrümmerbahnen." *Zeitschrift für Physik*, 42 (1927): 704–720.

Holoubek, Rudolf. "Der Nachweis von Atomtrümmern nach der Wilson-Methode." *Sitzungsberichte Akademie der Wissenschaften in Wien, Mathematisch-naturwisse nschaftliche Klasse*, Abteillung IIa, 136 (1927): 321–336.

Holsti, Lars. "How Radiation Oncology Emerged as a Discipline in Finland." *International Journal of Radiation Oncology, Biology, Physics*, 35 no. 40 (1996): 793–798.

Hönigschmid, Otto and Horovitz, Stefanie. "Revision des Atomgewichtes des Thoriums. Analyse des Thoriumbromids." *Sitzungsberichte der Kaiserl. Akademie der Wissenschaften in Wien, Mathematisch-naturwissenschaftliche Klasse*, Abteillung IIa, 125 no. 3 (1916): 149–178. Republished in *Monatshefte für Chemie*, 37 (1916) :335–45.

Hönigschmid, Otto and Horovitz, Stefanie. "Sur le poids atomique du plomb de la pechblende." *Comptes Rendus des Seanses de l' Academie des Sciences, Paris*, 158 (1914): 1796–1798.

Hönigschmid, Otto and Horovitz, Stefanie. "Über das Atomgewicht des Uranbleis." *Monatshefte für Chemie*, 35 (1914): 1557–1560.

Hönigschmid, Otto and Horovitz, Stefanie. "Über das Atomgewicht des Uranbleis." *Sitzungsberichte der Kaiserl. Akademie der Wissenschaften in Wien, Mathematisch -naturwissenschaftliche Klasse*, Abteillung IIa, Band cxxiii, (1914): 2407–2432.

Hönigschmid, Otto and Horovitz, Stefanie. "Über das Atomgewicht des Uranbleis II." *Monatshefte für Chemie*, 36 (1915): 335–380.

Hönigschmid, Otto and Horovitz, Stefanie. "Zur Kenntnis des Atomgewichtes des Urans." *Sitzungsberichte der Kaiserl. Akademie der Wissenschaften in Wien, Mathe matisch-naturwissenschaftliche Klasse*, Abteillung IIa, (1915): 1089–1094.

Hönigschmid, Otto and Horovitz, Stefanie. "Zur Kenntnis des Atomgewichtes des Ioniums." *Sitzungsberichte der Kaiserl. Akademie der Wissenschaften in Wien, Mat hematisch-naturwissenschaftliche Klasse*, Abteillung IIa, 125 (1916): 179–199. Republished in *Monatshefte für Chemie*, 37 (1916): 305–34.

Hornyak, Maria. "Über die Oberflächenionisation." *Sitzungsberichte Akademie der Wissenschaften in Wien, Mathematisch-naturwissenschaftliche Klasse*, Abteillung IIa, 130 (1921): 135–147.

Howes, Ruth and Herzenberg, Caroline. *Their Day in the Sun: Women of the Manhattan Project*. Philadelphia: Temple University Press, 1999.

Hubenstorf, Michael. "The Anatomical Science in Vienna." *The Lancet*, 355 (22 August 2000): 1385–1386.

Hughes, Bill. "U.S. Begins Compensating Workers Exposed to Toxic Substances." *The Journal News*, (August 20, 2001): 1.

Hughes, Jeffrey. "'Modernists with a Vengeance': Changing Cultures of Theory in Nuclear Science, 1920–1930." *Studies in the History and Philosophy of Modern Physics*, 29 no. 3 (1998): 339–367.

Hughes, Jeffrey. "Radioactivity and Nuclear Physics." In *The Cambridge History of Science: The Modern Physical and Mathematical Sciences* vol. 5 edited by Mary Jo Nye, Cambridge: Cambridge University Press, 2003.

Hughes, Jeffrey. *The Radioactivists: Community, Controversy and the Rise of Nuclear Physics*. Ph.D. Thesis: University of Cambridge, 1993.

Hunt, Felicity and Barker, Carol. *Women at Cambridge: A Brief History*. Cambridge: University of Cambridge Press, 1998.

Janik, Allan and Toulmin, Stephen. *Wittgenstein's Vienna*. New York: Touchstone, 1973.

Janik, Allan and Veigl, Hans. *Wittgenstein in Vienna: A Biographical Excursion Through the City and its History*. New York; Springer, 1998.

Jeffery, Charlie. *Social Democracy in the Austrian Provinces 1918–1934: Beyond Red Vienna*. London: Leicester University Press, 1995.

Johnson, Jeffrey. *The Kaiser's Chemists: Science and Modernization in Imperial Germany* Chapel Hill, NC: University of North Carolina Press, 1990.

Julian, Maureen. "The Early Days of the X-ray Revolution." *Chemical Heritage*, 4 no. 2 (1996): 38–39.

Julian, Maureen. "Women in Crystallography." In *Women of Science: Righting the Record* edited by G. Kass-Simon, and Patricia Farnes. 300–334. Bloomington: Indiana University Press, 1989.

Kant, Horst. "Von KWI für Chemie zum KWI für Radioaktivität" *Dahlemer Archivgespräche*, 8 (2002): 57–92.

Kara-Michailova, Elisabeth and Kamburov, K et al. "Radiological and Hydrological Research in the Area of Dolni Bogrov Village near Sofia in Respect to the Increased Cancer Disease." *Isvestiya na Fizicheskiya Institut s ANEB* 9 no. 1 (1961): 109.

Kara-Michailova, Elisabeth and Karlik, Berta. "Über die relative Helligkeit der Szintillationen von H-Strahlen bei verschiedenen Reichweiten." *Sitzungsberichte Akademie der Wissenschaften in Wien, Mathematisch-naturwissenschaftliche Klasse*, Abteillung IIa, 138 (1929): 581–587.

Kara-Michailova, Elisabeth and Przibram, Karl. "Orientierte Gleitbüschel auf Kristallflächen." *Zeitschrift für Physik*, 2 (1920): 297.

Kara-Michailova, Elisabeth and Przibram, Karl. "Über Radiolumineszenz und Radio-Photolumineszenz." *Sitzungsberichte Akademie der Wissenschaften in Wien, Mathematisch-naturwissenschaftliche Klasse*, Abteillung IIa, 131 (1922): 111–530.

Kara-Michailova, Elisabeth and Przibram, Karl. "Über Radiolumineszenz und Radio-Photolumineszenz." *Sitzungsberichte Akademie der Wissenschaften in Wien, Math ematisch-naturwissenschaftliche Klasse*, Abteillung IIa, 132 (1923): 285–298.

Kara-Michailova, Elisabeth. "Helligkeit und Zällbarkeit der Scintillationen von magnetische abgelenkten H-Strahlen verschiedener Geschwindigkeit." *Sitzungsberichte Akademie der Wissenschaften in Wien, Mathematisch-naturwissenschaftliche Klasse*, Abteillung IIa, 136 (1927): 357–368.

Kara-Michailova, Elisabeth. "The radioactivity of Water Sources in Villages with Endemic Nephritis." *Endemic Nephritis in Bulgaria*, (1960): 152–162.

Kara-Michailova, Elisabeth; Nikolov, K. and Doitchinova, K.. "The Radioactivity of Mineral Water-Springs in the Narechten Baths." *BKFLF*, 2 (1962): 6–10 and 3 (1962): 2–6.

Kara-Michailova, Elisasbeth and Pettersson, Hans. "Über die Messung der relativen Helligheit von Szintillationen." *Sitzungsberichte Akademie der Wissenschaften in Wien, Mathematisch-naturwissenschaftliche Klasse,* Abteillung IIa, 133 (1924): 163–168.

Kara-Michailova, Elizabeth. "Quantitative optische Unterscheidung von alpha- und H-Teilchen." *Physikalische Zeitschrift,* 25 (1924): 594–596.

Karlik, Berta and Bernert, Traude. "Das Element 85 in den natürlichen Zerfallsreihe." *Zeitschrift für Physik,* 123 (1943): 51.

Karlik, Berta and Bernert, Traude. "Das Element 85 in der Aktinium Reihe." *Naturwissenschaft,* 32 (1944): 44.

Karlik, Berta and Bernert, Traude. "Ein weiterer dualer Zerfall in der Thorium-Reihe." *Naturwissenschaft,* 31 (1943): 492.

Karlik, Berta and Bernert, Traude. "Eine neue natürliche Alpha-Strahlung." *Naturwissenschaft,* 31 (1943): 298.

Karlik, Berta and Bernert, Traude. "Entsehung des Ekajod-Isotops 218 in der Radiumreihe." *Sitzungsberichte Akademie der Wissenschaften in Wien, Mathematisch-naturwissenschaftliche Klasse,* Abteillung IIa, (1945): 34–35.

Karlik, Berta and Bernert, Traude. "Über die Entemanierung starker Radium-C-Präparate." *Sitzungsberichte Akademie der Wissenschaften in Wien, Mathematisch-naturwissenschaftliche Klasse,* Abteillung IIa, 151 (1942): 267–275.

Karlik, Berta and Bernert, Traude. "Über eine dem Element 85 zugeordnete alpha-Strahlung." *Sitzungsberichte Akademie der Wissenschaften in Wien, Mathematisch-naturwissenschaftliche Klasse,* Abteillung IIa, 152 (1943): 103–110.

Karlik, Berta and Bernert, Traude. "Über eine vermutete beta-Strahlung des Ra A und die natürliche Existenz des Elementes 85." *Naturwissenchaft,* 30 (1942): 685.

Karlik, Berta and Bernert, Traude. "Über zwei neue alpha-Strahluungen in der Thorium- und in der Actiniumreihe." *Sitzungsberichte Akademie der Wissenschaften in Wien, Mathematisch-naturwissenschaftliche Klasse,* Abteillung IIa, (1944): 2–3.

Karlik, Berta and Bernert, Traude. "Zur Frage eines dualen Zerfalls des Radium A." *Sitzungsberichte Akademie der Wissenschaften in Wien, Mathematisch-naturwissenschaftliche Klasse,* Abteillung IIa, 151 (1942): 255–265.

Karlik, Berta and Gilchrist H. S. "Separation of Normal Longchain Hydrocarbons Through High-Vacuum Distillation." *Journal of the Chemical Society,* 54 (1932): 1992.

Karlik, Berta and Kara-Michailova, Elisabeth. "Über die durch alpha-Strahlen erregte Lumineszenz und deren Zusammenhang mit der Teilchenenergie." *Sitzungsberichte Akademie der Wissenschaften in Wien, Mathematisch-naturwissenschaftliche Klasse,* Abteillung IIa, 137 (1928): 363–380.

Karlik, Berta and Kara-Michailova, Elisabeth. "Zur Kenntnis der Szintillationsmethode." *Zeitschrift für Physik,* 48 (1928): 765–783.

Karlik, Berta and Knaggs, I.. *Tables of Cubic Crystal Structures.* London: Adam Hilger, 1932

Karlik, Berta and Rona, Elisabeth. "Untersuchungen der Reichweite von alpha-Strahlen des Actinium X und seiner Folgeprodukte mit der Lumineszenzmethode." *Sitzungsberichte Akademie der Wissenschaften in Wien, Mathematisch-naturwissenschaftliche Klasse,* Abteillung IIa, 142 (1933): 121–126.

Karlik, Berta and Rona, Elisabeth. "Untersuchungen über die Reichweite von Alphastrahlen des Actinium X und seiner Folgeprodukte mit der Lumineszenzmethode." *Sitzungsberichte Akademie der Wissenschaften in Wien, Mathematisch-naturwisse nschaftliche Klasse,* Abteillung IIa, 143 (1934): 217–221.

Karlik, Berta and Schmid, Erich. *Franz Serafin Exner und sein Kreis: Ein Beitrag zur Geschichte der Physik in Österreich.* Wien: Österreichischen Akademie der Wissenschaften, 1982.

Karlik, Berta. "1938 bis 1950." *Festrift des Institutes für Radiumforschung Anlässlich seines 40jährigen Bestandes (1910–1950),* 35–41, Wien, 1950.

Karlik, Berta. "Eine Lumineszenzmethode zur Untersuchung der Reichweite von alpha-Strahlen." *Sitzungsberichte Akademie der Wissenschaften in Wien, Mathema tisch-naturwissenschaftliche Klasse,* Abteillung IIa, 142 (1933): 115–119.

Karlik, Berta. "Hans Pettersson Nachruf." *Almanach der Österreichischen Academie der Wissenschaften,* 119 (1970): 303–317.

Karlik, Berta. "Karl Przibram Nachruf." *Almanach der Österreichischen Academie der Wissenschaften*124 (1974): 379–387.

Karlik, Berta. "Über die Abhangigkeit der Szintillationen von der Beschaffenheit des Zinksulfids und das Wesen des Szintillationsvorganges." *Sitzungsberichte Akademie der Wissenschaften in Wien, Mathematisch-naturwissenschaftliche Klasse,* Abteillung IIa, 136 (1927): 531–561.

Karlik, Berta. "Uber die Szintillationsfähigeit von Kalziumwolframat." *Sitzungsberichte Akademie der Wissenschaften in Wien, Mathematisch-naturwissenschaftliche Klasse,* Abteillung IIa, 139 (1930): 319–326.

Karlik, Berta. "Untersuchungen zur Lumineszenz von Zinksulfid und Diamant unter Einwirkung von Alphastrahlen." *Sitzungsberichte Akademie der Wissenschaften in Wien, Mathematisch-naturwissenschaftliche Klasse,* Abteillung IIa, 139 (1930): 509–519.

Kautz, Theodora. "Ermittlung der Halbierungszeit von RaD mittels Warmemessung eines alten Ra-Präparates." *Sitzungsberichte Akademie der Wissenschaften in Wien, Mathematisch-naturwissenschaftliche Klasse,* Abteillung IIa, 135 (1926): 93–97.

Kindermann, Gottfried. *Hitler's Defeat in Austria, 1933–1934: Europe's First Containment of Nazi Expansionism.* Boulder, Colo.: Westview Press, 1988.

Kirby, H. W. "The Discovery of Actinium." *Isis,* 62 no. 3 (1971): 290–308.

Kirkham, Pat. *The Gendered Object.* Manchester: Manchester University Press, 1996.

Kirsch, Gerhard and Pettersson, Hans. "Experiments on the Artificial Disintegration of Atoms." *Philosophical Magazine,* 47 (1923): 500–512.

Kirsch, Gerhard and Pettersson, Hans. "Long-Range particles from Radium Active Deposit." *Nature,* 112 (1923): 394–395.

Kirsch, Gerhard and Pettersson, Hans. "Long-Range Particles from Radioactive Deposit." *Nature* 112 (1923): 687.

Kirsch, Gerhard and Pettersson, Hans. "Uber die Atomzentrümmerung durch alpha-Partikeln." *Sitzungsberichte Akademie der Wissenschaften in Wien, Mathematisch-naturwissenschaftliche Klasse,* Abteillung IIa, 132 (1923): 299–307.

Kirsch, Gerhard and Wambacher, Hertha. "Geschwindigkeit der Neutronen aus Beryllium." *Sitzungsberichte Akademie der Wissenschaften in Wien, Mathematisch -naturwissenschaftliche Klasse,* Abteillung IIa, 142 (1933): 241–249.

Kjäer, Illy. "Zehn Jahre parlamentarische Frauenarbeit in Österreich." *Die Östereicherin,* 2 no. 2 (1929): 2–4.

Klein, Ursula and Lefevre, Wolfgang. *Materials in Eighteenth-Century Science: A Historical Ontology,* Cambridge, Mass.: MIT Press, 2007.

Klein, Ursula and Lefevre, Wolfgang. *Shifting Ontologies: Materials in Eighteenth-Century Science.* Cambridge, Mass.: MIT Press (forthcoming).

Klein, Ursula. "Shifting Ontologies, Changing Classifications: Plant Materials from 1700 to 1830." *Studies in History and Philosophy of Science,* 36 (2005): 261–329.

Klein, Ursula. *The Making of Materials* (1500-1800) (forthcoming).

Koestler, Arthur. *The Case of a Midwife Toad.* New York: Random House, 1971.

Kogelnik, Dieter. "Inauguration of Radiotherapy as a New Scientific Specialty by Leopold Freund 100 Years Ago." *Radiotherapy and Oncology,* 42 (1997): 203–211.

Kogelnik, Dieter. "The History and Evolution of Radiotherapy and Radiation Oncology in Austria." *International Journal of Radiation Oncology, Biology, Physics,* 35 no. 2 (1996): 219–226.

Kohl, F. "Von den 'Becquerel-Strahlen' zur Radioactivität." *Deutsche medizinische Wochenschrift,* 122 no. 15 (1996): 488–491.

Kohler, Robert. *Lords of the Fly: Drosophila Genetics and the Experimental Life.* Chicago: Chicago University Press, 1994.

Kopytoff, Igor. "The Cultural Biography of Things: Commoditization as Process." In *The Social Life of Things: Commodities in Cultural Perspective* edited by Arjun Appadurai. 64–91. Cambridge: Cambridge University Press, 1986.

Korotin, Ilse. "Auf eisigen Firnen, Zur intellectuellen Tradition von Frauen." In *Wissenschafts als Kultur, Österreichs Beitrag zur Moderne* edited by Friedrich Stadler. 291–306, Bd 6, Wien: Veröffentlichungen des Wiener Kreises. 1997.

Kragh, Helge. *Introduction to Historiography of Science.* Cambridge: Cambridge University Press, 1987.

Krebs, Adolf. "Early History of the Scintillation Counter." *Science,* 122 (1955): 17–18.

Krebs, Adolf. "Ein Demonstrationsversuch zur Emanationsdiffusion." *Annalen der Physik,* 39 no. 5 (1941): 330–332.

Kreidl, Norbert."Zur Verwendberkeit des Geiger'schen Spitzenzählers für Versuche über Atomzetrümmerung." *Sitzungsberichte Akademie der Wissenschaften in Wien, Mathematisch-naturwissenschaftliche Klasse,* Abteillung IIa, 136 (1927): 589–602.

Krige, John. "Atoms for Peace, Scientific Internationalism and Scientific Intelligence." In *Global Power Knowledge. Science and Technology in International Affairs,* edited by John Krige and Kai Kenrik Barth. *Osiris,* 21 (2006).

Kurner, Charlotte. *Lise, Atomphysikerin: Die Lebensgeschichte der Lise Meitner.* Weinheim: Beltz & Gelberg, 1998.

LaFollette, M. "Eyes on the Stars: Images of Women Scientists in Popular Magazines." *Science, Technology and Human Values,* 13 no [[3Q]], (1988): 262–175.

Lamberg-Karlovsky, C. "The Biography of an Object: The Intellectual Style Vessels of the Third Millennium B.C." In *History from Things: Essays on Material Culture* edited by Steven Lubar and David Kingery. 270–292. Washington: Smithsonian Institution Press, 1993.

Lang, Daniel. "A Most Valuable Accident." *New Yorker,* (May 2, 1959): 49–92.

Lankford, John and Slavings, Rickey. "Gender and Science: Women in American Astronomy, 1859–1940." *Physics Today,* 43 no. 3 (1990): 58–65.

Larsen, Egon. *The Cavendish Laboratory: Nursery of Genius.* London: Edmund Ward, 1962.

Lawrence, Ernest. "The Evolution of the Cyclotron." In *Nobel Lectures, Physics 1922–1941* (1965): 425–43. Amsterdam: Elsevier.

Lawrence, Ernst. "The Origins of Big Science–Rutherford at McGill", in *Rutherford and Physics at the Turn of the Century* edited by M. Bunge and W.Shea. 23–41. New York, 1979.

Lawson, Robert. "Der Aggregatrücksto;sz als Sättigungsstromes für a-Strahlen im Plattenkondensator." *Sitzungsberichte Akademie der Wissenschaften in Wien, Mat hematisch-naturwissenschaftliche Klasse,* Abteillung IIa, 128 (1920): 795–830.

Lawson, Robert. "Prof. Stefan Meyer." *Nature,* 165 (1950): 549.

Lefebvre, Thierry and Raynal, Cécil, "De l' Institut Pasteur à Radio Luxemburg: L' histoire étonnante du Tho-Radia" *Revue d' histoire de la pharmacie,* 335 no. 3 (2002): 461–480.

Lenoir, Timothy. "Instrument Makers and Discipline Builders: The Case of Nuclear Magnetic Resonance." *Perspectives in Science,* 3 (1997): 276–345.

Lesky, Erna. *The Vienna Medical School of the 19th Century.* Baltimore: The John Hopkins University Press, 1976.

Levi, Hilde. *George de Hevesy: Life and Work.* Copemhagen: Rhodos, 1985.

Levy, Leonard and Willis, Herbert. *Radium and Other Radioactive Elements: A Popular Account Treated Experimentally.* London: Percival Marshall, 1904.

Lintner, Karl and Schmid, Erich. "Das II. Physikalisches Institut der Universität Wien." in *Aufgaben der Universität Wien in Gegenwart und Zukunft Aufsätze zur 600–Jahrerer.* Wien, 1965.

Lintner, Karl. "Berta Karlik, Nachruf." *Almanach der Österreichischen Akademie der Wissneschaften,* 140 (1990): 305–313.

Löffler, Heinz. "Limnology in Austria; A Condensed Synopsis." In *Groundwater Ecology: A Tool for Management of Water Resources* edited by Christian Griebler et al. Luxemburg: European Communities, 2001.

Luger, A. 70 *Jahre Krankenhaus der Stadt Wien-Lainz.* Wien, 1977.

Lykknes, Annette. *Ellen Gleditsch: Professor, Radiochemist, and Mentor.* Ph.D Thesis. Trondheim: Norwegian University of Science and Technology. 2005.

Mach, Ernst. *Popülar-wissenschaftliche Vorlesungen.* Leipzig, 1910.

Machamer, Peter; Pera, Marcello, Baltas Aristides. *Scientific Controversies: Philosophical and Historical Perspectives.* Oxford: Oxford University Press. 2000.

Mache, Heinrich and Meyer, Stefan. "Über die Radioaktivität der Quellen der böhmischen Bädergruppe: Karlsbad, Marienbad, Teplitz-Schünau-Dux, Franzensbad sowie von St. Joachimsthal." *Sitzungsberichte der Kaiserl. Akademie der Wissenschaften in Wien, Mathematisch-naturwissenschaftliche Klasse,* Abteillung IIa, 114 (1905): 355–385.

Mache, Heinrich and Suess, Edurad. "Über die Aufnahme von Radiumemanation in das menschliche Blut bei der Inhalations-und Trinkkur." *Sitzungsberichte der Kaiserl. Akademie der Wissenschaften in Wien, Mathematisch-naturwissenschaftliche Klasse,* Abteillung IIa. 121 (1912): 171–184.

Mache, Heinrich. "Über die im Gasteiner Wasser enthaltene radioactive Emanation." *Physikalische Zeitschrift,* 5 (1904a): 441–444.

Mache, Heinrich. "Über die im Gasteiner Wasser enthaltene radioactive Emanation." *Sitzungsberichte der Kaiserl. Akademie der Wissenschaften in Wien, Mathem atisch-naturwissenschaftliche Klasse,* Abteillung IIa, 113 (1904b): 1329–1352.

Macleod, Roy. "Of Medals and Men: a Reward System in Victorian Science, 1826–1914." *Notes and Records of the Royal Society of London* 26 (1971): 81–105.

Macrakis, Kristie. "The Rockefeller Foundation and German Physics under National Socialism." *Minerva,* 27 (1989): 33–57.

Macrakis, Kristie. *Surviving the Swastika: Scientific Research in Nazi Germany.* Oxford: Oxford University Press, 1993.

Maerker, Anna Katharina. *Model Experts: The Production and Uses of Anatomical Models at La Specola, Florence and the Josephinum, Vienna, 1775–1814.* Ph.D. Thesis. Cornel University, 2005.

Makower, Walter and Geiger, Hans. *Practical Measurements in Radioactivity.* London: Green, 1912.

Malley, Marjorie. "The Discovery of Atomic Transmutation: Scientific Styles and Philosophies in France and Britain." *Isis,* 70 (1979): 213–223.

Maracineanu, Stefania. "Researches on the Constant of Polonium." *Comptes Rendus des Seanses de l' Academie des Sciences, Paris,* 176 (1923): 1879–1881.

Marshall, Fitz-Hugh and Coltman J. W. "The Photo-Multiplier Radiation Detector." *Physical Review* 72 (1947): 528.

Martinez, Katharine and Ames, Kenneth. *The Material Culture of Gender, the Gender of Material Culture.* Winterthur Museum, 1997.

Massing, Jean Michel. "From Greek Proverb to Soap Advert: Washing the Ethiopian." *Journal of the Warburg and Courtauld Institutes,* 85 (1995): 180–201.

May, Arthur. *The Habsburg Monarchy 1867–1914.* New York: Norton, 1968.

Mayneord, W. V. and Belcher, E. H. "Scintillation Counting and its Medical Applications." *British Journal of Radiology,* Suppl. 2–3 (1950): 259.

Mazeron, Jean-Jacques and Gerbault, Alain. "The Centenary of Discovery of Radium." *Radiotherapy and Oncology,* 49 (1998): 205–216.

McCready, Ralph. "Bequerel's Legacy to Nuclear Medicine." *European Journal for Nuclear Medicine,* 23 (1996): 489–490.

McMillan, Edwin. "Early History of Particle Accelerators." In *Nuclear Physics in Retrospect: Proceedings on a Symposium on the 1930s* edited by Roger Stuewer. 111–155. Minneapolis: University of Minnesota Press, 1979.

Meitner, Lise. "Looking Back." *Bulletin of the Atomic Scientists*, 20 (1964): 2–7.

Merricks, Linda. *The World Made New: Frederick Soddy, Science, Politics, and Environment.* Oxford: Oxford University Press, 1996.

Meyer, Stefan "Magnetic Properties of Inorganic Compounds." *Wied. Annalen.*, 69 no. 1 (1899): 236–263.

Meyer, Stefan and Schweidler, Egon von. "Influence of Alternations of Temperature on Radioactive Substances." *Pyschicalische Zeitschrift*, 5 (1904): 319–320.

Meyer, Stefan and Suess, Erhard. "Zur Verwendung der Radiumemanation als Indicator für Diagnostik und Therapie. Untersuchungen an Atmungsvorgängen." *Sitzungsberichte Akademie der Wissenschaften in Wien, Mathematisch-naturwissensc haftliche Klasse*, Abteillung IIa, 139 (1930): 613–628.

Meyer, Stefan. "Das erste Jahrzehnt des Wiener Instituts für Radiumforschung." *Jahrbuch der Radioactivität und Electronik* XVII, Band, Heft 1, (1920): 1–29.

Meyer, Stefan. "Die Vorgeschichte der Gründung und das erste Jahrzehnt des Institutes für Radiumforschung." In *Festrift des Institutes für Radiumforschung Anlässlich seines 40jährigen Bestandes (1910–1950)*. 1–26. Wien, 1950.

Meyer, Stefan. "Magnetic Properties of the Elements" *Wied. Annalen.*, 68 no. 2 (1899): 325–334.

Meyer, Stefan. "Physikalische Grundlagen zur Radiumemenationstherapie." *Sitzungsberichte Akademie der Wissenschaften in Wien, Mathematisch-naturwissensc haftliche Klasse*, Abteillung IIa, 138 (1929): 557–580.

Meyer-Thurow, Georg. "The Industrialization of Invention: A Case Study from the German Chemical Industry." *Isis*, 73 (1982): 363–381.

Michl, Helga. *Geschichte des Studienfaches Chemie an der Universität Wien in der Letzten Hundert Jahren.* Ph.D Thesis. University of Vienna, 1950.

Mikoletzky, Juliane; Georgeacopol-Winischhofer, Ute; Pohl Margit. "'Dem Zuge der Zeit entsprechend . . .' Zur Geschichte des Frauenstudiums in Össtereich am Beispiel der Technischen Universität Wien" *Schriftenreiche des Universitätsarchivs der Technischen Universität Wien* 1 (1997).

Molisch, Hans. "Über das Treiben der Pflanzen mittels Radium." *Sitzungsberichte der Kaiserl. Akademie der Wissenschaften in Wien, Mathematisch-naturwissenschaftlic he Klasse*, Abteillung IIa. 121 (1912): 121–139.

Molisch, Hans. "Über der Einfluss der Radiumemanation auf die höhere Pflanzen." *Sitzungsberichte der Kaiserl. Akademie der Wissenschaften in Wien, Mathematisch -naturwissenschaftliche Klasse*, Abteillung IIa. 121 (1912): 833–857.

Moore, Walter. *A Life of Erwin Schrödinger.* Cambridge: Cambridge University Press, 1994.

Mould, Richard. "The Discovery of Radium in 1898 by Maria Sklodowska-Curie (1867–1934) and Pierre Curie (1859–1906) with Commentary on their Life and Times." *The British Journal of Radiology*, 71 no. 852 (1998): 1229–1254.

Mould, Richard. *A Century of X-rays and Radioactivity in Medicine: With Emphasis on Photographic Records of the Early Years.* Philadelphia: Institute of Physics Publishing, 1993.

Mühlleitner, Elke and Reichmayr Johannes. "The Exodus of Psychoanalysis from Vienna." In *The Cultural Exodus from Austria* edited by Friedrich Stadler and Peter Weibel. 98–121. New York: Springer-Verlag, 1995.

Mullner, Ross. *Deadly Glow: The Radium Dial Painters.* Washington: American Public Health Association, 1999.

Myers, Fred. *The Empire of Things: Regimes of Value and Material Culture.* Santa Fe: School of American Research Press, 2001.

Nazarska, Georgeta. "Elissaveta Karamichailova." In *A Biographical Dictionary of Women's Movements and Feminisms. Central, Eastern, and Southeastern Europe: 19th 20th Centuries* edited by Francisca De Haan, Krassimira Daskalova, and Anna Loutfi. 222–225. Budapest, New York: CEU Press, 2006.

Nelkin, Dorothy. *Controversy: Politics of Technical Decisions.* Newbury Park Calif.: Sage, 1992.

Neuburger, Max. *British Medicine and the Vienna School: Contacts and Parallels.* London, 1943.

Nornvall, Fredrik and Svansson, Artur. "Bornö Oceanographic Station: The Foundation of a Marine Station in Sweden." *History of Oceanography*, 10 (1998): 5–8.

Nye, Mary Jo. *The Question of the Atom: From Karlsruhe Congress to the First Solvay Conference, 1860–1911.* Los Angeles: Tomash Publishers, 1986.

Oberhummer, Klaus. "The Leading Chemical Company in the Habsburg Empire: Der Österreichische Verein für chemische und metallurgische Produktion zu Aussig a.d. Elbe." In *Pioneering Ideas for the Physical Sciences: Josef Loschmidt's Contributions and Modern Developments in Structural Organic Chemistry, Atomistics, and Statistical Mechanics* edited by W. Fleischhacker and T. Schönfeld. New York: Plenum Press, 1997.

Obrist, Ulrich and Vanderlinden, Barbara. eds. *Laboratorium.* Dumont: Antwerpen Open, Roomade, 1999.

Okey, Robin. *The Habsburg Monarchy: From Enlightenment to Eclipse.* New York: St. Martin's Press, 2001.

Oliphant, Mark. "Working with Rutherford." In *Cambridge Physics in the Thirties* edited by John Hendry. Bristol: Adam Hilger, 1984.

Ortner, Gustav and Stetter, Georg. "Die Hörbarmachung von H-Strahlen." *Physikalische Zeitschrift*, 28 (1927): 70–72.

Page, Benjamin. "The Rockefeller Foundation and Central Europe: A Reconsideration." *Minerva*, 40 (2002): 265–287.

Pais, Abraham. *Inward Bound: of Matter and Forces in the Physical World.* Oxford: Oxford University Press, 1986.

Pais, Abraham. *Niels Bohr's Times in Physics, Philosophy and Polity.* Oxford: Clarendon Press, 1991.

Paneth, Friedrich. "Prof. Stefan Meyer." *Nature*, 165 (1950): 548.

Paneth, Friedrich. "Über die Arbeit des Instituts für Radiumforschung." *Die Naturwissenschaften*, 34 (1915): 437–43.

Pawkowicz, Rainer. *Die Österreichische Akademie der Wissenschaften und die Bauten für ihre Forschungsinstitute.* Ph.D. Thesis. Vienna: University of Vienna, 1978.

Peh, W. C. "The Discovery of Radioactivity and Radium." *Singapore Medical Journal*, 6 (1996): 627–630.

Perlmutter, Arlond. "Marietta Blau's Work After World War II" (unpublished manuscript).

Petterson, Hans and Angström, A. "Ein neues Totalimmersions-Aräometer mit Kettenbelastung." Z. *Instrumentenkunde*, 37 no. 9 (1917): 177–180.

Pettersson, Dagmar. "Long-Range Particles from Radium-Active Deposit." *Nature*, 113 (1924): 641–642.

Pettersson, Dagmar. "Über die maximale Reichweite der von Radium C ausgeschleuderten Partikeln." *Sitzungsberichte Akademie der Wissenschaften in Wien, Mathematisch-naturwissenschaftliche Klasse*, Abteillung IIa, 133 (1924): 149–162.

Pettersson, Hans. "A New Microbalance and its Use." (Diss. Stockholms Högskola) *Göteborgs K. Vetensk. –o. Vitterh. Samh. Handl.* 16 no. 4 (1914): 1–95.

Pettersson, Hans. "Contributions à la connaissance du dégagement de chaleur du radium." *Arkiv Mat. Astr. Fys.*, 6 no. 26 (1910): 1–9.

Pettersson, Hans. "Recent Oceanographic Research in Sweden." *Högskolas arsskrift*, 39 no. 2 (1933): 207–218.

Pettersson, Hans. "Some New Measurements for Oceanographic Research." *Monthly Weather Review*, 45 no. 4 (1917): 159–164.

Pettersson, Hans. "Zur Herstellung von RaC." *Sitzungsberichte Akademie der Wissenschaften in Wien, Mathematisch-naturwissenschaftliche Klasse*, Abteillung IIa, 132 (1923): 55–57.

Pettersson, Hans. "Zur Technik der Dichtigkeitsbestimmung von Meerwasser." *Ann. Hydrogr. Berl.*, 45 (1917): 87–93.

Pflaum, Rosalynd. *Grand Obsession: Madame Curie and her World*. New York: Doubleday, 1989.

Pinch, Trevor and Bijker, Wiebe. "The Social Construction of Facts and Artifacts: Or How the Sociology of Science and the Sociology of Technology Might Benefit each Other." In *The Social Construction of Technological Systems* edited by W. Bijker, T. Hughes, and T. Pinch, 19–50. Cambridge, Mass.: MIT Press, 1987.

Pintaric, Horvat. *Vienna 1900: The Architecture of Otto Wagner*. London: Studio Editions, 1989.

Plaschka, Richard Georg. "Universität 1884- Haus mit neuen Weichenstellungen. Die Universität Wien in den Herausforderungen der zweiten Hälfte des 19. Jahrhunderts." In *100 Jahre Universität am Ring: Wissenschaft und Forschung an der Universität Wien seit 1884* edited by G. Hamann, K. Mühlberger, F. Skacel. 9–25. Schriftenreihe des Universitätsarchivs, Band 3, Universität Wien, 1986.

Plassmeyer, Peter. "Nineteenth Century Architecture: From Classicism to the Era of the Ringstrasse, 1790–1890." In *Vienna: Art and Architecture* edited by Rolf Toman. 144–215. Cologne: Könemann, 1999.

Plech, Pia Maria. "Das frauenstudium an den philosochischen Fakultäten" in Forkl, Martha and Koffmahn, Elizabeth. Frauenstudium und Akademische Frauenarbeit in Österreich Vienna: Universitäts;szVerlagsbuchandlung Ges. M. b. H, 1968.

Popescu, Miruna; Rayner-Canham, Marelene and Rayner-Canham, Geoffrey. "Stefania Maracineanu: Ignored Romanian Scientist." In *A Devotion to Their Science: Pioneer Women in Radioactivity* edited by Marelene Rayner-Canham and Geoffrey Rayner-Canham. 87–91. London: McGill-Queen's University Press, 1997.

Powell, Cecill. "Applications of the Photographic Method to Problems in Nuclear Physics." *Proceedings of the Royal Society of London*, 181 no. 987 (1943): 344–359.

Pringle, R. W. "The Scintillation Counter." *Nature*, 166 (July 1, 1950): 11–14.

Probst, Edith. "Emigration und Exil österreichischer Wissenschafterinnen." In *Vertriebene Vernunft I, Emigration und Exil österreichischer Wissenschafter* 1930–1940 edited by Friedrich Stadler. 444–470. München: Jugend und Volk, 1987.

Przibram, Karl and Belar, Maria. "Die Verfärbungen durch Becquerelstrahlen und die Frage des blauen Steinsalzes." *Sitzungsberichte Akademie der Wissenschaften in Wien, Mathematisch-naturwissenschaftliche Klasse*, Abteillung IIa, 132 (1923): 261–277.

Przibram, Karl. "1920 bis 1938." In *Festrift des Institutes für Radiumforschung Anlässlich seines 40jährigen Bestandes (1910–1950)*. 27–34. Wien, 1950.

Przibram, Karl. "Errinerungen an ein altes physikalisches Institut." In *Beiträge zur Physik und Chemie des 20.Jahrhunderts* edited by O. R. Frisch, F. A.Paneth, F. Laves, and P. Rosbaud. 1–6. Braunschweig: Friedrich Vieweg & Sohn, 1959.

Przibram, Karl. *Irradiation Colours and Luminescence: A Contribution to Mineral Physics*. London, 1956.

Pycior, Helena. "Marie Curie: Time only for Science and Family." In *A Devotion to their Science: Pioneer Women of Radioactivity* edited by Marelene Rayner-Canham, and Geoffrey Rayner-Canham. 31–50. Philadelphia: Chemical Heritage Foundation, 1997.

Pycior, Helena. "Marie Curie's 'Anti-Natural Path': Time Only for Science and Family." In Abir-Am, Pnina and Outram, Dorinda (eds.). *Uneasy Careers and Intimate Lives: Women in Science* 1789–1979, 191–215. New Brunswick: Rutgers University Press, 1987.

Pycior, Helena. "Reaping the Benefits of Collaboration While Avoiding its Pitfalls; Marie Curie's Rise to Scientific Prominence." *Social Studies of Science*, 23 (1993): 301–323.

Quinn, Susan. *Marie Curie: A Life*, New York, 1995.

Rabinbach, Anson. "Politics and Pedagogy: The Austrian Social Democratic Youth Movement 1931–1934." *Journal of Contemporary History*, 13 no. 2 (1978): 337–356.

Rabinbach, Anson. *The Crisis of Austrian Socialism: From Red Vienna to Civil War* 1927–1934. Chicago: The University of Chicago Press, 1983.

Rayner-Canham, Marelene and Rayner-Canham Geoffrey. *Harriet Brooks: Pioneer Nuclear Scientist*. Montreal: McGill-Queen's University Press, 1992.

Rayner-Canham, Marelene and Rayner-Canham, Geoffrey. "Elizabeth Rona: The Polonium Woman." In *A Devotion to their Science: Pioneer Women of Radioactivity* edited by Marelene Rayner-Canham and Geoffrey Rayner-Canham. 209–216, Philadelphia: Chemical Heritage Foundation and Montreal: McGill-Queen's University Press, 1997.

Rayner-Canham, Marelene and Rayner-Canham, Geoffrey. "Stefanie Horovitz: A Crucial Role in the Discovery of Isotopes." In *A Devotion to their Science: Pioneer Women of Radioactivity* edited by Marelene Rayner-Canham and Geoffrey Rayner-Canham, 192–195. Philadelphia: Chemical Heritage Foundation and Montreal: McGill-Queen's University Press, 1997.

Rayner-Canham, Marelene and Rayner-Canham, Geoffrey. "Stefanie Horovitz, Ellen Gleditsch, Ada Hitchins, and the Discovery of Isotopes." *Bulletin for the History of Chemistry*, 25 no. 2 (2000): 103–109.

Rayner-Canham, Marelene and Rayner-Canham, Geoffrey. *A Devotion to Their Science: Pioneer Women in Radioactivity*. Montreal: McGill-Queen's University Press and Philadelphia: Chemical Heritage Foundation, 1997.

Rayner-Canham, Marelene and Rayner-Canham, Geoffrey. *Women in Chemistry: Their Changing Roles From Alchemical Times to the Mid-Twentieth Century*. Philadelphia: Chemical Heritage Foundation, 1998.

Regener, Erich. "Über Zählung der a-Teilchen durch die Szintillation und die Grö;sze des elektrischen Elementarquantums." *Verhandlungen der deutschen physikalischen Geselschaft*, 10 (1908): 78–83.

Reiter, Wolfgang. "Stefan Meyer: Pioneer of Radioactivity." *Physics in Perspective* 3 (2001): 106–127.

Reiter, Wolfgang. "The Year 1938 and its Consequences for the Sciences in Austria." In *The Cultural Exodus from Austria* edited by Friedrich Stadler and Peter Weibel. 188–205, New York: Springer-Verlag, 1995.

Reiter, Wolfgang. "Vienna: A Random Walk in Science." *Physics in Perspective*, 3 no. 4 (2001): 462–489.

Reiter, Wolfgang. "Zerstört und vergessen: Die Biologische Versuchsanstalt und ihre Wissenschaftler/innen." *Österreichische Zeitschrift für Geschichtswissenschaften*, 10 no. 4 (1999): 585–614.

Renneberg, Monika and Walker, Mark. eds. *Science, Technology and National Socialism*. Cambridge, Cambridge University Press, 1994.

Rentetzi, Maria (coordinator), "Women Pioneers in Radiaoctivity Research." In *Women Scholars and Institutions: Proceedings of the International Conference Prague, June 8–11, 2003* edited by S. štrbáň$ová, I. H. Stamhuis and K. Mojsejová, 581–725. Prague, 2004.

Rentetzi, Maria. "The City as a Context of Scientific Activity: Creating the *Mediziner-Viertel* in *fin-de-siècle* Vienna." *Endeavour* 28 (2004): 39–44.

Rentetzi, Maria. "Women Radium Dial Painters as Experimental Subjects or What Counts as Human Experimentation." In *Twentieth Century Ethics of Human Subjects Research: Historical Perspectives on Values, Practices, and Regulations* edited by Volker Roelcke and Giovanni Maio. 275–291. Stuttgart: Steiner, 2004. (republished in the *Internationale Zeitschrift für Geschichte und Ethik der Naturwissenschaften, Technik und Medizin-NTM*, , 12 (2004): 233–248)

Rheinberger, Hans-Jörg. "Putting Isotopes to Work: Liquid Scintillation Counters, 1950–1970." preprint 121, *Max Planck Institute for the History of Science*, Berlin, 1999.

Rheinberger, Hans-Jörg. *Toward a History of Epistemic Things: Synthesizing Proteins in the Test Tube.* Stanford, CA: Stanford University Press, 1997.

Richard, Thomas. *The Commodity Culture of Victorian England: Advertising and Spectacle, 1851–1914.* New York: Verso, 1990.

Richter, Grete. "Messungen im Schutzringplattenkondesator mit RaF nebst eingehender Diskussion der Verwendung des Binanten-oder Quadranten-Elektrometers als Strommessinstrument. " *Sitzungsberichte Akademie der Wissenschaften in Wien, Mathematisch-naturwissenschaftliche Klasse,* Abteillung IIa, 128 (1919): 539–569.

Rife, Patricia. *Lise Meitner and the Dawn of the Nuclear Age.* Boston: Birkhäuser, 1999.

Riordan, Teresa. *Inventing Beauty: A History of the Innovations that Have Made Us Beautiful,* Broadway Books, 2004.

Robinson, Gabrielle and Keen, Mike. "Café kultur: the coffeehouses of Vienna." *Contemporary Review,* 256 (1996.): 24–29.

Rocke, Alan. "Origins and Spread of the 'Giessen Model' in University Science," *Ambix,* 50 (2003): 90–115.

Rona, Elisabeth and Neuninger, Elisabeth."Beiträge zur Frage der künstlichen Aktivität des Thoriums." *Sitzungsberichte Akademie der Wissenschaften in Wien, Mathe matisch-naturwissenschaftliche Klasse,* Abteillung IIa, 145 (1936): 479–482.

Rona, Elisabeth and Schmidt, Ewald. "Eine Methode zur Herstellung von hochkonzentrierten Poloniumpräparaten." *Sitzungsberichte Akademie der Wissenschaften in Wien, Mathematisch-naturwissenschaftliche Klasse,* Abteillung IIa, 137 (1928): 103–105.

Rona, Elisabeth and Schmidt, Ewald. "Herstellung von hochkonzentrierten Poloniumpräparaten durch Destillation." *Zeitschrift für Physik,* 48 (1928): 784–789.

Rona, Elisabeth and Schmidt, Ewald. "Untersuchungen über das Eindringen des Poloniums in Metale." *Sitzungsberichte Akademie der Wissenschaften in Wien, Ma thematisch-naturwissenschaftliche Klasse,* Abteillung IIa, 136 (1927): 65–73.

Rona, Elisabeth. "Absorptions- und Reichweitebestimmung an 'natürlichen' H-Strahlen." *Sitzungsberichte Akademie der Wissenschaften in Wien, Mathematisc h-naturwissenschaftliche Klasse,* Abteillung IIa, 135 (1926): 117–126.

Rona, Elisabeth. "Diffusionsgrösse und Atomdurchmesse der Radiumemanation." *Zeitschrift für Physikalische Chemie,* 92 (1917): 213–218.

Rona, Elisabeth. "Über die Reduktion des Zimtaldehyds durch Hefe. I; Vergärung von Benzylbrenztraubensäure." *Biochemische Zeitschrift,* 67 (1914): 137–142.

Rona, Elisabeth.. "Über die Wirksamkeit der Fermente unter abnormen Bedingungen über die angebliche Aldehydnatur der Enzyme." *Biochemische Zeitschrift,* 109 (1920): 279–89.

Rona, Elizabeth and Schintlmeister, Josef. "Untersuchung der Alpha-Stralung des Methor 2." *Sitzungsberichte Akademie der Wissenschaften in Wien, Mathematisch-naturwissenschaftliche Klasse,* Abteillung IIa, 147 (1938): 49–62.

Rona, Elizabeth. "Laboratory Contamination in the Early Period of Radiation Research." *Health Physics,* 37 (1979): 723–727.

Rona, Elizabeth. *How it Came About: Radioactivity, Nuclear Physics, Atomic Energy.* Oak Ridge Associated Universities, 1978.

Roque, Xavier. "Displacing radioactivity" In *Instrumentation Between Science, State and the Industry* edited by B. Joerges and T. Shinn. 51–68. Boston, Mass., 2001.

Roque, Xavier. "Marie Curie and the Radium Industry: A Preliminary Sketch." *History and Technology,* 13 (1997): 267–291.

Rosner, Robert and Strohmaier, Brigitte. eds. *Marietta Blau – Sterne der Zetrümmerung.* Wien Böhlau Verlag, 2003.

Rossiter, Margaret, "Sexual Segregation in the Sciences: Some Data and a Model" *Signs,* 4 no. 1 (1978): 146–151.

Rossiter, Margaret. "Women's Work in Science, 1880–1910." *Isis,* 71 (1980): 287–304.

Rossiter, Margaret. *Women Scientists in America: Before Affirmative Action 1940–1972.* Baltimore: John Hopkins University Press, 1995.

Rossiter, Margaret. *Women Scientists in America: Struggles and Strategies to 1940.* (Volume I) Baltimore: Johns Hopkins University Press, 1984.

Ruthenberg, Klaus. "Friedrich Adolf Paneth (1887–1958)." *International Journal for Philosophy of Chemistry,* 3 (1997): 103–106.

Rutherford, Ernest and Chadwick, James and Ellis, C. *Radiations from Radioactive Substances.* New York: Macmillan, 1930.

Rutherford, Ernest. "Collision of alpha-Particles with Light Atoms. I. Hydrogen." *Philosophical Magazine,* 36 (1919): 537–561. [In *The Collected Papers of Lord Rutherford* edited by James Chadwick vol. II. 547–567. New York: Interscience Publishers, 1963.]

Rutherford, Ernest. "Collisions of alpha-Particles with Light Atoms. IV. An Anomalous Effect in Nitrogen." *Philosophical Magazine,* 37 (1919): 581–587. [In *The Collected Papers of Lord Rutherford* edited by James Chadwick, vol. II. 585–590. New York: Interscience Publishers, 1963.]

Rutherford, Ernest. "Nuclear Constitution of Atoms (Bakerian Lecture)." *Proceedings of the Royal Society,* A 97 (1920): 374–400. [In *The Collected Papers of Lord Rutherford* edited by James Chadwick. vol. III, 14–38. New York: Interscience Publishers, 1963.]

Rutherford, Ernest. "The Chemical Nature of the alpha-Particles from Radioactive Substances." Before the Royal Academy of Science at Stockholm. (1908) In *The Collected Papers of Lord Rutherford* edited by James Chadwick. vol. II. 137–146. New York: Interscience Publishers, 1963.

Rutherford, Ernset and Chadwick, James. "The Artificial Disintegration of Light Elements." *Philosophical Magazine,* 42 (1921): 809–825. [In *The Collected Papers of Lord Rutherford* edited by James Chadwick. vol. III. 48–61. New York: Interscience Publishers, 1963.]

Sabine, Ernst. *Lise Meitner an Otto Hahn: Briefe aus den Jahren 1912 bis 1924.* Stuttgart: Wissenschaftliche Verlagsgesellschaft mbH, 1992.

Sablik, Karl. *Julius Tandler. Mediziner und Sozialreformer.* Wien, 1983.

Schmidt, Ewald and Stetter, Georg. "Die Anwendung des Röhrenelektrometers zur Untersuchungen von Protonstrahlen." *Sitzungsberichte Akademie der Wissenschaften in Wien, Mathematisch-naturwissenschaftliche Klasse,* Abteillung IIa, 138 (1929): 271–287.

Schmidt, Ewald and Stetter, Georg. "Die Ionisastion einzelner alpha- und H- Strahlen am Ende der reichweite." *Sitzungsberichte Akademie der Wissenschaften in Wien, Mathematisch-naturwissenschaftliche Klasse*, Abteillung IIa, 139 (1930): 123–138.

Schmidt, Ewald and Stetter, Georg. "Untersuchungen mit dem Röhrenelektrometer über die alpha Reflexion und den Zertrümmerungseffekte an Leichtelementen." *Sitzungsberichte Akademie der Wissenschaften in Wien, Mathematisch-naturwissenschaftliche Klasse*, Abteillung IIa, 139 (1930): 139–150.

Schmidt, Ewald. "Über die Zertrümmerung des Aluminiums durch a-Strahlen, I." *Sitzungsberichte Akademie der Wissenschaften in Wien, Mathematisch-naturwissenschaftliche Klasse*, Abteillung IIa, 134 (1925): 385–404.

Schmidt, Ewald. "Über die Zertrümmerung des Aluminiums durch a-Teilchen." *Zeitschrift für Physik*, 42 (1927): 721–740.

Schneidt Selma. "Das elektrochemische Verhalten von Polonium in Lösungen verschidener H-Ionenkonzentration." *Sitzungsberichte Akademie der Wissenschaften in Wien, Mathematisch-naturwissenschaftliche Klasse*, Abteillung IIa, 138 (1929): 754–765.

Schönfeld, Thomas and Ipser Herbert. *Die Geschichte der Chemie an der Universität Wien und des Instituts für Anorganische Chemie*. Wien: Institut für Anorganische Chemie, Universität Wien, 1996.

Schorske, Carl. *Fin-de-siècle Vienna: Politics and Culture*. New York: Alfred Knopf, 1980.

Schürmann, Astrid. "Promoting International Women's Research in Radioactivity: The Curie Laboratory in Paris." In *Women Scholars and Institutions: Proceedings of the International Conference Prague, June 8–11, 2003* edited by S. štrbáňová, I. H. Stamhuis and K. Mojsejová. 591–609. Prague, 2004.

Schweidler, Egon von. "Bericht des Generalsekretärs." In *Akademie der Wissenschaften in Wien, Almanach für das Jahr* 1927. 175–178, Wien, 1927.

Segre, Emilio. "Nuclear Physics in Rome." In *Nuclear Physics in Retrospect: Proceedings on a Symposium on the* 1930s edited by Roger Stuewer. 35–62. Minneapolis: University of Minnesota Press, 1979.

Seidel, Robert. "The Origins of the Lawrence Berkley Laboratory." In *Big Science: The Growth of Large-Scale Research* edited by Peter Galison and Bruce Hevly. 21–45. Stanford, Calif.: Stanford University Press, 1992.

Seidler, K. "Das Frauenstudium an der Wiener Technischen Hochschule." In *30 jahre Frauenstudium in Österreich, Festschrift*, Wien, 1927.

Sell, Arne. "The Development of Radiotherapy in Denmark during 100 Years: From Radiology to Oncology." *Acta Oncologica*, 34 no 8 (1995): 1005–1010.

Sengoopta, Chandak. "'Dr Steinach coming to make old young!': Sex glands, Vasectomy and the Quest for Rejuvenation in the Roaring Twenties." *Endeavour*, 27 no. 3 (2003): 122–126.

Sengoopta, Chandak. "Glandular Politics: Experimental Biology, Clinical Medicine, and Homosexual Emancipation in Fin-de-Siecle Central Europe." *Isis*, 89 no. 3 (1998): 445–473.

Shackelford, Jole. "Tycho Brahe, Laboratory Design, and the Aim of Science: Reading Plans in Context." *Isis*, 84 no. 2 (1993): 211–230.

Shapin, Steven and Schaffer, Simon. *Leviathan and the Air-Pump: Hobbes, Boyle and the Experimental Life*. Princeton: Princeton University Press, 1985.

Shapiro, Maurice. "Tracks of Nuclear Particles in Photographic Emulsions." *Reviews of Modern Physics*, 13 (1941): 58–71.

Shea, William. *Otto Hahn and the Rise of nuclear Physics*. Boston: Kluwer Academic Publishers, 1983.

Shell, Kurt. *The Transformation of Austrian Socialism*. New York: State University of New York, 1962.

Siegmund-Schultze, Reinhard. *Rockefeller and the Institutionalization of Mathematics Between the Two World Wars: Documents and Studies for the Social History of Mathematics in the 20th Century*. Birkhäuser Verlag, 2001.

Sime, Ruth. *Lise Meitner: A Life in Physics*. Berkley: University of California Press, 1996.

Smereker, Hilda and Juris, K. "Messung der Beta-Strahlung des Radiums in r-Einheiten." *Strahlentherapie*, 52 (1935): 327–337.

Smereker, Hilda and Juris, K. "Versuche über die nicht direkte Ionisierung der Gammastrahlen." *Strahlentherapie*, 61 (1938): 161–166.

Smereker, Hilda and Schloss, W. "Zur Radiumbehandlung der sogenannten strahlenresistenten Epitheliome." *Strahlentherapie*, 55 (1936): 102–113.

Smereker, Hilda. "Dosimetrische Fragen der Radium-und Röntgentherapie." *Strahlentherapie* 58 (1937): 676.

Smereker, Hilda. "Untersuchungen der Strahlungsintesität in der Nähe radioaktiver Präparate mit dünnwandigen Ionisationskammern." *Strahlentherapie*, 58 (1937): 267.

Soddy, Frederick. "Radium as a Therapeutic Agent." *Medical Record* (New York), 64 (1903): 660.

Souczek, Helene. "Messungen des Radiumgehalts der bei der Verarbeitung von St. Joachimstaler Uranpecherzückständen resultierenden radiumarmen Produkte." *Sitzungsberichte der Kaiserl. Akademie der Wissenschaften in Wien, Mathematisch -naturwissenschaftliche Klasse*, Abteillung IIa, 119 (1910): 371–376.

Sretenova, Nikolina. "Elizaveta Karamihailova (1897–1968)—the First Woman Physicist, Who Managed to Cross the 'Safety Curtain' of Sofia University." In *Philosophy between Old and New Values Proceedings of XXII Varna International Philosophical School 29 May–01 June 2002*, 202–208. Sofia: IPhR-BAS, 2003.

Stadler, Friedrich and Weibel, Peter. eds. *The Cultural Exodus from Austria*. Wien, New York: Springer Verlag, 1995.

Stadler, Friedrich. "The Emigration and Exile of Austrian Intellectuals." In *The Cultural Exodus from Austria* edited by Friedrich Stadler and Peter Weibel. 14–26. Wien, New York: Springer Verlag, 1995.

Stadler, Friedrich. "The Vienna Circle and the University of Vienna." In *The Cultural Exodus from Austria* edited by Friedrich Stadler and Peter Weibel. 44–55. Wien, New York: Springer Verlag, 1995.

Stadler, Friedrich. Ed. *Vertriebene Vernunft I: Emigration und Exil österreichischer Wissenschaft 1930–1940*. Wien and München: Jugend und Volk, 1987.

Stadler, Friedrich. Ed. *Vertriebene Vernunft II: Emigration und Exil österreichischer Wissenschaft 1930–1940*. Wien and München: Jugend und Volk, 1988.

Stadler, Karl. "Austria." In *European Fascism* edited by Woolf S. J. New York: Vintage Books, 1968.

Stadler, Karl. *Austria*. New York: Praeger Publishers, 1971.

Star, Susan Leigh and Griesemer, John. "Institutional Ecology, 'Translations' and Boundary Objects: Amateurs and Professionals in Berkley's Museum of Vertebrate Zoology, 1907–1939." *Social Studies of Science*, 19 (1989): 387–420

Stawström, Carl-Olov. "Relative Acceptance: How Einstein's Theories of Relativity Were Received in Sweden, 1905–1965." In *Center on the Periphery: Historical Aspects of 20th Century Swedish Physics* edited by Svante Lindqvist, Marika Hedin, and Thomas Kaiserfeld. Canton. Mass.: Science History, 1993.

Stetter, Georg and Wambacher, Hertha. "Neuere Ergebnisse von Untersuchungen über die Mehrfachzetrümmerung von Atomkernen durch Höhenstrahlen." *Physikalische Zeitschrift*, 40 (1939): 702–6.

Stetter, Georg. "Die Bestimmung des Quotienten Ladung/Masse für natürlich H-Strahlen und Atomtrümmer aus Aluminium." *Sitzungsberichte Akademie der Wissenschaften in Wien, Mathematisch-naturwissenschaftliche Klasse*, Abteillung IIa, 135 (1926): 61–69.

Stetter, Georg. "Die Massenbestimmung von Atomtrümmern aus Aluminium, Kohlenstoff, Bor und Eisen." *Zeitschrift für Physik*, 42 (1927): 741–758.

Stetter, Georg. "Die Massenbestimung von H-Partiklen." *Zeitschrift für Physik*, 34 (1925): 158–177.

Stetter, Georg. "Die neueren Untersuchungen über Atomzetrümmerung." *Physikalische Zeitschrift*, 28 (1927): 712–723.

Stetter, Georg. "Massenbestimmung von Atomtrümmern." *Physikalische Zeitschrift*, 27 (1926): 735–738.

Stewart, Larry. "Other centers of calculation, or, where the Royal Society did not count: commerce, coffee-houses and natural philosophy in early modern London." *British Journal for the History of Science*, 32 (1999): 133–153.

Stuewer, Roger. "Artificial Disintegration and the Vienna-Cambridge Controversy." In *Observation, Experiment, and Hypothesis in Modern Physical Science* edited by Peter Achinstein and Owen Hannaway. 239–307. Cambridge, Mass.: MIT Press, 1985.

Stuewer, Roger. "The Origin of the Liquid-Drop Model and the Interpretation of the Nuclear Fission." *Perspectives on Science*, 2 (1994): 76–129.

Svansson, Artur. *Otto Pettersson: the Oceanographer, the Chemist, the Inventor*. Gothenburg: Tre Boecker, 2006.

Sviedrys, Romualda. "The Rise of Physical Science at Victorian Cambridge." *Historical Studies in the Physical Sciences*, 2 (1970): 127–52.

Tausky-Todd, Olga. "Olga Tausky-Todd: An Autobiographical Essay." In *Mathematical People: Profiles and Interviews* edited by Donald Albers and G. Alexaderson. 309–336. Boston: Birkhäuser, 1985.

Tesinska, Emilie. "Women in the Field of Radioactivity: The Case of the Czech Physical Chemist and Radiobiologist Jarmila Petrova." In *Women Scholars and*

Institutions: Proceedings of the International Conference Prague, June 8–11, 2003 edited by S. štrbáňová, I. H. Stamhuis and K. Mojsejová. 659–691. Prague, 2004.

Thomas, Nicholas. *Entangled Objects: Exchange, Material Culture and Colonialism in the Pacific*. Cambridge, Mass.: Harvard University Press, 1991.

Trenn, Thaddeus. *Transmutation Natural and Artificial*. Philadelphia: Heyden, 1981.

Trevor, Levere. "Public and private science at home and abroad: networks among chemists, physicians and industrialists in the 1780s." *Endeavour*, 26 no. 3 (2002): 107–112.

Tsoneva-Mathewson, Snezha; Rayner-Canham, Marelene and Rayner-Canham, Geoffrey. "Elizaveta KaraMichailova: Bulgarian Pioneer of Radioactivity." In *A Devotion to Their Science: Pioneer Women in Radioactivity* edited by Marelene Rayner-Canham, and Geoffrey Rayner-Canham. 205–208. London: McGill-Queen's University Press, 1997.

Tubiana, Maurice; Dutreix, Jean and Pierquin, Bernard. "One Century of Radiotherapy in France 1896–1996." *International Journal of Radiation Oncology, Biology, Physics*, 35 (1996): 227–242.

Tuma, Renate. "Die österreichischen Studentinnen der Universität Wien (ab 1897)." In *"Durch Erkenntnis zu Freiheit und Glück" Frauen an der Universität Wien* edited by Waltraud Heindl and Marina Tichy. 79–107. Schriftenreihe des Universitätsarchivs, Band 5, Wien: Universität Wien, 1990.

Urbach, Franz. "Einiges aus dem Physikalischen Laboratorium eines Krankenhauses." *Zeitschrift für das gesamte Krankenhauswesen*, 25 (1933): 537–541.

Urbach, Franz. "Über eine spontane Veränderung des latenten Bildes." *Sitzungsberichte Akademie der Wissenschaften in Wien, Mathematisch-naturwissenschaftliche Klasse*, Abteillung IIa. 72 (1935): 222–223.

Vincente, Bénédicte. "Genesis of the Pavillion Pasteur of the Institut du radium in Paris" *History and Technology*, 13 (1997): 293–305.

Viol, Charles. "History and Development of Radium Therapy." *Journal of Radiology*, 2 (1921): 29–34.

Vogt, Annette. "The Timofeeff-Ressovsky's–A Couple in Science." *Max Planck Institute for the History of Science*, reprint 157, 2000.

Vogt, Annette. "Von der Ausnahme zur Normalität? Wissenschaftlerinnen in Akademien und in der Kaiser-Wilhelm-Gesellschaft (1912–1945)." In *Zwischen Vorderbühne und Hinterbühne: Beiträge zum Wandel der Geschlechterbeziehungen in der Wissenschaft vom 17. Jahrhundert bis zur Gegenwart* edited by Theresa Wobbe. 159–188. Berlin: Berlin-Braderburgischen Akademie der Wissenschaften, Band 12, 2003.

Vogt, Annette. "Women Members of the Academies of Science-A Comparative Study with Special Consideration of the Kaiser Wilhelm Society (1912–1945)." *Max-Planck Institute for the History of Science*, reprint 155, 2000.

Wagner, Otto. *Die Baukunst unserer Zeit*. Wien: Löcker Verlag, 1979.

Wajcman, Judy. "Reflections on Gender and Technology Studies: In What State is the Art?" *Social Studies of Science*, 30 no. 3 (2000): 447–464.

Walker, Mark and Sachse, Carola. eds. *Politics and Science in Wartime: Comparative International Perspectives on Kaiser Wilhelm Institute*, Osiris 20, Chicago: University of Chicago Press, 2005.

Walkhoff, F. "Unsichtbare, photographisch wirksame Strahlen." *Photographische Rundschau*, 14 (1900): 189–191.

Walzel, Karin. "Ärztinnen in Wien." In *Töchter des Hippokrates: 100 Jahre akademische Ärztinnen in Österreich* edited by Birgit Bolognese-Leuchtenmüller and Sonia Horn. 113–116. Wien: Pressestelle und Verl. Der Österr. Ärztkammer, 2000.

Wambacher, Hertha. "Höhenstrahlung und Atomkernbau." *Österreichische Chemikerzeitung*, 40 (1940): 116–121.

Wambacher, Hertha. "Mehrfachzertrümmerung von Atomkernen durch komische Strahlung." *Zeitschrift für technische Physik*, 19 (1938): 569–576.

Wambacher, Hertha. "Mehrfachzertrümmerung von Atomkernen durch kosmische Strahlung; Ergebnisse aus 154 Zertrümmerungssternen in photographischen Platten." *Physikalische Zeitschrift*, 39 (1938): 883.

Wambacher, Hertha. "Untersuchungen der photographischen Wirkung radioaktiver Strahlen auf mit Chromsäure und Pinakryptolgelb vorbehandelten Filmen und Platten." *Sitzungsberichte Akademie der Wissenschaften in Wien, Mathematisch-naturwissenschaftliche Klasse*, Abteillung IIa, 141 (1931): 271–292.

Wambacher, Hertha. "Kernzertrümmerung durch Höhenstrahlung in der photographischen Emulsion." *Sitzungsberichte Akademie der Wissenschaften in Wien, Mathematisch-naturwissenschaftliche Klasse*, Abteillung IIa, 149 (1940): 157–211.

Wambacher, Hertha. "Wirkung der Korpuskularstrahlen auf die photographische Emulsion." *Zeitschrift für wissenschaftliche Photographie*, 38 (1939): 38–61.

Weidler Kubanek, Anne-Marie and Grzegorek Grete. "Ellen Gleditsch: Professor and Humanist." In *A Devotion to their Science: Pioneer Women of Radioactivity* edited by Marelene Rayner-Canham and Geoffrey Rayner-Canham. 51–75. Philadelphia: Chemical Heritage Foundation, 1997.

Weindling, Paul. "Public Health and Political Stability: The Rockefeller Foundation in Central and Eastern Europe between the two World Wars." *Minerva*, 31 no. 3 (1993): 253–267.

Weiss-Tessbach, Felicitas. "Mikrokalorimetrische Messung der Absorption der γ-Strahlung von RaC." *Sitzungsberichte Akademie der Wissenschaften in Wien, Mathematisch-naturwissenschaftliche Klasse*, Abteillung IIa. 137 (1928): 551–562.

Wettstein, R. "Karl Kupelwieser." In *Neue Österreichische Biographie 1815–1918* edited by A. Bettelheim et al. 9–15. Vienna, 1928.

Wibiral, Norbert and Mikula, Renata. "Heinrich von Ferstel." In *Die Wiener Ringstrasse*, edited by Wagner-Riger, Band VIII. 44–75. Wiesbaden: Franz Steiner Verlag, 1974.

Wick, Frances. "Über Triboluminieszenz." *Sitzungsberichte Akademie der Wissenschaften in Wien, Mathematisch-naturwissenschaftliche Klasse*, Abteillung IIa. 145 (1936): 689–705.

Wick, Frances. "Verschuche über Radiothermolumineszenz." *Sitzungsberichte Akademie der Wissenschaften in Wien, Mathematisch-naturwissenschaftliche Klasse*, Abteillung IIa. 139 (1930): 497–508.

Wiesloch, Hegar. "Sterilization in the interest of Race Betterment." *American Journal of Radiology*, 3 (1916): 46.

Wyklicky, Helmut. "Zur Entwicklung der Medizinishen Forschung in Wien." In *100 Jahre Universität am Ring: Wissenschaft und Forschung an der Universität Wien seit 1884* edited by G. Hamann, K. Mühlberger, F. Skacel. 35–41. Schriftenreihe des Universitätsarchivs, Band 3, Wien: Universität Wien, 1986.

Wyklicky, Helmut. "Zur Geschichte der Strahlentherapie in Österreich." *Wiener klinische Wochenschrift*, 5 (1980): 165–171.

Wyklicky, Helmut. *Das Institut für Geschichte der Medizin der Universität.* Wien, Wien: Universität Wien, 1990.

Zintl, E. "Otto Hönigschmid zum 60 Geburtstag" Zeitschrift für anorganische und allgemeine Chemie, 236 no. 1–4 (1938): 12.

TABLES AND CHARTS

Table 01/3: Enrollment of female and male students at the University of Vienna, 1897–1914						
Acad Year	**Women (phil. faculty)**	**Women (phil. faculty)**	**Women (med. faculty)**	**Women (med. faculty)**	**Women (all faculty)**	**Men (all faculty)**
	Matric.	Non- matric.	Matric.	Non- matric.		
1897/98	3	34	-	-	37	6775
1900/01	31	87	10	25	153	6975
1904/05	75	235	32	5	347	7886
1908/09	157	333	62	16	568	8340
1912/13	285	330	152	12	779	9535
1913/14	314	292	184	6	796	9645

Source: The figures in this table are based on Renate Tuma's work on Austrian women students in the University of Vienna, in Tuma, "Die Österreichischen Studentinnen der Universität Wien" (1990), 80.

Table 02/3: Number of women enrolled in courses in the philosophical faculty of the University of Vienna, by field.

Fields	**1897/98**	**1904/05**	**1908/09**	**1913/14**
Philosophy	26	178	329	404
Psychology	17	123	244	204
German Philology	18	186	322	305
Pedagogy	13	172	310	323
History	15	86	141	157
History of Art	6	70	135	156
Physics	10	72	116	145
Chemistry	9	70	108	133
Mathematics	12	55	97	128
Total	32	218	400	520

Source: Tuma, "Die Österreichischen Studentinnen" (1990), 87. The total in the table refers to the actual number of women enrolled at the philosophical faculty of the University of Vienna. Those were able to attend several courses in different fields. This is why the total does not match the numbers given by field.

Table 01/4: Enrollment of female and male students, University of Vienna, 1918-1934

Academic Year	Women	Men	Total
1918/19	1569	8946	10515
1919/20	1749	9693	9693
1923/24	1863	8042	9905
1925/26	1669	7631	9300
1928/29	2459	8685	11144
1933/34	3144	8801	11945

Source: The figures in this table are based on Renate Tuma's work on the female Austrian students in the University of Vienna (Tuma, "Die österreichischen Studentinnen" (1993), 81).

Table 02/4: Number of women enrolled in the philosophical faculty of the University of Vienna by field, 1918-1934

Fields	**1918/19**	**1928/29**	**1933/34**
Philosophy	432	402	383
Psychology	122	424	431
German Philology	255	479	618
Pedagogy	51	357	404
History	218	176	323
History of Art	193	180	160
Physics	172	236	318
Chemistry	223	307	365

Source: The figures in this table are based on Renate Tuma's work on the female Austrian students in the University of Vienna (Tuma, "Die österreichischen Studentinnen" (1993), 88).

Table 03/4: Professional and career information of the director and the assistants of the institute, 1919-1938			
Name	**Position at RI**	**Years at RI**	**Position at U of V**
Meyer, Stefan	Director	1919-1938	1900 Privatdozent 1908 ausserodentlicher Professor 1911 wirklicher Extraordinarius Prof. 1915 ordentlicher Professor 1920 wirklicher ordentlicher Prof. 1938 dismissed
Przibram, Karl	Ordentlicher Assistent	1919-1938	1905 Privatdozent 1916 ausserodentlicher Professor 1927 wircklicher ausserodentlicher Professor 1938 dismissed
Fonovits, Hilda	Ausserordentlicher Assistent	1919-1922	none
Geiger, Sebastian	Ausserordentlicher Assistent	1922-1924	none
Ortner, Gustav	Ausserordentlicher Assistent	1924-1938	1939 ausserordentlicher Professor

Source: Data from *Almanac of the Austrian Academy of Sciences*; *Mitteilungen* of the Radium Institute; and correspondence.

Abbreviations

aus. A. ausserordentlicher Assistent

Dir. Director

Dr. Doctor

Ing. ingenieur

o. A. ordentlicher Assistent

Pr. Practicum student

Pr. Dr. Professor Doctor

w. H. wissenschaftlicher Hilfscraft

° individual was listed as a collaborator of the institute but without a title; probably was a practicum student

Table 04/4: Researchers who remained at the Radium Institute for longer than four years from the academic year 1919/20 to 1933/34

Names	1919/20	1920/21	1921/22	1922/23	1923/24	1924/25	1925/26	1926/27	1927/28	1928/29	1929/30	1930/31	1931/32	1932/33	1933/34	no. of years
Baiersdorf, G												°	°	°	°	4
Belar, Maria		°	Dr.	Dr.	Dr.	Dr.	Dr.	Dr.	Dr.	Dr.	Dr.					10
Blau, Marietta		Dr.			Dr.	Dr.	Dr.	Dr.	Dr.	Dr.	Dr.	Dr.	Dr.	Dr.	Dr.	12
Czapek, A.										°	°	Dr.	Dr.			4
Fonovits-Smereker, Hilda	aus. A.	aus. A.	aus. A.	aus. A.												4
Hernegger, F												°	°	°	°	4
Holesch, Elsa											°	°	°		Dr.	4
Kailan, Anton	Pr.	Pr.	Pr.	Pr.												4
Kara-Michailova, Elisabeth		°	Dr.	Dr.	Dr.	Dr.	Dr.	Dr.	Dr.	Dr.-w.H.	Dr.-w.H.	Dr.-w.H.	Dr.-w.H.	Dr.-w.H.	Dr.-w.H.	14
Karl, A										°	°	°	°			4
Karlik, Berta												Dr.	Dr.	Dr.	Dr.-w.H.	4
Kindinger, Max						°	°	°	°	°	Dr.	Dr.	Dr.	Dr.	Dr.	10
Kirsch, Gerhard	°	Dr.		Dr.	Dr.	Dr.	Dr.	Dr.	Dr.	Dr.	Dr.	Dr.	Dr.	Dr.	Dr.	14
Krafft, P.								Ing.	Ing.	Ing.	Ing.	Ing.	Ing.			6
Kreidl, N						°	°	°	°							4
Kürti, G												Dr.	Dr.	Dr.	Dr.	4
Mayer, Gertrud									°	°	°	°	Dr.			5
Meyer, Stefan	Dir.	Dir.	Dir.	Dir.	Dir.	Dir.	Dir.	Dir.	Dir.	Dir.	Dir.	Dir.	Dir.	Dir.	Dir.	15

Names	1919/ 20	1920/ 21	1921/ 22	1922/ 23	1923/ 24	1924/ 25	1925/ 26	1926/ 27	1927/ 28	1928/ 29	1929/ 30	1930/ 31	1931/ 32	1932/ 33	1933/ 34	no. of years
Ortner, Gustav					Dr.	aus. A.	aus. A.	aus. A.	aus. A.	aus. A.	aus. A.	aus. A.	aus. A.	aus. A.	aus. A.	11
Pelz, St										°	°	°	°	°	°	6
Pettersson, Dagmar			Dr.	Dr.	Dr.	Dr.	Dr.									5
Pettersson, Hans			Dr.	Dr.	Dr.	Dr.	Dr.	Dr.	Dr.	Dr.	Dr.	Pr.	Pr.	Pr.	Pr.	13
Przibram, Karl		o. A.	o. A.	o. A.	o. A.	o. A.	o. A.	o. A.	o. A.	o. A.	o. A.	o. A.	o. A.	o. A.	o. A.	14
Rona, Elizabeth						Dr.	Dr.	Dr.	Dr.	Dr.- w.H.	Dr.	Dr.	Dr.	Dr.	Dr.	10
Schmid, Ewald						Dr.	Dr.		w. H.	w. H.	Dr.	Dr.	Dr.			7
Stetter, Georg											Dr.	Dr.	Dr.	Dr.	Dr.	5
Ulrich, Karl	Dr.	Dr.	Dr.	Dr.	Dr.	Dr.										6
Urbach, Franz				°	°	°	°	Dr.	Dr.	Dr.	Dr.	Dr.	Dr.			10
Weiß-Tessbach, Felizitas								°	°	Dr.	Dr.	Dr.				5
Wolf, St							°	Dr.	Dr.	Dr.	Dr.					5

Source: *Almanac of the Austrian Academy of Sciences*, 1933–1939.

Table 05/4: Number of publications of those researchers who published more than ten papers from 1920 to 1934

Names	**1920**	**1921**	**1922**	**1923**	**1924**	**1925**	**1926**	**1927**	**1928**	**1929**	**1930**	**1931**	**1932**	**1933**	**1934**	**no. of publicatio**
Belar, Maria		1		2			1									4
Blau, Marietta					1	1	1	1	1	1	2	2	3		2	15
Gräven, Heinrich											1		2			3
Haberlandt, Herbert													2	5	5	12
Hoffman, Josef											2	1	1	2	1	7
Kailan, Anton	1	2	1		1		1			1		1			1	9
Kara-Michailova, Elisabeth			2	1	1			1	1	1		1		1	3	12
Karlik, Berta								1	1	1	2			3	4	12
Kirsch, Gerhard	1		1	1	2	2		1	1				4	2		15
Meyer, Stefan	1		1	1	1				2	3	1		1		1	12
Ortner, Gustav					1		1	1	1	1	1	1		3	2	12
Pelz, St												1		2		3
Pettersson, Hans				2	6	3		2	2	1			1	2	4	23
Przibram, Karl		2	3	2		2	4	4	2	5	3		6	5	5	43
Rieder, Fritz													5	1		6
Rona, Elizabeth							2	1	2	1	1		3	1	1	12
Schintlmeister, Josepf													2	3	4	9
Schmid, Ewald						1		1	2	1	2					7
Stetter, Georg							1	1	1	1	2	1	6	1		14
Urbach, Franz							1		3	3	5					12
Wambacher, Hertha												1	3	1	2	7

Source: *Mitteilungen* of the Radium Institute.

Table 06/4: Collaborations at the Institute for Radium Research, 1920-1934			
year	**men - men**	**women - women**	**men - women**
1920			• Hess - Hornyak
1922	• Przibram - Meyer		• Przibram - Kara-Michailova
1923	• Kirsch - Pettersson • Meyer - Urlich		• Przibram - Belar • Przibram - Kara-Michailova
1924	• Kirsch - Pettersson • Ortner - Pettersson		• Przibram - Kara-Michailova
1925	• Kirsch - Pettersson • Michel - Pettersson		
1926		• Blau - Rona	
1927	• Kirsch - Pettersson		• Rona - Schmidt
1928	• Ortner - Stetter	• Karlik - Kara-Michailova	• Rona - Schmidt
1929	• Blank - Urbach • Schmidt - Stetter	• Blau - Rona • Karlik - Kara-Michailova	
1930	• Schmidt - Stetter • Meyer - Suess • Urbach - Schwarz	• Rona - Blau	
1931		• Blau - Kara-Michailova	
1932	• Kirsch - Rieder • Kirsch - Graeven • Pettersson - Schintlmeister	• Blau - Wambacher	• Rona - Rieder • Deseyve - Kirsch - Rieder
1933	• Pettersson - Schintlmeister • Kirsch - Trattner • Przibram - Haberlant • Ottner - Stetter • Stetter - Schintlmeister • Koehler - Haberlant	• Rona - Karlik	• Wallner - Merhaut • Wambacher - Kirsch • Hoffman - Karlik - Przibram
1934	• Ortner - Schintlmeister • Pettersson - Schintlmeister • Haberlandt - Przibram • Stetter - Schintlmeister	• Rona - Karlik • Blau - Wambacher	• Haberlandt - Karlik - Przibram • Kara-Michailova - Pettersson • Karlik - Pettersson

Source: *Mitteilungen* of the Radium Institute.

Abbreviations

aus. A.	ausserordentlicher Assistent
Dir.	Director
Dr.	Doctor
Ing.	ingenieur
o. A.	ordentlicher Assistent
Pr.	Practicum student
Pr. Dr.	Professor Doctor
w. H.	wissenschaftlicher Hilfscraft
°	individual was listed as a collaborator of the institute but without a title; probably was a practicum student

Table 01/6: Men and women, members of the Radium Institute from 1933/34 to 1938/39						
Names	**1933/ 34**	**1934/ 35**	**1935/ 36**	**1936/ 37**	**1937/ 38**	**1938/ 39**
Adler, Elfrieda			°	°	°	°
Baiersdorf, G	°	Dr.	Dr.			
Belar, Maria					Dr.	
Blau, Marietta	Dr.	Dr.	Dr.	Dr.	Dr.	
Böhm, F				°	°	
Brück, Lilly	Dr.	Dr.				
Buchgraber, Dora	°	°	°			
Deinlein, Maria	Dr.	Dr.	Dr.			
Dostal, H.	Dr.	Dr.				
Eckstein, H			°	°	°	
Eysank, Elfrida	°	°	°			
Föyan, E		Dr.	Dr.			
Friedrich, W	°					
Fuchs, G			Dr.	Dr.	Dr.	
Gleditsch, Ellen					Pr. Dr.	
Großmann, R		°	°	Dr.	Dr.	Dr.
Haberfeld, Magda	°					
Haberlandt, H	Dr.	Dr.	Dr.	Dr.	Dr.	Dr.
Hernegger, F	°	°	°	°	Dr.	o. A.
Hiesinger, D					°	Dr.
Hoffer, Margarete	°	°	Dr.			
Hoffmann, J	Dr.					Hofrat
Holesch, Elsa	Dr.	Dr.	Dr.			
Jagitsch, R			°			
Jentschke, W		°	Dr.			Dr.
Kara-Michailova, Elisabeth	Dr.-w.H.	Dr.	Dr.			
Karlik, Berta	Dr.-w.H.	w. H.	w. H.	w. H.	w. H.	w. H.
Kellermann, W			°	°	°	
Kindinger, Max	Dr.	Dr.	Dr.	Dr.	Dr.	Dr.

Source: *Almanac of the Austrian Academy of Sciences*, 1933–1939.

Printed in the United States
210209BV00001BA/63/P